Crop Ecology: Productivity and Management in Agricultural Systems

Crop Ecology: Productivity and Management in Agricultural Systems

Editor: Corey Aiken

RCALLISTO
REFERENCE

www.callistoreference.com

Callisto Reference,
118-35 Queens Blvd., Suite 400,
Forest Hills, NY 11375, USA

Visit us on the World Wide Web at:
www.callistoreference.com

ISBN: 978-1-63239-951-9 (Hardback)

Cataloging-in-Publication Data

Crop ecology : productivity and management in agricultural systems / edited by Corey Aiken.
    p. cm.
Includes bibliographical references and index.
ISBN 978-1-63239-951-9
1. Crops--Ecology. 2. Agricultural productivity. 3. Crop improvement. I. Aiken, Corey.
SB106.E25 C76 2018
577.55--dc23

Table of Contents

Preface

This book has been a concerted effort by a group of academicians, researchers and scientists, who have contributed their research works for the realization of the book. This book has materialized in the wake of emerging advancements and innovations in this field. Therefore, the need of the hour was to compile all the required researches and disseminate the knowledge to a broad spectrum of people comprising of students, researchers and specialists of the field.

Crop ecology is an emerging field of study. It studies the methods of farming and assesses the use of technology in agriculture. This book elucidates new techniques and their applications in a multidisciplinary approach. The research done in this field focuses on the techniques and practices that can maximize the profits produced by cropping systems. This book is an essential guide for both academicians and those who wish to pursue this discipline further.

At the end of the preface, I would like to thank the authors for their brilliant chapters and the publisher for guiding us all-through the making of the book till its final stage. Also, I would like to thank my family for providing the support and encouragement throughout my academic career and research projects.

Editor

Screening wheat genotypes for coleoptile length: A trait for drought tolerance

Md. Farhad[1], Md. Abdul Hakim[1], Md. Ashraful Alam[1], N. C. D. Barma[2]

[1]Wheat Research Centre, BARI, Nashipur, Dinajpur-5200, Bangladesh
[2]Regional Wheat Research Centre, BARI, Joydebpur, Gazipur-1701, Bangladesh

Email address:
farhadnabin@bari.gov.bd (Md. Farhad)

Abstract: The study was conducted during Rabi season of 2013-14 at the Wheat Research Centre (WRC), Bangladesh Agricultural Research Institute (BARI), Dinajpur. Thirty wheat genotypes including local control BARI Gom 26 were evaluated in split-split plot design having two replications with irrigation in the main plot, seeding depth in a sub-plot and genotype was in sub-sub-plot. The main objective of this study was to evaluate new exotic lines against drought, with emphasis on coleoptile length under Bangladeshi conditions, and to identify drought tolerant germplasm. To measure potential coleoptile length (CL), disease free, healthy, uniform seeds were sown in wooden trays with sandy soil in a temperature controlled room at 200 degree days ($20°$ c X 10 days). The genotypes were evaluated for yield, and yield components i.e., plant establishment, plant height (cm), spikes per m^2, grains per Spike, 1000-grain weight (g) and visual grain quality. Selection of genotypes was based on Schneider's stress severity index (SSSI), yield under drought condition and coleoptiles length. Deep seeding over normal seeding had a significant effect on yield and the yield components, as did water stress. The interaction of the two factors showed that seeding depth causes more yield loss than irrigation. More traits showed significant relationships in deep seeding conditions than normal conditions, meaning that there is greater scope for screening wheat using sowing depth. Based on higher negative value of SSSI and higher yield in deep sowing conditions the genotypes G 16, G 13, G 12, G 24, G 2, G 18, G 19 and G 3 were primarily selected for drought tolerance and will be evaluated further for advanced studies. These genotypes also have longer coleoptiles ranging from 7.4 to 10.5 cm.

Keywords: Wheat, Drought Tolerance, Deep Sowing, Coleoptile Length, Irrigation, Index

1. Introduction

Climate change in Bangladesh is expected to aggravate the situation along with the withdrawal of upstream water, resulting in increasing droughts and depletion of the water table. Crop production becomes impossible especially in drier northern and western regions of the country. Bangladesh already faces drought in the northwestern region [10] and it is expected that the moderately drought affected areas will become severely drought prone areas within the next two decades. The intensity and frequency of climatic hazards has brought to light the necessity for introducing stress tolerant varieties into breeding programs in Bangladesh, and for quick extension to growers. In addition, global climate change is negatively affecting crop yields under the current climate and is predicted to have a more severe impact on food production in future climate scenarios [22].

Wheat (*Triticum aestivum* L) is the second most important cereal crop in Bangladesh in respect of area and production cultivated in the winter season. However, due to light rainfall and scarcity of available irrigation facilities in the winter season, it suffers from soil moisture stress during the growing period. Being adapted to a wide range of moisture conditions, wheat is grown on more land area worldwide than any other crop, including in drought prone areas. In these marginal rain-fed environments where at least 60 m ha of wheat is grown, amount and distribution of rainfall are the predominant factors influencing yield variability [35]. Exposure of plants to drought led to a noticeable reduction in yield and yield contributing characters such as plant height, number of spikes per plant, total dry matter, number of seeds per spike, and 100- grain weight and grain yield [20]. There is an ever increasing interest in improving drought tolerance of wheat to attain a yield substantial enough to meet the increasing demand of a rapidly increasing population through breeding for drought tolerant and high yielding cultivars [54].

1.1. Significance of Long Coleoptile

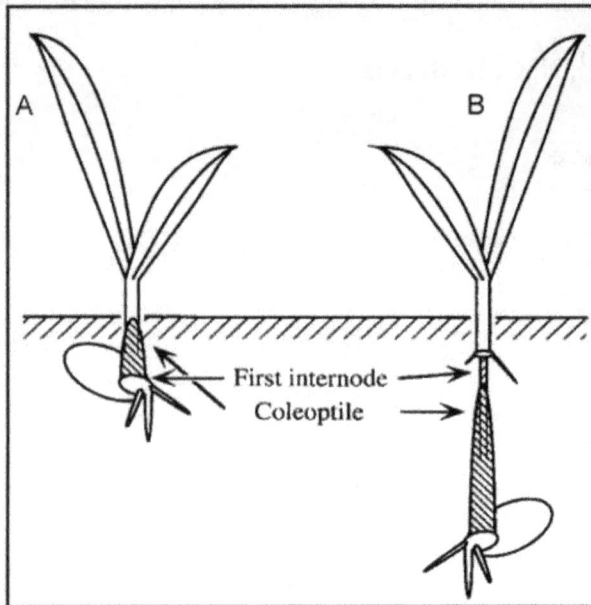

Fig 1. Schematic presentation of seedling growth under normal (A) and deep-seeding (B) conditions.

The coleoptile is an outer covering that protects the first leaf of the developing wheat plant as it pushes its way toward the surface of the soil during germination. If the coleoptile is shorter than the sowing depth, the first leaf must push through the soil and emerge in a dark environment. The longer it takes the first leaf to reach the surface, the more vulnerable the seedling is to soil crusting and diseases [21]. The coleoptile is essential for successful emergence and early plant vigour. Drought could promote the increase of Coleoptile Length (CL) and inhibit seedling height [21].

Wheat with long coleoptiles emerges with higher frequency than those with short coleoptiles especially when sown deep. [45, 46 and 47] When sown deep, wheat seedlings with short coleoptiles do emerge but much later and lack seedling vigour [19 and 47]. Deep sowing allows growers to exploit soil moisture lying below the drying topsoil and is an option considered by growers in Australia [11]. In some genotypes, shoots are able to reach the soil surface from the deep soil by elongating basal organs, e.g. coleoptile and first internode in wheat and barley [64] (Figure 1). Deeper sowing also assists in reducing removal of seeds by birds and rodents [8] and in avoiding phytotoxicity associated with some pre-emergent herbicides [40]. Deep sown seedlings with short coleoptile have smaller relative growth rates and slower leaf area development resulting in smaller leaf area early in the season which reduces the competitiveness of wheat crop against weeds and increases water loss through evaporation from the soil surface thus

reducing crop water use efficiency, biomass and finally yield [30]. Deep seed placement and reduced coleoptile elongation in the predominantly hot soil can have a potentially devastating impact on stand establishment [55]. It was reported that seed size had little impact on coleoptiles lengths in barley but not in wheat and oats [9]. Studies have demonstrated a strong association between coleoptiles length and seedling emergence with shallow [14] and deep sowing [19, 29, 32, 50 and 61]. Surveys of Australian farmers' fields show that grain yield is reduced by a minimum 10% when short coleoptiles wheats are sown deeper than 5 cm [49]. Reduced establishment and lower grain yields associated with deep sowing have been reported for shorter coleoptile wheats [31, 32, 37, 44 and 50]. A recombinant inbred line (RIL) population showed considerable variation, normal distribution and transgressive segregation for CL under field and controlled environment conditions [26].

Wheat with longer coleoptiles emerged sooner, produced more plants and had greater early vigour particularly with deep sowing. Semi dwarf wheat has CL 30–40% shorter than non-semi dwarf wheat [2]. Selection for CL usually occurs in either a greenhouse in controlled environment condition or in field plots through deep planting. Selection under the controlled conditions of a growth chamber has the advantage of being quicker, cheaper and possibly more effective for increasing CL than field selection [25]. Selection of wheat cultivars with long coleoptile is an important component of improving emergence, weed suppression and grain yield in low rainfall regions. This would be useful tool in managing climate variability and would assist wheat growers to sow closer to the optimum sowing time in situations where moisture is present at depth but not on the soil surface. Considerable genetic variation has been observed for this trait and several studies have reported relationship between CL and seedling emergence and subsequent effect on early growth, biomass and grain yield.

2. Materials and Methods

The experimental material of the study was consisted of thirty genotypes of spring wheat (*Triticum aestivum* L.) including local control variety BARI Gom 26, collected from "4th CSISA drought yield trial 2011-12" [63]. To measure potential CL of the materials, disease free, healthy uniform seeds were sown in wooden trays with sandy soil in a temperature controlled room at 200 degree days (20° c X 10 days). The field study was conducted at the experimental field of the Wheat Research Centre (WRC), Bangladesh Agricultural Research Institute (BARI), Dinajpur during 2013-14 cropping season. The genotypes were grown in rain fed, irrigated, deep seeding (around 5 cm below the soil surface) and normal seeding conditions (around 10 cm below the soil surface). Three irrigations were applied in the irrigated plots whereas no irrigation was applied in non irrigated plots to induce water stress. Root depths were between 5 cm below soil surface and around 10 cm below soil surface. The experiment was laid out in split-split plot design with two replications with irrigation in the main plot, seeding depth in the sub-plot and genotype in the sub-sub-plot. Randomization for the trial was generated by Cropstat 7.2 software [23]. Seeds were sown continuously in 2.5m long, 4-row plots with a row spacing of 20 cm with 1 meter between replicates. A two meter distance was maintained between the irrigated and non irrigated plots to control the unexpected water movement from irrigated plots to non irrigated plots. Recommended fertilizers and cultivation practices were followed in every plot. Data were recorded on yield (Y) and other yield contributing characters e.g., plant establishment (EPP), plant height in cm (PH), number of spikes per m^2 (SPN), grains per Spike (GPS) and 1000-grain weight in grams (TGW). At maturity, the whole plot was harvested to estimate grain yield measured in kg per hectare. Data were analyzed by ANOVA using Cropstat 7.2 software. A correlation matrix was calculated by Mltibase_2015 add-in to MS Excel [36]. Duncan's multiple range test was performed by using DSAASTAT add-in to MS Excel [41].

Selection of genotypes was based on index proposed by a modified formula of Stress Susceptibility Index (SSI) [15]. According to the formula (Yi)s denote the yield of the ith genotype under stress, (Yi)ns the yield of the ith genotype under nonstress (i.e., irrigated) conditions and Ys and Yns is the mean yields of all genotypes evaluated under stress and nonstress conditions, respectively. SSI is expressed by $SSI = \frac{1 - \frac{(Yi)s}{(Yi)ns}}{SI}$, SI, the stress intensity is estimated as $SI = 1 - \frac{Ys}{Yns}$. Lower SSI values indicate lower differences in yield across stress levels, in other words, more resistance to drought. The modified formula for Schneider's stress severity index [51,

52] is $SSSI = (1 - \frac{(Yi)s}{(Yi)ns}) - (1 - \frac{Ys}{Yns})$. The SSSI estimates the relative tolerance for yield reduction of a genotype relative to the population mean reduction in grain yield response due to stress. Selections based on these indices were carried out by many authors [18, 28, 39, 42, 43, 52, 53, and 57].

3. Results and Discussion

The average coleoptiles lengths of the 30 genotypes varied in range, presented in table 1. The CL varied from 3.5 cm (G6) to 12.10 cm (G12) in different genotypes and the highest mean CL was 10.9 cm in G 25. The mean CL was recorded above 10 cm in G26, G30, G12 and G20, and 5 genotypes (G22, G 11, G 17, G 24 and G 21) were produced from 9.0 to 9.9 cm. Mean CL from 8.2 cm to 8.8 cm was in 8 genotypes. Another group having 7 to 7.8 cm CL consist of 8 genotypes. The remaining 4 genotypes vary for mean CL between 5.5 and 6.8 cm. CL variation among the CIMMYT lines in this collection was very narrow with the majority having a CL between 7-10.9cm. This narrow variation is probably due to the semi-dwarf nature of all the collected lines and/or due to a common genetic background and having been previously selected from 4th CSISA drought trial. A number of non-genetic factors have been reported to affect seedling vigour and CL including variation in grain size [27, 13], grain positions in the ear [56] and the environment from which the grain was harvested [37, 44]. Seed source is obviously a very important determinant of CL [24].

Table 1. Average CL with range of 30 genotypes.

Genotype	Average CL (cm)	Range (cm)	Genotype	Average CL (cm)	Range (cm)	Genotype	Average CL (cm)	Range (cm)
G-1	7.5	5.5-9.5	G-11	9.8	8.8-10.2	G-21	9.3	6.7-10.8
G-2	8.2	6.1-11.0	G-12	10.5	6.2-12.1	G-22	9.9	7.2-11.5
G-3	7.5	5.0-10.5	G-13	7.4	5.2-11.5	G-23	6.1	4.9-8.1
G-4	7.6	7.1-7.9	G-14	7.7	5.9-9.5	G-24	9.5	8.0-10.9
G-5	7.8	5.8-9.6	G-15	8.5	7.0-10.5	G-25	10.9	9.0-11.9
G-6	5.5	3.5-9.5	G-16	8.7	6.9-11.5	G-26	10.8	10.1-11.3
G-7	8.5	5.0-11.4	G-17	9.8	8.5-11.0	G-27	7.3	5.9-10.1
G-8	8.8	6.6-12.0	G-18	8.7	7.5-10.6	G-28	6.8	5.2-9.3
G-9	6.8	5.4-9.0	G-19	8.6	6.2-10.8	G-29	7.0	5.0-9.0
G-10	8.6	7.0-10.8	G-20	10.2	8.3-11.6	G-30	10.8	9.0-11.6

Table 2. Analysis of variance and coefficient of variation (CV) of each trait with different treatments in the field condition.

Variation Source	df	Mean sum of squares					
		EPP	PH	SPN	GPS	TGW	Y
Sowing depth (D)	1	111208**	817.705**	656993**	5219.2**	136.33**	77535400**
Irrigation (IR)	1	291.233	24.7042	187042**	448.814**	3.37542	100255000**
D X IR	1	304.601	30.1041	8449.07*	52.8282	13.1552	1368520
Genotype (G)	29	337.558**	41.6323	4080.67**	83.695**	26.6389**	2345680**
IR X G	29	121.021	44.1956	5408.37**.	43.8669	12.3041	951139**
D X G	29	460.024**	53.6955	5599.47**	66.3746**	9.84239	1724130**
D X IR X G	29	135.765	31.7507	3498.87**	39.1607	11.7532	831609*
CV		18.5	7.5	14.9	10.4	9.1	14.5

The mean sums of squares for the characters studied are presented in table 2. The mean sums of squares due to genotypes were significant for five characters studied while PH shows no significance for genotypes. The mean sum of

squares suggests that the genotypes selected were genetically variable and a considerable amount of variability existed among them. This indicates selection for different quantitative characters for wheat improvement. These findings are in accordance with the findings of [3, 6 and 12] who also observed significant variability in wheat germplasm. Analysis of variance also revealed significant differences between treatments and among the genotypes. All the traits studied varied significantly for sowing depth while only SPN, GPS and Y showed a significant variation for irrigation. The sowing depth × genotype interactions were also significant for all the characters except PH and TGW. On the other hand Irrigation X Genotype interaction is significant for SPN and Y. This interaction revealed that genotypes performed inconsistently over the stress conditions. This significant variation in water stress conditions may serve as good indicator of drought tolerance.

The correlation coefficients of yield and yield contributing characters indicated some significant relation exist among the character studied. More traits showed significant relationship in deep seeding condition than normal condition, meaning that there is greater scope for screening wheat using sowing depth. CL showed a non-significant relationship except with number of spikes in irrigated deep seeding condition. As in [4 and 16] we found that plant height and

CL were not significantly correlated. There were no differences in the mean plant height for entries with coleoptiles longer than 90 mm in wheat as in [34]. They also indicated that coleoptiles longer than 90 mm showed no advantage for emergence from deep planting and might even have a negative effect. Again, it was also found that thousand kernel weight did not affect emergence on any days after planting nor was it associated with plant height at maturity [34]. It was revealed that wheat seedling emergence did not have linear relationship with CL as many other factors have been implicated in the process [48, 7 and 38]. Emergence and CL is reported to be influenced both by genetic background and environmental factors including soil texture, seed-zone water content, temperature, light penetration, and crop residue [58]. As we do not know the nature of dwarfing genes in the world collection, these could possibly be gibberellic acid (GA) sensitive genotypes that are reported to have no adverse effect on CL but reduce plant height, or possibly have favorable alleles for CL in semi-dwarf background [60].

Table 3. Simple Correlation coefficient between yield and the traits associated with yield.

Table 3a. Correlation Coefficient Matrix for normal sowing in irrigated and non-irrigated condition

	CL	EPP	PH	SPN	G	TGW	Y
CL		0.18	-0.04	0.30	-0.07	-0.01	-0.08
EPP	0.01		0.34	0.31	-0.25	-0.23	0.43*
PH	-0.09	0.23		0.17	0.15	-0.23	0.20
SPN	-0.28	0.09	0.28		-0.41*	-0.23	0.15
GPS	0.12	-0.34	0.02	0.12		0.14	0.02
TGW	0.15	-0.24	-0.26	-0.25	0.10		0.04
Y	0.22	-0.08	0.19	0.16	0.25	-0.06	

Parentheses are the r^2 of non-irrigated condition

Table 3b. Correlation Coefficient Matrix for deep sowing in irrigated and non-irrigated condition

	CL	EPP	PH	SPN	GPS	TGW	Y
CL		-0.09	0.25	0.43*	-0.18	-0.01	-0.05
EPP	-0.11		0.21	0.23	-0.24	-0.10	0.45**
PH	-0.30	0.28		0.47**	0.09	-0.08	0.55**
SPN	-0.19	0.26	0.57**		-0.09	-0.09	0.68**
GPS	-0.06	-0.32	-0.46**	-0.32		-0.36*	0.25
TGW	-0.27	-0.04	0.23	0.21	-0.36*		-0.14
Y	-0.15	0.32	0.36*	0.58**	-0.12	-0.10	

Parentheses are the r^2 of non-irrigated condition

Table 4. Mean effect of depth of seeding on yield and yield contributing characters of wheat.

Treatment	EPP	PH	SPN	GPS	TGW	Y
Normal Seeding	78	97.4	328	49	46.11	5359
Deep Seeding	33	93.4	223	59	44.70	4151
F- value	**	**	**	**	**	**
LSD (5%)	2.63	1.82	10.49	1.43	1.05	176.39
CV	14.7	7.6	16.3	10.2	9.2	13.3

Table 4 shows that seeding depth had a significant effect on yield and yield components in wheat. Higher seedling emergence, taller plant, more spikes per square meter, increased thousand grain weight and ultimately higher grain yield was found in normal seeding condition compared to deep seeding condition. Fewer grains per spike were observed in normal seeding condition than deep seeding. As in [1] semi dwarf varieties of wheat produced more grains per spike when planted at 10 cm deep where moisture was not a limiting factor. An increased number of grains per spike as well as grain yield per spike was noted in a local wheat variety had sown under 9 cm deep below the soil surface [33].

Table 5. *Mean effect of water stress on yield and different yield parameters of wheat*

Treatment	EPP	PH	SPN	GPS	TGW	Y
Irrigated	55	95.7	303	56	46	5419
Water stressed	57	95.1	247	53	45	4091
F- value	ns	ns	**	**	ns	**
LSD (5%)	2.63	1.82	10.49	1.43	1.05	176.39
CV	14.7	7.6	16.3	10.2	9.2	13.3

Table 6. *Interaction effect of yield and yield parameters imposed by depth and irrigation.*

Treatment		EPP	PH	SPN	GPS	TGW	Y
Normal Seeding	Irrigated	78	97.2	349	50	46.5	5960
	Water stressed	78	97.6	306	48	45.7	4758
Deep Seeding	Irrigated	31	94.2	257	61	44.7	4878
	Water stressed	36	92.6	189	57	44.7	3424
F- value		**	ns	**	**	ns	**
LSD (5%)		3.73	2.58	14.83	2.03	1.49	259.46
CV		14.7	7.6	16.3	10.2	9.2	13.3

Table 7. *Yield of wheat genotypes in different treatments and Schneider's stress severity index (SSSI) with average CL.*

Genotype	Yield (Kg/h)*								SSSI		CL
	Normal Seeding				Deep Seeding				Normal seeding	Deep Seeding	
	Irrigated		Dryland		Irrigated		Dryland				
G 1(BARI Gom 26)	5080	bc	4700	abc	3960	efghi	4770	abc	-0.125	-0.505	7.5
G 2	5710	abc	5535	ab	5084	abcdef	4114	abcdef	-0.169	-0.109	8.2
G 3	6547	ab	4050	bc	4326	defgh	3140	cdefghij	0.181	-0.026	7.5
G 4	7355	a	5305	ab	3391	ghi	2266	ij	0.079	0.032	7.6
G 5	5965	abc	3870	bc	3324	hi	2825	efghij	0.151	-0.150	7.8
G 6	6280	abc	3059	c	3380	ghi	2010	j	0.313	0.105	5.5
G 7	5310	bc	5000	ab	2290	i	2416	fghij	-0.142	-0.355	8.5
G 8	6105	abc	4280	bc	5860	abcd	3705	abcdefghij	0.099	0.068	8.8
G 9	5640	abc	4880	ab	3165	hi	2425	fghij	-0.065	-0.066	6.8
G 10	6085	abc	5000	ab	3805	fghi	2695	fghij	-0.022	-0.008	8.6
G 11	6390	abc	4135	bc	4505	cdefgh	2285	hij	0.153	0.193	9.8
G 12	6760	ab	4225	bc	4330	defgh	3630	abcdefghij	0.175	-0.138	10.5
G 13	6050	abc	4810	ab	5505	abcde	4885	ab	0.005	-0.187	7.4
G 14	5960	abc	5095	ab	6580	a	4505	abcde	-0.055	0.015	7.7
G 15	6245	abc	5240	ab	5215	abcdef	3295	bcdefghij	-0.039	0.068	8.5
G 16	5965	abc	4640	abc	5650	abcd	5295	a	0.022	-0.237	8.7
G 17	6485	ab	5150	ab	5400	abcdef	2690	fghij	0.006	0.202	9.8
G 18	6105	abc	4360	bc	5035	abcdefg	3920	abcdefghi	0.086	-0.079	8.7
G 19	5860	abc	4435	abc	5950	abcd	4535	abcd	0.043	-0.062	8.6
G 20	5315	bc	4910	ab	3890	efghi	2330	hij	-0.124	0.101	10.2
G 21	6585	ab	5125	ab	5875	abcd	2935	defghij	0.022	0.200	9.3
G 22	5345	bc	5395	ab	5850	abcd	3860	abcdefghi	-0.209	0.040	9.9
G 23	5875	abc	5380	ab	6450	ab	2990	defghij	-0.116	0.236	6.1
G 24	5940	abc	4580	abc	5950	abcd	4865	ab	0.029	-0.118	9.5
G 25	4655	c	3925	bc	3850	efghi	2645	fghij	-0.043	0.013	10.9
G 26	5195	bc	6140	a	5045	abcdefg	3400	bcdefghij	-0.382	0.026	10.8
G 27	5555	bc	4060	bc	5805	abcd	3855	abcdefghi	0.069	0.036	7.3
G 28	6695	ab	5045	ab	6210	abc	4065	abcdefg	0.046	0.045	6.8
G 29	5920	abc	5115	ab	5905	abcd	4005	abcdefgh	-0.064	0.022	7
G 30	5835	abc	5290	ab	4755	bcdefgh	2365	ghij	-0.107	0.203	10.8
Mean	5960		4758		4878		3424				

*Duncan's multiple range test (p= 0.05)

Plant establishment was counted before CRI stage when no irrigation was applied, thus table 5 revealed no variation in plant establishment in both irrigated and non-irrigated plots. Irrigation had also no effect on plant height. These non-significant variations might be due to the semi-dwarf stature of the genotypes having very little or no variation for plant height. As we do not know the nature of dwarfing genes in the genotype collection, these could possibly be GA sensitive genotypes that are reported to have no adverse effect of irrigation but reduce plant height, or possibly have favorable alleles for drought effect in a semi-dwarf background [60]. Spikes per square meter, grains per spike, thousand grain weight and yield varied due to irrigation in wheat. Similar results were recorded as in [58] where a significant effect of irrigation on 1000-grain weight was reported. Application of five irrigations at different wheat growth stages resulted in higher spike length, higher number of grains and wheat grain yield [5].

Mean interaction effect of seeding depth and irrigation in table 6 revealed that highest yield was obtained from irrigated normal seeding condition whereas lowest yield was from the non-irrigated deep sowing condition. No significant difference in yield was observed between non irrigated deep seeding conditions and irrigated normal seeding conditions. Spikes per square meter and number of grains per spike were highest (mean) in irrigated deep seeding condition. As several authors indicate that CL has a significant effect on yield and yield controlling characters, and a negative but non-significant correlation of CL found in this experiment (table 3), screening of wheat genotypes for drought and CL through deep planting might be fruitful.

G 4 was the highest yielder (7355 kg) followed by G 12, G 28, G 21 and G 3, respectively whereas genotype 25 (4655 kg) was the lowest yielder in irrigated normal seeding condition (table 7). G 26 (6140 kg) was the maximum yielder followed by G 2, G 22, G 23 and G 4 in normal seeding in the dryland condition, whereas G 6 (3059 kg) was the lowest yielder for this environment. The highest yield under deep sowing with regular irrigation was found in G 14 (6580 kg) followed by G 23, G 28, G24 and G 19 respectively and the lowest yielder was G 7 in this condition.. Moreover, the table revealed that G 16 (5295 kg) was the maximum yielder in the non-irrigated deep sowing condition followed by G 13, G 24 and G 19. Yield was lowest in G 6 (2010 kg) in non irrigated deep sowing condition.

Sixteen genotypes showed positive SSSI values for normal seeding condition, suggesting that they suffered high stress and high grain yield loss for irrigation in normal seeding. Fourteen genotypes had negative SSSI values, indicating that they experienced low stress and low grain yield loss in the same condition. Similarly, positive SSSI values for 17 genotypes in deep seeding condition indicated that they are prone to stress caused by drought induction, and also suffered greater grain yield loss. Negative SSSI values for 13 genotypes in the deep seeding condition indicate that they experience low stress in drought conditions caused by no irrigation in deep sowing.

No irrigation in the deep sowing condition imposed drought and allowed the genotype to survive in a deeper layer of soil by absorbing deep soil moisture. The survival within a deeper layer of soil must have some genotypic basis for drought tolerance. This stress tolerance is also related to coleoptiles length [11, 19, 29, 32, 37, 45, 46, 47, 48, 61 and 62]. Based on the higher negative value of SSSI and higher yield in deep sowing conditions the genotypes G 16, G 13, G 12, G 24, G 2, G 18, G 19 and G 3 were primarily selected for drought tolerance and will be evaluated further for advanced studies. These genotypes also have longer coleoptiles ranged from 7.4 to 10.5 cm.

4. Conclusion

Considering the overall yield and other characteristics, eight genotypes have been provisionally selected at WRC, Dinajpur for further evaluation for drought tolerance and can be included in crossing blocks. The variability for CL may be partly accounted for by the unexplained variation in the relationship between CL and sowing depth. High gain from selection for increased CL requires that screening conditions are repeatable and that phenotypic differences in CL largely reflect underlying genetic factors. Large variability in seed depth arising from non uniform seed placement demonstrates the need for better sowing equipment if screening directly for CL in the field. Further testing is required to adapt the method for a wider range of crop types and soil conditions and testing for crops grown to maturity.

Authors' Contributions

Md. Farhad was the principle investigator (PI) who planed and set-up the experiment. Md. Abdul Hakim and Dr. Md. Ashraful Alam helped the PI to collect data, analysis the data and manuscript preparation. Dr. N. C. D. Barma had technical contribution by suggesting the plan of experiment and data collection and interpretation.

Acknowledgements

Authors thank to Dr. David Bonnett, Senior Scientist and Leader of Yield Potential Program, CIMMYT, Mexico for delivering the indoor protocol for coleoptile length study. Authors are also thankful to Harriet Benbow, a PhD student at Bristol University for language correction of the manuscript.

References

[1] Ahmad S., S. Yasmin, N. I. Hashmi and A. Qayyum (1988) Influence of seed size and Seeding depth on Performance of tall and semidwarf wheats under limited soil moisture. Pakistan J. Agric. Res. 9(1): 300-304

[2]   Allan R. E., O. A. Vogel and C. J. Peterson. (1962) Seedling emergence rate of fall sown wheat and its association with plant height and coleoptile length. *Agron J.* 54: 347-350. 73: 153–168

[3]   Asif M., M. Y. Mujahid, M. S. Kisana, S. Z. Mustafa and I. Ahmad, (2004) Heritability, genetic variability and path analysis of traits of spring wheat. Sarhad Journal of Agriculture, 20(1):87-91.

[4]   Awan S.I., S. Niaz, M. Faisal, A. Malik and S. Ali (2007). Analysis of variability and relationship among seedling traits and plant height in semi-dwarf wheat (*Triticum aestivum* L.) J. Agri. Soc. Sci. 3(2):59-62 (http://www.fspublishers.org)

[5]   Badaruddin Khokhar, Imtiaz Hussain and Zafar Khokhar (2010) Effect of different irrigation frequiencies on growth and yield of different wheat genotypes in sindh. Pakistan J. Agric. Res. 23(3-4):108-112

[6]   Bergale S., M. Billore, A. S. Halkar, K. N. Ruwali, S. V. S. Prasad and B. Mridulla (2001) Genetic variability, diversity and association of quantitative traits with grain yield in bread wheat. Madras Agril. Journal, 88 (7-9): 457-461.

[7]   Botwright, T. L., G. J. Rebetzke, A. G. Condon and R. A. Richards (2001) Influence of variety, seed position and seed source on screening for coleoptile length in bread wheat (*Triticum aestivum* L.). Euphytica 119: 349–356.

[8]   Brown, P. R., G. R. Singleton, C. R. Tann and I. Mock (2003) Increasing sowing depth to reduce mouse damage to winter crops; Crop Prot 22 : 653–660.

[9]   Ceccarelli, S., M. T. Pegiati and F. Simeoni (1980) Relationship between coleoptile length and culm length in barley; Can J Plant Sci 60 : 687–693.

[10]   Dey, N. C., M. S. Alam, A. K. Sajjan, M. A. Bhuiyan, L. Ghose, Y. Ibaraki and F. Karim, (2011). Assessing environmental and health impact of drought in the northwest Bangladesh, J. Environ. Sci. & Natural Resources, 4(2): 89-97.

[11]   Donald, C. M. and D. W. Puckridge (1975) The ecology of the wheat crop; in wheat and other temperate cereals (eds.) A Lazenby.

[12]   Dwivedi, A. N., I. S. Pawar, and S. Madan (2004) Studies on variability parameters and characters association among yield and quality attributing traits in wheat. J. Crop Res., 32: 77-80.

[13]   Evans, L. E. and G. M. Bhatt. (1977) Influence of seed size, protein content and cultivar on early seedling vigour in wheat. Can. J. Plant Sci., 57: 929-935.

[14]   Fick, G. N. and C. O. Qualset (1976) Seedling emergence, coleoptile length, and plant height relationships in crosses of dwarf and standard-height wheats. Euphytica. 25, 679–684.

[15]   Fischer, R.A. and R. Maurer, (1978) Drought resistance in spring wheat cultivars. I. Grain yield response. Aust. J. Agric. Res., 29: 897–907.

[16]   Gabriela Şerban (2012) Identification of longer coleoptile mutants in an rht-b1b semidwarf wheat population. Romanian. Agril. Res. 29: 17-21.

[17]   Gan, Y., E. H. Stobbe and J. Moes (1992) Relative date of wheat seedling emergence and its impact on grain yield; Crop Sci., 32 : 1275–1281.

[18]   Golabadi, M., A. Arzani, and S. A. M. Mirmohammadi Maibody (2006) Assessment of drought tolerance in segregating populations in durum wheat. Afr J Agric Res., 1:162–171.

[19]   Hadjichristodoulou, A., A. Della and J. Photiades (1977) Effect of sowing depth on plant establishment, tillering capacity and other agronomic characters of cereals; J Agric Sci 89 : 161–167.

[20]   Haque, M. R., M. A. Aziz, M. T. Rahman, B. Ahmed and F. Saberin. (2010) Screening of wheat genotypes for drought tolerance at vegetative stage. J. Agrofor. Environ. 4 (2): 189-192.

[21]   Hong Zhang and Honggang Wang. (2012) Evaluation of drought tolerance from a wheat recombination inbred line population at the early seedling growth stage. African J. Agril Res.7 (46): 6167-6172 (Available online at http://www.academicjournals.org/AJAR).

[22]   IPCC (2013) Working Group I Contribution to the IPCC Fifth Assessment Report Climate Change 2013: The Physical Science Basis, Summary for Policymakers. www.climatechange2013.org/images/uploads/WGIAR5-SPM_Approved27Sep2013.pdf.

[23]   IRRI (2007) CropStat for windows version 7.2, International Rice Research Institute, Metro Manila, Philippines.

[24]   Jennifer Pumpa, P. Martin, F. McRae and Neil Coombes (2013) Coleoptile length of wheat varieties. NSW Department of Primary Industries. Australia.REF: INT 13/1516.

[25]   Kalpana Singh and R. K. Chopra. (2012) Physiology and QTL analysis of coleoptile length, a trait for drought tolerance in wheat. J. Plant Biol. 37 (2): 1-9.

[26]   Kalpana Singh, S. Shukla, S. Kadam, V. K. Semwal, N. K. Singh, R. K. Chopra. (2014) Genomic regions and underlying candidate genes associated with coleoptile length under deep sowing conditions in a wheat RIL population. J. Plant Biochem. Biotechnol. 10: 1-7.

[27]   Kaufmann, M. L. (1968) Coleoptile length and emergence in varieties of barley oats and wheat. Can. J. Plant Sci. 48: 357-361.

[28]   Khayatnezhad, M., R. Gholamin, S. Jamaati-e-Somarin, R. Zaibhi-e-Mahmoodabad (2010) Study of drought tolerance of maize genotypes using the stress tolerance index. Am-Eurasian J Agric Environ Sci 9:359–363.

[29]   Loeppky, H., G. P. Lafond and D. B. Fowler (1989) Seeding depth in relation to plant development, winter survival, and yield of no-till winter wheat. Agron. Journal 81: 125–129.

[30]   Lopez-Castaneda, C. and R. A. Richards (1994) Variation in temperate cereals in rain fed environment: III. Water use and water use efficiency; Field Crops Res. 39: 85–98.

[31]   Mahdi, L., C. J. Bell, J. Ryan (1998) Establishment and yield of wheat (*Triticum turgidum* L.) after early sowing at various depths in a semi-arid Mediterranean environment. Field Crops Res. 58: 187–196. doi: 10.1016/S0378-4290(98)00094-X.

[32]   Matsui, T., S. Inanaga, T. Shimotashiro, P. An and Y. Sugimoto (2002) Morphological characters related to varietal differences in tolerance to deep sowing in wheat. Plant Prod. Sci. 5: 169–174.

[33] Mehmet Yagmur and Digdem Kaydan (2009) The effects of different sowing depth on grain yield and some grain yield components in wheat (*Triticum aestivum* L.) cultivars under dryland conditions. African J. Biotechnol.8 (2): 196-201

[34] Mohan, A., W. F. Schillinger, K. S. Gill (2013) Wheat seedling emergence from deep planting depths and its relationship with coleoptile length. PLoS ONE 8(9): e73314. doi:10.1371/journal.pone.0073314.

[35] Monneveux, P., and S. C. Jing Rand Misra (2012) Phenotyping for drought adaptation in whea tusing physiological traits. Front. Physio. 3:429.doi: 10.3389/fphys.2012.00429.

[36] Multibase_2015: Excel Add-ins for PCA and PLS, (2014) NumericalDynamics.com/DownLoad2.html.

[37] Murray, G. M. and J. Kuiper (1988) Emergence of wheat may bereduced by seed weather damage and azole fungicide and is related to coleoptile length; Aust. J. Exp. Agric. 28 : 253–261.

[38] Nebreda, I. M. and P. C. Parodi (1977) Effect of seed type on coleoptile length and weight in triticale, *X Triticosecale Wittmack*. Cereal Res. Communications 5: 387–398.

[39] Nouri, A., A. Etminan, A. Jaime, T. da Silva and R. Mohammadi (2011) Assessment of yield, yield-related traits and drought tolerance of durum wheat genotypes (*Triticum turjidum* var. *durum Desf.*). Aust. J. Crop Sci. 5:8–16.

[40] O'Sullivan, P. A., G. M. Weiss and D. Friesen (1985) Tolerance of spring wheat (*Triticum aestivum* L.) to trifluralin deep-incorporated in the autumn or spring; Weed Res 25 : 275–280.

[41] Onofri A. (2007) Routine statistical analyses of field experiments by using an Excel extension. Proceedings 6th National Conference Italian Biometric Society: "La statistica nelle scienze della vita e dell'ambiente", Pisa, 20-22 June 2007, 93-96.

[42] Ouk, M., J. Basnayake, M. Tsubo, S. Fukai, K. Fischer, M. Cooper and H. Nesbitt (2006) Use of drought response index for identification of drought tolerant genotypes in rainfed lowland rice. Field Crops Res. 99:48–58.

[43] Pantuwan, G., S. Fukai, M. Cooper, S. Rajatasereekul and J. C. O'Toole (2002) Yield response of rice (*Oryza sativa* L.) genotypes to different types of drought under rainfed lowlands. I. Grain yield and yield components. Field Crops Res.

[44] Radford, B. J. (1987) Effect of constant and fluctuating temperature regimes and seed source on the coleoptile length of tall and semi dwarf wheats. Aust. J. Exp. Agric. 27: 113-117.

[45] Rebetzke, G. J., S. E. Bruce and J. A. Kirkegarrd (2005) Longer coleoptiles improve emergence through crop residues to increase seedling number and biomass in wheat; Plant Soil 272 : 87–100.

[46] Rebetzke, G. J., M. H. Ellis, D. G. Bonnett and R. A. Richards (2007a) Molecular mapping of genes for coleoptile growth in bread wheat (*Triticum aestivum* L.); Theor Appl Genet 114 : 1173–1183.

[47] Rebetzke, G. J., R.. A. Richards, N. A. Fettell, M. Long, A. G. Condon, R. I. Forrester and T. L. Botwright. (2007b)

[48] Rebetzke, G. J., R. A. Richards, V. M. Ficher and B. J. Mickelson. (1999) Breeding long coleoptile, reduced height wheats. Euphytica 106: 159-168.

[49] Reithmuller, G. P. (1990) Machinery for improved crop establishment in Western Australia. In 'Agricultural Engineering Conference1990' pp. 40–45. (Institute of Engineers: Australia).

[50] Schillinger, W. F., E. Donaldson, R. E. Allan and S. S. Jones (1998) Winter wheat seedling emergence from deep sowing depth; Agron J 90 : 582–586.

[51] Schneider, K. A., R. Rosales-Serna, F. Ibarra-Perez, B. Cazares-Enriquez, J. A. Acosta-Gallegos, P. Ramirez-Vallejo, N. Wassini and J. D. Kelly (1997) Improving common bean performance under drought stress. Crop Sci 37:43–50.

[52] Singh, B. U., K. V. Rao and H. C. Sharma (2011) Comparison of selection indices to identify sorghum genotypes resistant to the spotted stemborer *Chilo partellus* (Lepidoptera: Noctuidae). Int. J. Trop. Insect Sci. 31:38–51.

[53] Sio-Se Mardeh, A., A. Ahmadi, K. Poustini and V. Mohammadi (2006) Evaluation of drought resistance indices under various environmental conditions. Field Crops Res. 98:222–229.

[54] Slafer, G. A., J. L. Araus, C. Royo and L. F. G. DelMoral (2005) Promising eco-physiological traits for genetic improvement of cereal yields in Mediterranean environments; Ann. App. Biol. 146 : 61–70.

[55] Stocktom, R. D., E. G. Krenzer, J. Solie and M. E. Payton (1996) Stand establishment of winter wheat in Oklahoma; A survey; J Prod Agric 9: 571–575.

[56] Stoddard, F. L. (1999). Variation in grain mass, grain nitrogen and starch B-granule content within wheat heads. Cereals Chem. 76: 139-144.

[57] Talebi, R., F. Fayazl and A. M. Naji (2009) Effective selection criteria for assessing drought stress tolerance in durum wheat (*Triticum durum* Desf.). Gen. Appl. Plant Physiol. 35:64–74.

[58] Trethowan, R., R. Singh, J. Huerta-Espino, Crossa, and M. Van Ginkel (2001) Coleoptile length variation of near-isogenic Rht lines of modern CIMMYT bread and durum wheats. Field Crops Res. 70: 167–176.

[59] Wajid, A., A. Hussain, M. M. Ahmed and M. Waris (2002) Influence of sowing date and irrigation levels on growth and grain yield of wheat. Pak. J. Agric. Sci. 39 (1): 22-24.

[60] Wang, J., S. C. Chapman, D. G. Bonnett and G. J. Rebetzke (2009) Simultaneous selection of major and minor genes: use of QTL to increase selection efficiency of coleoptile length of wheat (*Triticum aestivum* L.). Theoretical and Applied Genetics 119: 65–74. doi:10.1007/s00122–009–1017–2.

[61] Whan, B. R. (1976a) The emergence of semidwarf and standard wheats, and its association with coleoptile length. Aust. J Exp. Agril. Animal Husb. 16: 411–416.

[62] Whan, B. R. (1976b) The association between coleoptile length and culm length in semidwarf and standard wheats; J. Aust .Inst. Agri. Sci. 42 : 194–196.

Genotypic increases in coleoptile length improves stand establishment, vigour and grain yield of deep-sown wheat; Field Crops Res. 1: 10-23

[63] WRC Annual Report 2011-2012 (2012), Wheat Research Centre, Bangladesh Agricultural Research Institute (BARI), Nashipur, Dinajpur, Bangladesh.

[64] Yashu, T. and K. Fujii (1979) Studies on the elongation of mesocotyle and coleoptile length in gramineouss crops. I. On the elongation ratio of mesocotyle and coleoptile; Jpn J. Crop Sci. 48: 356–364.

2

Assessment of the profitability and the effects of three maize-based cropping systems on soil health in Western Africa

Kodjovi Sotomè Detchinli, Jean Mianikpo Sogbedji

Ecole Supérieure d'Agronomie, Université de Lomé, Lomé, Togo

Email address:

skdetch@gmail.com (K. S. Detchinli), mianikpo@yahoo.com (J. M. Sogbedji)

Abstract: Enhanced livelihoods for populations, especially smallholder farmers in sub-Saharan Africa may be achieved through improved cropping systems. We assessed the economic returns from maize grain yield and the effects of three cropping systems on soil properties in an eight-year study segmented in cycles of two years each: continuous maize (*Zea mays* L.), maize-mucuna (*Mucuna pruriens* var. utilis), and maize-pigeon pea (*Cajanus cajan*). The rainfall pattern in the study region allows for two growing seasons per year, leading to four growing seasons per cycle. Nitrogen (N) and phosphorus (P) fertilizer rates were imposed on maize in each system and maize grain yields and associated cash values as well as soil properties were measured. Seeding mucuna and pigeon pea crops into maize crop in the first year did not result in maize grain yield increases from N and P fertilizers in the subsequent year. Continuous maize system increased mean maize grain yields by 6.2 to 60.3% in the fallow year of the 2002-2003 and 2006-2007 cycles and by 5.1 to 8.2% on a cycle basis in the 2002-2003 cycles. For the remaining periods of the study, mucuna and pigeon pea based maize cropping increased grain yields by 28.6 to 47.6%, 22 to 260% and 28.3 to 136.1% in fallow year, non-fallow years and on a cycle basis, respectively, compared to yields under continuous maize. On a cycle basis, economic returns for maize-mucuna and maize-pigeon pea based systems were 105.1 and 66.5%, respectively, higher than that for continuous maize. The mucuna and pigeon pea based systems increased the initial soil total carbon (C) content by 55 and 69%, respectively, resulted in increases of 110 to117%, 33 to 63%, 29%, and 16-17% for exchangeable Ca^{2+}, Mg^{2+}, K^+ and total cation exchange capacity (CEC), respectively, and enhanced water stable macroaggregates stability, compared to continuous maize. Maize mucuna and pigeon pea-based maize cropping systems with mucuna and pigeon crops in alternate years should be advised towards sustaining enhanced profitability and improved soil physical and chemical properties.

Keywords: Maize, Mucuna, Pigeon Pea, Fertilizer, Soil Properties, Profitability

1. Introduction

There is still an increasingly growing concern about the issue of food shortages in Africa which has become a major obstacle to the development of the continent, especially the sub-Saharan region. During the last three decades the region has experienced a population growth of 3.1% against a 2.1% food production growth rate [1]. Therefore, a major challenge for scientists, governments and other stakeholders in the region is that food production should increase by 70% by 2050 to meet the necessary caloric requirements [2]. The agricultural intensification is recognized as the main opportunity to meet rising food needs [3]. In Sub-Saharan

Africa (SSA), smallholder farmers have experienced declining yields, increasing costs of production and growing uncertainty of producing the food needed by their families. Major factors contributing to such uncertainty and decline in productivity are: soil degradation, dry spells, erratic availability of inputs particularly mineral fertilizers, inefficient use of soil and water resources and high cost for soil fertility improvement [4]. In addition, compounded factors, such as poor access to financing, innovation and markets, have caused soil mining. This situation is affecting the livelihood of smallholder farmers in SSA. Efforts towards improving agricultural productions to enhancing food security in the region should address major constraints with focus on reversing nutrient depletion from soils, mitigating the effect of drought spells and erosion, increasing nutrient and water use efficiency and adaptation of improved crop varieties. These constraints contribute to the fact that SSA is the only continent that has grown poorer in the past 35 years [5] and may be expected to remain primary concerns during the coming decades with increasingly negative consequences, unless technological, economical and socio-political measures are taken to curtail further soil degradation and to accelerate agricultural growth.

It is well established that soil fertility depletion in smallholder farms is the fundamental biophysical cause for declining per capita food production in SSA [4, 6]. There is ample evidence that the most significant biophysical constraint to increased production of both crops and livestock in SSA is the poor mineral and organic content of the soils. This constraint leads to inadequate availability of assimilated energy, protein and phosphorus for livestock production and not enough nitrogen, phosphorus and organic matter for crop production [7]. Hence, there is no way out of the poverty cycle for SSA farmers unless strong emphasis is placed on reversing nutrient depletion and increasing nutrient and water use efficiency for each particular farming system.

The use of low external input sustainable agriculture (LEISA), promoted by many donors and NGOs, presumes that organic resources are efficient in sustaining production and the natural resource base. In most cases, however, the use of organic inputs such as manure and composting is part of an internal flow of nutrients within the farm and, therefore, does not add nutrients to infertile soils. Their production is further constrained by the same limitation as food crops (poor soils and limited water). Also, the low availability of manure in Africa is inadequate to meet nutrient demand over a large area. Moreover, the low nutrient content and high labour demands for processing and application are negative factors limiting organic matter-based soil management. Several studies in West Arica [4, 8-10] have reported that cropping systems involving legume crops or short duration planted tree fallow as a means of organic matter input improved soil fertility and maize yields. However, such cropping systems result in a land use based competition between the cereal and legume crops leading in some cases to a complete loss of the cereal cropping season. Furthermore, questions remain about the potential of the organic matter technology alone to sustain high maize yields [10, 11].

Several other studies [8, 12, 13] concluded that the combined application of mineral and organic fertilizers, together with methods to conserve organic matter may be the most promising strategies for improving soil fertility and sustaining maize yields. The sustainability of a cropping system is primarily a function of both crop yield expressed in terms of economic returns and the associated soil health status. A quantitative characterization of complex cropping systems that include organic inputs in terms of profitability and soil health status is poorly established in the West Africa

sub-region.

The objectives of this research were 1) to quantitatively evaluate three cropping systems including various organic and inorganic nutrient inputs with regard to maize grain yields and associated economic returns and 2) to determine and compare the soil health status under the three systems. The ultimate aim was to identify appropriate cropping systems that sustain maize production and mitigate the degradation of the resource base in coastal West Africa.

2. Materials and Methods

2.1. Experimental Site

The study was conducted at the University of Lomé Research Station near Lomé, Togo (6°22'N, 1°13'E; altitude = 50 m). The soil type was a rhodic Ferralsol locally called "Terres de Barre" that developed from a continental deposit [14]. This soil type covers part of the arable lands in Togo, Bénin, Ghana, and Nigeria [15] and is commonly used for maize production in coastal Western Africa. It is a well-drained soil, very low in organic matter (< 10 g kg^{-1}) and K (< 0.2 meq 100g^{-1}), and has total P contents ranging from 250 to 300 mg kg^{-1}, cation exchange capacity of 3 to 4 ceq kg^{-1}, and pH of 5.2 to 6.8 [15, 16]. Sand content is approximately 80% at the 0 to 0.20 m depth, and decreases to less than 60% at the 0.50 to 1.20 m depth [17]. The experimental site has a slope of less than 1%. Annual precipitation typically ranges from 800 to 1100 mm and allows for two maize growing seasons, one from April to July and another from September to December. At the onset of this experiment, the site, which has usually been used by farmers for unfertilized continuous maize cropping, was under a 1-year grass fallow.

2.2. Crop and Soil Management

An eight-year period (2002-2009) split-plot experiment was established with three replicates (Fig. 1). The eight-year period was segmented in 4 cycles of 2 years (2002-2003, 2004-2005, 2006-2007, and 2008-2009) with 4 growing seasons per cycle. Three cropping systems were the main plot effects and four fertilizer levels were at the subplot level.

The site was manually plowed and 12 main plots (16 x 16 m) and 48 subplots (8 x 8 m) were laid out in a spatially-balanced complete block design [18]. Spatially-balanced complete block (SBCB) designs are a model-based approach that guarantees that the experiment is insensitive to trends, spatial correlation, or periodicity in the research domain [19]. It aims to equalize variances among treatment contrasts and allows for conventional statistical analysis methods. The cropping system scenarios include: (i) maize monoculture for the four growing seasons (MaMaMaMa) of each cycle, (ii) relay (interseeding) of a mucuna crop into the first maize crop so that it grew from June to December for the first year; in the second year, both the first and the second seasons were grown to maize (MaMuMaMa and (iii) relay of a pigeon pea crop into the first maize crop so that it grew from June to April for the first year; in the second year, both the first and the second seasons were grown to maize (MaPpMaMa). The maize cowpea-based cropping system (Fig. 1) is not discussed in this paper because cowpea growth was hampered by pests during the period of study.

Fertilizer treatments were applied to subplots only when maize was grown in all three cropping systems. Four subplots were treated with combinations of three levels of N (0, 40, and 80 kg ha^{-1}) and two levels of P (0 and 30 kg ha^{-1}): N_0-P_0, N_{40}-P_0, N_{40}-P_{30}, and N_{80}-$P_{30.}$ All maize plots were fertilized with 60 kg K ha^{-1}. Fertilizer P and K rates were manually broadcast as P_2O_5 and K_2O, respectively, at maize planting while N rates were manually point-placed as urea three

weeks after planting at approximately 8 cm depth. Maize (IKENNE, the most commonly used improved variety) was planted in April and harvested in July during the first growing season, and was planted in September and harvested in December during the second season at a density of 50,000 plants ha^{-1}. The crop was manually weeded three times during each growing season. Pigeon pea and mucuna were planted at a density of 42,000 and 35,000 plants ha^{-1}, respectively. Crop residues from pigeon pea (after grain harvesting) and mucuna fallow (after seed harvesting) were incorporated into the soil during land preparation for the subsequent maize crop.

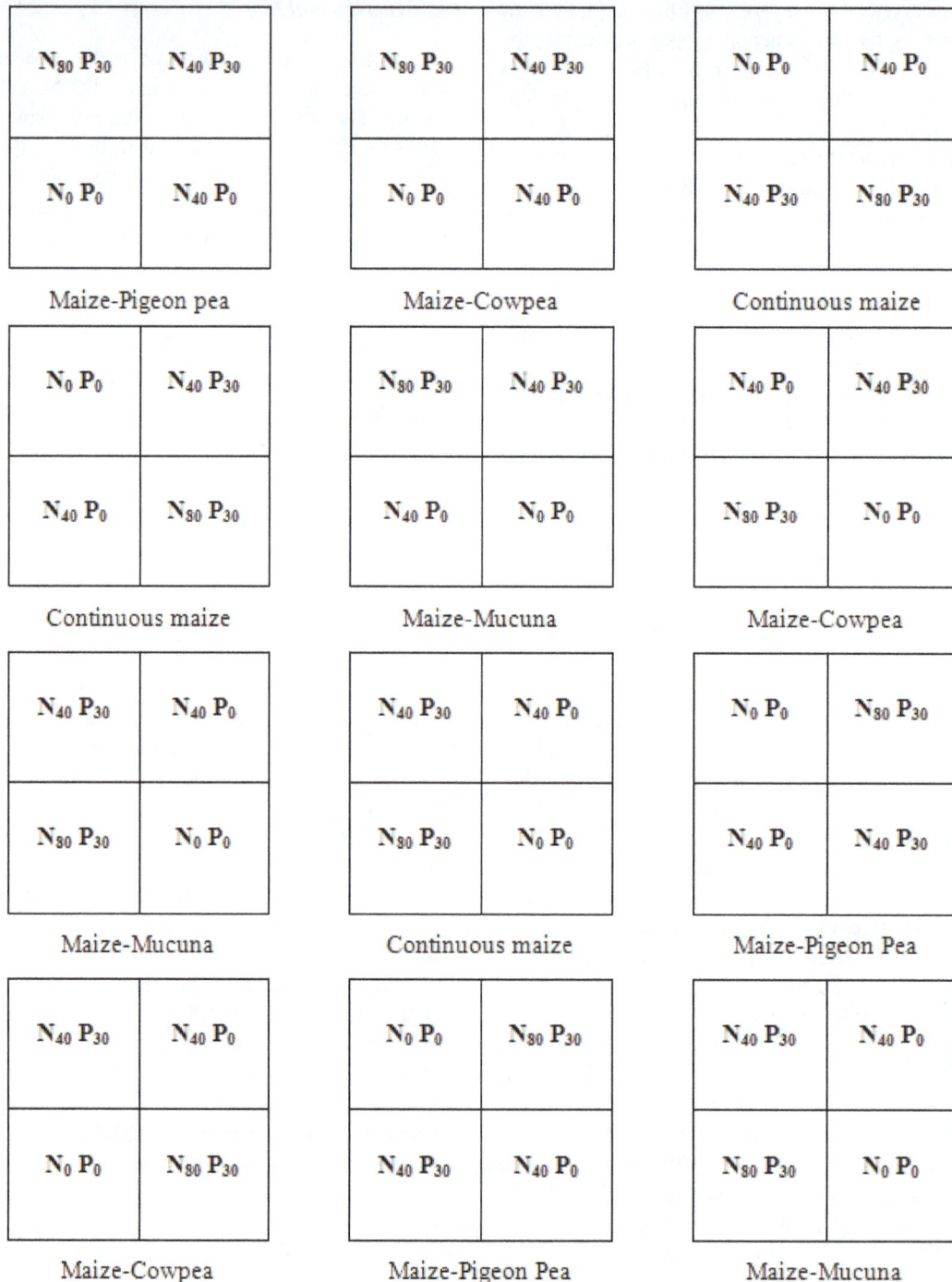

Figure 1. Plot layout and experimental design.

2.3. Data Collection

At the onset of the experiment in 2002 (at maize planting in April), initial soil properties including total C and N contents, exchangeable bases (Ca++, Mg++, Na+ and K+), pH and total cation exchange capacity (CEC) were measured for the first 20 cm soil layer (0-20 cm depth) on the experiment site from twenty four composite soil samples using the standard methods of the International Institute for Tropical Agriculture [20]. At the end of the experiment in 2009 (at maize harvest in December) the same soil properties were measured on each main plot from twelve composite soil samples as described above. In addition, at the end of the experiment the water-stable aggregates (WSA) for the 0-10 cm soil depth from twelve composite soil samples was measured on each main plot. In preparation for the WSA measurement, soil samples were crushed by hand and passed through 2000, 500, 250 and 50 μm sieve meshes. The coarse fraction and plant residues that remained on the 2000 μm sieve were discarded along with the fraction that passed through the 50 μm sieve. Three fractions of soil aggregate sizes remained: the 500–2000 μm fraction, referred to as macroaggregates, the 250–500 μm fraction, referred to as mesoaggregates and the 50–250 μm fraction, referred to as microaggregates. Samples were moistened with distilled water using a fine sprayer. A wet sieving apparatus (Eijkelkamp Giesbeek, the Netherlands) was used to determine the aggregate stability following the procedure described by [21]. Wet sieving was carried out by placing the pre-wetted soil on 500 μm mesh size for the macroaggregates, 250 μm mesh size for the mesoaggregates and on 50 μm mesh size for the microaggregates. The sieving times were fixed at 5, 15, 30, 60, 120 and 240 min, except that the 5 min period was not used for the microaggregates. The aggregate stability was expressed as the percentage of sand-free aggregates retained on the sieve after sieving, with the initial sample also being corrected for sand content [22]. Analysis of variance (ANOVA) was performed on the gathered soil chemical and physical data sets using the MSTAT-C software, and the Student Newman-Keuls test was used to discriminate among cropping systems.

Maize grain yield was determined under each cropping system scenario from four 6-m long rows of maize from the center of each subplot that were harvested and adjusted to 14% moisture content. Due to management problems no data were collected for 2004-2005 cycle. Maize grain yield data were analyzed using the general linear mixed model with rep and rep*cropping system as random, and fertilizer level and cropping system as fixed effects. Significant effects were followed by multiple comparisons adjusted with a Bonferoni correction. The MIXED procedure in Statistical Analytical System [23] was used to run the analysis.

2.4. Economic Analysis

The profitability of the MaMaMaMa, MaMuMaMa and MaPpMaMa treatments was estimated through a partial budget analysis. Output consisted of the amount of cash corresponding to the maize mean grain yield for the three cycles, which was assumed to be sold at 160 F CFA (US$0.32) kg^{-1}, the average sale price in the country. For continuous maize cropping (MaMaMaMa), grain yield under the $N_{80}P_{30}$ fertilization was used, and average yield values for the four mineral fertilization treatments were used for the cropping systems involving mucuna and pigeon pea (MaMuMaMa and MaPpMaMa). The inputs consisted of the costs associated with each cropping system, including those for soil preparation, seed, crop planting and related tasks, fertilizer purchase and application, crop weeding and crop harvesting and associated tasks. Mucuna and pigeon pea grain yield sale values and harvesting costs were not included in the budget because mucuna grain is a non-food product and has no sale value as seed at the farmers' level in the country. Pigeon pea grain is used as food mainly in rural areas, but its sale value is not well established. No weeding costs were associated with the mucuna and pigeon pea crops as they were relayed into maize crops and because of their competitive growth and ability to provide soil cover. Labor costs were determined to be 1500 FCFA (US$3.0) per person day, and fertilizer costs were based on prices used by the Direction Régionale de l'Agriculture, de l'Elevage et de la Pêche (DRAEP) (pers. comm.) Estimates of labor for maize, mucuna and pigeon pea crops in a growing season as defined in the MaMaMaMa, MaMuMaMa and MaPpMaMa systems are presented in Table 1, and are based on labor records from the experiment.

Table 1. Estimated labor associated with a season of maize, mucuna and pigeon crop under a cycle of continuous maize, maize mucuna-based, and maize pigeon pea-based cropping systems.

	MaMaMaMa	MaMuMaMa	MaPpMaMa
	person day ha^{-1}		
Soil preparation	30	0	0
Planting and related tasks	35	12	12
Weeding	90	0	0
Fertilizer application	20	0	0
Harvesting and related tasks	70	0	0
Total labor	245	12	12
Total labor cost¶ (F CFA§)	367500	18000	18000

3. Results and Discussion

3.1. Maize Grain Production

Maize grain yield was not responsive to cropping system and fertilization pattern in the first year of the study (Table 2).

Table 2. Mean maize grain yields (Mg ha^{-1}) for each growing season, year and the 2-years cycle period.

Cropping systems	Year 1			Year 2			Year 1 + Year 2
	GS^{T}1	GS2	Total	GS1	GS2	Total	Total
Cycle 2002-2003							
MaMaMaMa$^{\pm}$							
N_0P_0	6.1	3.7a	9.8a	4.5a	2.5a	7.0a	16.8a
$N_{40}P_0$	6.3	3.8a	10.1a	5.7b	2.7a	8.4a	18.5ab
$N_{40}P_{30}$	6.3	3.9a	10.2a	5.6b	2.8a	8.4a	18.6ab
$N_{80}P_{30}$	6.5	4.0a	10.5a	5.9b	3.7b	9.6b	20.1b
Mean	6.3	3.8	10.1	5.4	2.9	8.3	18.5
MaMuMaMa							
N_0P_0	6.1	§	6.1b	6.9b	4.4b	11.3b	17.5ab
$N_{40}P_0$	6.3	§	6.3b	6.8b	4.6b	11.4b	17.9ab
$N_{40}P_{30}$	6.3	§	6.3b	7.0b	4.3b	11.3b	17.5ab
$N_{80}P_{30}$	6.5	§	6.5b	7.0b	4.4b	11.4b	17.9ab
Mean	6.3		6.3	6.9	4.4	11.3	17.6
MaPpMaMu							
N_0P_0	6.1	§	6.1b	6.6b	3.8b	10.4b	16.7a
$N_{40}P_0$	6.3	§	6.3b	6.7b	4.1b	10.8b	17.2a
$N_{40}P_{30}$	6.3	§	6.3b	6.6b	3.8b	10.4b	16.7a
$N_{80}P_{30}$	6.5	§	6.5b	6.7b	4.2b	10.9b	17.1a
Mean	6.3		6.3	6.6	4.0	10.6	17.1
Cycle 2006-2007							
MaMaMaMa							
N_0P_0	2.9a	1.9b	4.8c	2.2c	1.4c	3.6c	8.4c
$N_{40}P_0$	4.6b	1.8b	6.4b	3.4a	1.8c	5.2d	11.6d
$N_{40}P_{30}$	5.0b	2.1b	7.1d	3.8a	1.7c	5.5d	12.6d
$N_{80}P_{30}$	6.2	2.6c	8.8a	4.0a	2.5a	6.5d	15.3e
Mean	4.7	2.1	6.8	3.4	1.9	5.2	12.0
MaMuMaMa							
N_0P_0	6.3	§	6.3b	6.6b	4.0b	10.6b	16.9a
$N_{40}P_0$	6.2	§	6.2b	6.8b	4.2b	11.0b	17.2a
$N_{40}P_{30}$	6.5	§	6.5b	7.0b	4.2b	11.2b	17.7ab
$N_{80}P_{30}$	6.6	§	6.6b	6.9b	4.4b	11.3b	17.9ab
Mean	6.4		6.4	6.8	4.2	11.0	17.4
MaPpMaMa							
N_0P_0	5.4	§	5.4b	5.6b	3.3b	8.9a	14.3e
$N_{40}P_0$	5.5	§	5.5b	5.6b	3.7b	9.3a	14.8e
$N_{40}P_{30}$	6.0	§	6.0b	6.3b	3.7b	10.0b	16.0a
$N_{80}P_{30}$	6.2	§	6.2b	6.4b	3.9b	10.3b	16.5a
Mean	5.8		5.8	6.0	3.7	9.6	15.4
Cycle 2008-2009							
MaMaMaMa							
N_0P_0	1.8c	1.0d	2.8e	1.2d	0.8d	2.0e	4.8f
$N_{40}P_0$	2.8a	1.2d	4.0c	1.5c	0.9d	2.5e	6.5g
$N_{40}P_{30}$	3.0a	1.1d	4.1c	1.7c	0.9d	2.6e	6.7g
$N_{80}P_{30}$	4.2b	1.8b	6.0b	3.2a	1.6c	4.8d	10.8h
Mean	3.0	1.3	4.2	1.9	1.1	3.0	7.2
MaMuMaMa							
N_0P_0	6.1	§	6.1b	6.4b	4.2b	10.6b	16.7a
$N_{40}P_0$	6.2	§	6.2b	6.3b	4.3b	10.6b	16.8a
$N_{40}P_{30}$	6.3	§	6.3b	6.8b	4.0b	10.8b	17.1a
$N_{80}P_{30}$	6.2	§	6.2b	6.8b	4.5b	11.3b	17.5a
Mean	6.2		6.2	6.6	4.3	10.8	17.0
MaPpMaMa							
N_0P_0	4.8b	§	4.8c	5.2b	3.0b	8.2a	13.0e
$N_{40}P_0$	5.0b	§	5.0c	5.2b	3.5b	8.7a	13.7e
$N_{40}P_{30}$	5.7	§	5.7b	5.9b	3.3b	9.2a	14.9e
$N_{80}P_{30}$	6.2	§	6.2b	6.2b	3.7b	9.9b	16.1a
Mean	5.4		5.4	5.6	3.4	9.0	14.4

Grain yield from all cropping system scenarios ranged from 6.1 to 6.5 and 3.7 to 4.0 Mg ha^{-1} during the first and the second growing seasons, respectively. The yield depression in the second growing season as compared with the first growing season, which was also observed during the whole period of the study, presumably resulted from lower rainfall (154.1 mm) compared with the first growing season (529.6 mm), similar to previous research [24]. The limited yield response to N and P occurred primarily as a result of the high initial soil NO_3-N content (46.1 kg ha^{-1}) and labile P content (368.9 kg ha^{-1}). In addition, the lack of yield response suggests that mucuna and pigeon pea crops that were relayed 50 to 60 days after maize planting did not significantly reduce maize nutrient use and growth. Reference [25] found that relay of mucuna into maize 30 days after maize planting resulted in maize yield depression due to competition, and suggested a longer time period between the planting times of the two crops.

In the second year, the effects of fertilizer and cropping system and their interaction were significant. During the first growing season under continuous maize (MaMaMaMa), grain yield was significantly lower under N_0P_0 fertilization compared with those for others ($N_{40}P_0$, $N_{40}P_{30}$ and $N_{80}P_{30}$, Table 2). The lack of response to P fertilization and the interaction between N and P presumably resulted from the high (368.9 kg P ha^{-1}) April 2002 soil P content. Except for the N_0P_0 fertilization level under MaMaMaMa, grain yield was similar for all fertilization levels under the three cropping systems (Table 2). This demonstrates that the interaction of fertilizer rate*cropping system was significant and that nutrient restitution to soil through incorporation of the cover crops prevented the need for additional fertilizer. During the second growing season of the second year, grain yields for the highest fertilization level ($N_{80}P_{30}$) under MaMaMaMa and all fertilization levels under MaMuMaMa and MaPpMaMa were similar (3.7 to 4.6 Mg ha^{-1}), but higher than the three other fertilization levels (N_0P_0, $N_{40}P_0$ and $N_{40}P_{30}$, 2.5 to 2.8 Mg ha^{-1}) under MaMaMaMa. This, again, indicates that the effects of fertilization level on grain yield varied with cropping system. In each of the two growing seasons of the second year of the study, maize grain yields were similar or slightly higher for MaMuMaMa and MaPpMaMa and lower for MaMaMaMa compared to those in the corresponding seasons of the first year (Table 2). These results indicate that MaMuMaMa and MaPpMaMa sustained higher maize yields at minimal mineral fertilizer rates.

In the first year of the study, two-season cumulative grain yields for MaMaMaMa were higher (9.8 to 10.5 Mg ha^{-1}, Table 2) than those for MaMuMaMa and MaPpMaMa (6.2 to 6.5 Mg ha^{-1}) because the latter did not allow for a second maize crop. In the second year, however, yearly cumulative grain yields were higher (10.4 to 11.4 Mg ha^{-1}) for MaMuMaMa and MaPpMaMa than those for MaMaMaMa (7.0 to 9.6 Mg ha^{-1}). On a cycle basis (2-years cumulative value) grain yield data showed that the highest fertilization level ($N_{80}P_{30}$) under MaMaMaMa resulted in higher yield

(20.1 Mg ha^{-1}) than the N_0P_0 (16.8 Mg ha^{-1}) and all fertilization levels under MaPpMaMa (16.7 to 17.2 Mg ha^{-1}, Table 2). Except for the $N_{80}P_{30}$ under MaMaMaMa, all fertilization levels under MaMaMaMa, MaMuMaMa and MaPpMaMa provided similar cycle-based grain yields (16.7 to 18.6 Mg ha^{-1}). Only significant additional fertilizer allowed for higher yields (20.1 Mg ha^{-1} under $N_{80}P_{30}$) for MaMaMaMa. On average (mean value for all fertilization levels), annual maize grain yield in the fallow year increased by 60.3% under MaMaMaMa as compared with yields under MaMuMaMa and MaPpMaMa, but in the non-fallow year yield increased by 28 and 22% under MaMuMaMa and MaPpMaMa, respectively, as compared with yield under MaMaMaMa. On a cycle basis, mean yield value was 5.1 and 8.2% higher than those for MaMuMaMa and MaPpMaMa, respectively, indicating that in short term continuous maize cropping proved superiority over maize-cover cropping based systems.

During the first year of the 2006-2007 cycles, maize grain yields were lowest, intermediate and highest for the N_0P_0, $N_{40}P_0$ and $N_{40}P_{30}$, and $N_{80}P_{30}$, respectively, for the MaMaMaMa system (Table 2), indicating that the soil fertility has decreased and N and P effects were measurable. However, the fertilization level did not affect grain yields under the MaMuMaMa and MaPpMaMa systems which were similar to the yield for the highest fertilization rate for the continuous maize system. Unlike the 2002-2003 cycle where the first year based cumulative yields for all fertilization levels under the continuous maize system were systematically higher than those under the MaMuMaMa and MaPpMaMa systems, yearly cumulative yields were lowest and highest under the N_0P_0 and $N_{80}P_{30}$ fertilization levels for MaMaMaMa and intermediate under all levels for the mucuna and pigeon pea based systems (Table 2). This indicates that even with the loss of the second growing season the latter systems challenged the continuous maize system. During the second year of the cycle, seasonal and annual grain yields were in general similar for all fertilization levels under MaMuMaMa and MaPpMaMa systems, but systematically higher than those for all fertilization levels under the MaMaMaMa system. This suggests that continuous cultivation contributed yield depression even at a high mineral fertilization level. Annual mean (average value for all fertilization levels) maize grain yield in the fallow year increased by 6.2 and 17.2% under MaMaMaMa as compared with yields under MaMuMaMa and MaPpMaMa, respectively, but in the non-fallow year yields were 111.5 and 84.6% higher under MaMuMaMa and MaPpMaMa, respectively, than mean yield under MaMaMaMa. On a cycle basis, mean yield values increased by 45 and 28.3% under MaMuMaMa and MaPpMaMa, respectively, as compared with value under MaMaMaMa.

The yield results for the first year of the 2008-2009 cycle followed similar trends as those for the second year of the 2006-207 cycle (Table 2) as described above. But during the second year of the 2008-2009 cycles, yield depression was very accentuated leading to seasonal and annual values ranging from 0.8 to 4.8 and from 3.0 to 11.3 Mg ha^{-1} for MaMaMaMa and, MaMuMaMa and MaPpMaMa, respectively. Annual mean (average value for all fertilization levels) maize grain yield in the fallow year increased by 47.6 and 28.6% under MaMuMaMa and MaPpMaMa, respectively, as compared with yield under MaMaMaMa, and in the non-fallow year yields were 260 and 200% higher under MaMuMaMa and MaPpMaMa, respectively, than mean yield under MaMaMaMa. On a cycle basis, mean yield values increased by 136.1 and 100% under MaMuMaMa and MaPpMaMa, respectively, as compared with value under MaMaMaMa.

Annual mean maize grain yield results from this study (except the first year of the experiment) agreed with those of [26, 27] in that a mucuna cover crop may allow for similar or higher yearly maize grain yields even if it causes the loss of the second maize crop of the year. Such a yield increase in the fallow year occurred during the 2008-2009 cycle of this study at a magnitude of 47.6 and 28.6% under mucuna and pigeon pea fallow, respectively. The magnitude of the mean yield increase under MaMuMaMa and MaPpMaMa in the non fallow year and on a cycle basis ranged from 27.7 to 260%, which corroborate reasonably well values ranging

from 24 to 220% published by [28, 29].

3.2. Partial Budget Analysis

Results of the budget of inputs (total costs associated with MaMaMaMa, MaMuMaMa and MaPpMaMa) and corresponding outputs (cash values of maize grain yield for the four growing seasons) are presented in Table 3.

The outputs from MaMuMaMa (2,768,000 FCFA) and MaPpMaMa (2,496,000 FCFA) were 12.3 and 1.3% higher, respectively, than the 2,464,000 FCFA output from MaMaMaMa with high fertilization level ($N_{80}P_{30}$). However, the input associated with MaMaMaMa was 28.9 and 30.1% higher than those for MaMuMaMa and MaPpMaMa, respectively. The balance was positive in all cases, but was on a per hectare basis 105.1% (1,377,871 FCFA = US$2,756) and 66.5% (1,118,871 F CFA = US$2,238) higher for MaMuMaMa and MaPpMaMa, respectively, compared to that (671,868 FCFA = US$1,344) of MaMaMaMa with $N_{80}P_{30}$ mineral fertilization (Table 3). The cash value superiority of MaMuMaMa and MaPpMaMa over MaMaMaMa may be accentuated if other benefits such as mucuna and pigeon pea grain values are accounted for.

Table 3. Partial budget analysis for continuous maize, maize mucuna-based and maize pigeon pea-based cropping systems.

	MaMaMaMa	MaMuMaMa	MaPpMaMa
		F CFA ha^{-1}	
Output (Maize grain value)	+2,464,000	2,768,000	2,496,000
Input (labor +seeds + fertilizer)	-1,792,192	1,390,129	1,377,129
Labor	(1,470,000)	(1,120,500)	(1,120,500)
Seeds	(76,000)	(85,000)	(72,000)
Fertilizer	(246,172)	(184,629)	(184,629)
Balance	+ 671,828 (US$1,344)	+ 1,377,871 (US$2,756)	+ 1,118,871 (US$2,238)

3.3. Soil Physical and Chemical Properties

Soil pH and stored total N in the soil were not responsive

to cropping system (Table 4).

Table 4. Soil properties at the onset (2002) and at the end (2009) of the experiment.

Soil Properties	Year 2002	Year 2009		
		MaMaMaMa	MaMuMaMa	MaPpMaMa
Chemical Properties				
pH (H$_2$O)	7.22	7.19	7.35	7.10
Total C (%)	0.71a	0.83a	1.10b	1.20b
Total N (%)	0.06	0.08	0.11	0.09
Exchangeable bases (cmol kg^{-1})				
Ca++	30.75a	38.37a	64.75b	66.63b
Mg++	7.75a	7.12a	10.44b	12.62b
Na+	6.75a	5.0b	7.37a	6.75a
K+	5.63a	3.38b	7.25c	4.40b
Total CEC (cmol kg^{-1})	2.35a	2.00b	2.73c	2.76c
Physical Properties				
WSA$_{240}$ min (%)				
Macroaggregates		65.60a	80.40b	71.50c
Mesoaggregates		73.30	74.20	74.30
Microaggregates		97.60	97.60	97.50

Unlike continuous maize which did not improve the soil C stock, mucuna and pigeon pea based cropping systems enhanced carbon sequestration, leading to an increase in the initial soil total C content by 55 and 69%, respectively. Similarly, mucuna and pigeon pea based systems increased soil exchangeable Ca^{2+} by 110 and 117%, respectively, and Mg^{2+} by 33 and 63%, respectively, whiles no improvement was observed under the continuous maize cropping (Table 4). Continuous maize and pigeon pea based cropping systems resulted in soil exchangeable K^+ depletion by 40 and 22%, respectively, but the mucuna based system increased exchangeable K^+ by 29%. Exchangeable Na^+ was maintained in the soil by mucuma and pigeon pea based systems, but was depleted by 35% under continuous maize cropping. A decrease of total CEC by 17.5% occurred under the MaMaMaMa, but MaMuMaMa and MaPpMaMa increased total CEC by 16 and 17%, respectively (Table 4). In a 2-years study to assess the effect of several cover crops including pigeon pea on soil physical and chemical properties in Burkina Faso, [30] found that soil exchangeable Ca^{2+}, Mg^{2+}, and Na^+, total CEC and total C were not affected by cover crop. This disagrees with the findings of our study which however reasonably corroborated research results published by [31] in that mucuna cover crop raised soil total C, exchangeable Ca^{2+} and Mg^{2+} by 81, 14, and 28%, respectively. Results of this study were also largely similar to those published by [32] who used tithonia green manure and water hyacinth compost as organic sources to restore soil fertility and found increases in soil exchangeable Ca^{2+}, Mg^{2+}, K^+, Na^+, and total CEC in the range of 61 to 74, 127 to 149, 172 to 187, 79 to 83 and 78 to 94%, respectively.

The stability of the mesoaggregates and microaggregates was not affected by cropping system (Table 4). However, 65.60, 80.40 and 71.50% of the macroaggregates were water stable under MaMaMaMa, MaMuMaMa and MaPpMaMa, respectively. This indicates that the mucuna and pigeon based cropping systems raised the macroaggregates stability by 22.6 and 9.0%, respectively, as compared with the continuous maize cropping, and mucuna based system was superior to pigeon pea based system by 12.44%. These results were comparable to the over 60% water stable macroaggregates found by [33] as a result in part of a mucuna cover crop, and reasonably agreed with [34] who reported a 26% increase in water stable macroaggregate stability due in part to the use of compost.

4. Conclusions

A threshold of 60 days after maize planting appeared to be an appropriate timing to relaying mucuna and pigeon pea into a maize crop. Relay of mucuna and pigeon pea into maize in alternate years sustained higher maize yields with minimal mineral fertilizer rates compared to the continuous maize system, but such a superiority of the cover cropping based system was more evident in non-fallow years. In a short term (over the first two to four years), continuous maize system may

provide higher grain yields when using high levels of mineral fertilization. Maize cropping with mucuna and pigeon pea as cover crops in alternate years proved largely more profitable in terms of economic returns compared to continuous maize cropping even with high mineral fertilization levels, with the profit substantially increasing over time. Continuous maize practice systematically induced soil degradation, but the maize mucuna and pigeon pea-based maize cropping systems enhanced soil physical and chemical properties, with a greater performance of the mucuna-based system.

References

[1]   Henao S, Baanante C. Agricultural production and soil nutrient mining in Africa: Implication for resource conservation and policy development. IFDC, Muscle shoals, AL 35662, USA, 75p, 2006.

[2]   Liniger HP, Mekdaschi Studer R, Hauert C, Gurtner M. La pratique de la gestion durable des terres. Directives et bonnes pratiques en Afrique subsaharienne. TerrAfrica, Panorama mondial des approches et technologies de conservation (WOCAT) et Organisation des Nations Unies pour l'alimentation et l'agriculture (FAO). Rome, Italie, 243p, 2011.

[3]   Kihara J, Fatondji D, Jones JW, Hoogenboom G, Tabo R, Bationo A. Improving Soil Fertility Recommendations in Africa using the Decision Support System for Agrotechnology Transfer (DSSAT). VIII, 187, 2012.

[4]   IFDC (International Fertilizer Development Center). Mainstreaming pro-poor fertilizer access and innovative practices in West Africa. IFAD Technical Assistance Grant No. 1174 report. Muscle Shoals, Alabama, USA, 2013.

[5]   IFPRI. Reaching Sustainable Food Security for All by 2020. PDF file and Powerpoint presentation available at www.ifpri.org/ 2020 vision. Rome, 2002.

[6]   Bationo A, Hartemink A, Lungu O, Naimi M, Okoth P, Smaling E, Thiombiano L, Waswa B. Knowing the African Soils to Improve Fertilizer Recommendations. P 19-42. VIII, 187p, 2012.

[7]   IFDC (International Fertilizer Development Center). Development and Dissemination of Sustainable Integrated Soil Fertility Management Practices for Smallholder Farmers in Sub-Saharan Africa. Technical Bulletin IFDC – T-71. Muscle Shoals, Alabama, USA, 2009.

[8]   Detchinli KS. Analyses multidimensionnelles des effets de trois systèmes culturaux sur le rendement du maïs (Zea mays L.) et le sol : bilan d'une expérimentation sur sols ferralitiques au Togo méridional. Diplôme d'Etudes Approfondies, Sciences des Agroressources et Génie de l'Environnement, option : Sciences des Agroressources, février 2013. Ecole Supérieure d'Agronomie de l'Université de Lomé, Togo. 41p.

[9]   SARI. Savanna Agricultural Research Institute (SARI) annual Report for 2005, Nyakpala, Ghana.105p, 2005.

[10] Sanchez PA, Jawa BBA. Soil fertility replenishment takes off in East and Southern Africa. p. 23-45. In B. Vanlauwe et al. (ed.). Integrated plant nutrient management in Sub-Saharan Africa: from concept to practice. CABI, Wallingford, UK, 2002.

[11] Place F, Christopher B, Barett H, de Freeman JA, Ramisch J, Vaulauwe B. Prospect for integrated soil fertility management using organic and inorganic inputs: evidence from smallholder African agricultural systems. Nairobi, Kenya, 24 p, 2003.

[12] Kombiok JM, Buah JSS, Sogbedji JM. Enhancing Soil Fertility for Cereal Crop Production Through Biological Practices and the Integration of Organic and In-Organic Fertilizers in Northern Savanna Zone of Ghana. In: Soil Fertility. R. Issaka (ed). INTECH free online publication, Croatia, pp 3-31, 2012.

[13] Adjei-Nsiah S, Kuyper TW, Leeuwis C, Abekoe MK, Giller KE. Evaluating sustainable and profitable cropping sequences with cassava and four legume crops: effects on soil fertility and maize yields in the forest/savannah transitional agro-ecological zone of Ghana. Field Crop Res. 103, 87–97, 2007.

[14] Saragoni H, Olivier R, Poss R. Dynamique et lixiviation des éléments minéraux. Agron. Trop. 45 : 259-273, 1991.

[15] Louette D. Synthese des travaux de recherche sur la fertilité des terres de barre au Bénin et au Togo. CIRAD-DSA, 34p. Montpellier, France, 1988.

[16] Tossah BK. Influence of soil properties and organic inputs on phosphorus cycling in herbaceous legume-based cropping systems in the West African derived savanna. Ph.D. Thesis No. 428, K.U. Leuven, Belgium, 2000.

[17] Sogbedji JM. Maize nitrogen utilization and nitrate leaching modeling in Togo and New York. Ph.D. Thesis, Cornell University, New York, USA, 1999.

[18] van Es HM, van Es CL. Spatial nature of randomization and its effect on the outcome of field experiments. Agron. J. 85: 420-428, 1993.

[19] van Es HM, Gomes C, Sellmann M, van Es CL. Spatially-balanced designs for experiments on autocorrelated fields. In: 2004 Proc. Am. Statistical Assoc., Statistics & the Environment Section [CDROM], Alexandria, VA, 2004.

[20] IITA (International Institute for Tropical Agriculture). Automated and Semi-automated Methods for soil and plant analysis. IITA, Ibadan, Nigeria, 2014.

[21] Mathieu C, Pieltain F. Analyse physique des sols : méthodes choisies. Lavoisier, Paris, 1998.

[22] Whalen, JK, Hu Q, Liu A. Compost applications increase water-stable aggrgates in conventional and no-tillage systems. Soil Science Society of America Journal, 67, 1842 – 1847, 2003.

[23] SAS Institute. Base SAS 9.4 Procedures Guide. SAS Institute, Cary, NC, 2014.

[24] Sogbedji JM, van Es HM, Tamelokpo FA. Optimizing N fertilizer use for maize on ferralsols in Western Africa. Revue Togolaise des Sciences. (2) 2-18, 2006.

[25] Traoré K, Bado BV, Hien V. Effet du mucuna sur la productivité du maïs et du coton. INERA, Bobo Dioulasso, Burkina Faso, 1999.

[26] Galiba M, Vissoh P, Dagbenonbakin G, Fagbahon F. 1998. Réactions et craintes des paysans à la vulgarisation du pois mascate (Mucuna pruriens var. utilis). pp 55-65 In : D. Buckles, et al. (eds.) Cover crops in West Africa contributing to sustainable agriculture. IDRC, Ottawa, Canada; IITA, Ibadan, Nigeria; Sasakawa Global 2000, Cotonou, Bénin.

[27] Lamboni D. Effet de l'amélioration par le mucuna sur l'efficacité des engrais azotés et phosphatés sur le rendement en grain du maïs : Cas de l'association maïs-mucuna dans la Région Maritime. Mémoire d'Ingénieur Agronome, Université du Bénin, Lomé, Togo, 106p, 2000.

[28] Ngome, AFE, Becker M, Mtei KM. Leguminous cover crops differentially affect maize yields in three contrasting soil types of Kakamega, Western Kenya. Journal of Agricultural and Rural Development in the Tropics and Subtropics. Vol. 112 No. 1(2011)1-10, 2011.

[29] Chabi-Olaye A, Nolte C, Schulthess F, Borgemeister C. Effects of grain legumes and cover crops on maize yield and plant damage by Busseola fusca (Fuller) (Lepidoptera: Noctuidae) in the humid forest of southern Cameroon. Agriculture, Ecosystems and Environment. 108(1) 17-28, 2005.

[30] Hulugalle NR. Effect of cover crop on soil physical and chemical properties of an alfisol in the Sudan savannah of Burkina Faso. Arid Soil Research and Rehabilitation. Vol. 2 (4)251-267, 2009.

[31] Adediran JA, Akande MO, Oluwatoyinbo FI. Effect of mucuna intercropped with maize on soil fertility and yield of maize. Ghana Jnl Agric. Sci. 37, 15-22, 2003.

[32] Omotayo OE, Chukwuka KS. Soil fertility restoration techniques in sub-Saharan Africa using organic resources. African Journal of Agricultural Research Vol. 4 (3), pp. 144-150, 2009.

[33] Taboada-Castr MM, Alves MC, Whalen J. Effect of tillage practices on aggregate size distribution in a Latossolo Vermelho (Oxisol) of Sp-Brazil. 13th International Soil Conservation Organisation Conference – Brisbane, July 2004.

[34] Ouattara K, Ouattara B, Nyberg G, Sedogo MP, Malmer A. Effects of ploughing frequency and compost on soil aggregate stability in a cotton–maize (Gossypium hirsutum-Zea mays) rotation in Burkina Faso. Soil Use and Management. 24, 19–28, 2008.

3

Influence of NPK 15-15-15 fertilizer and pig manure on nutrient dynamics and production of cowpea, *Vigna unguiculata* L. Walp

Omotoso Solomon Olusegun

Department of Crop, Soil and Environmental Sciences, Faculty of Agricultural Sciences, Ekiti State University, Ado-Ekiti, Nigeria

Email address:

solomon.omotoso@eksu.edu.ng

Abstract: A constant challenge for farmers in Nigeria is how to increase crop production in the face of low inherent nutrient status and rapid soil fertility depletion. This has attracted studies on how to build up nutrient capital in soil. Influence of NPK 15-15-15 fertilizer and pig manure on nutrient dynamics and production of cowpea *vigna unguiculata* L. Walp were evaluated at the Teaching and Research Farm, Ekiti State University, Ado-Ekiti, Nigeria in experiments consisting of six treatments laid out in a randomized complete block design with three replicates. The treatments consisted of 60kg NPK 15-15-15, 4t/ha Pig manure (PM), 8t/ha Pig manure, 4t/ha PM+60kg NPK 15-15-15, 8t/ha PM+60kg NPK 15-15-15 and no fertilizer as control. Data on plant height, no of branches, no of leaves, no of nodules/plant, dry matter yield taken at 50% flowering, number of pods/plant, number of seeds/pod, 100 seed weight and seed yield were collected. The result showed that 8t/haPM+60kgNPK gave significantly (p<0.05) higher number of nodules.plant^{-1}(13.7), dry matter (40.3g.plant^{-1}), number of pods.plant^{-1} (23.7), number of seeds.pod^{-1} (12.3) and 100 seed weight (25.5g) respectively. Maximum seed yield of 1.40t/ha was obtained with application of 8t/haPM + 60kgNPK. Sole application of pig manure and its combination with NPK significantly increased soil N, P, K, Ca and Mg. It can be concluded that for maximum production, the amount of pig manure required can reduce the chemical fertilizer that would be needed for cowpea.

Keywords: Cowpea, Yield Attributes, Nutrient Dynamics, NPK Fertilizer, Organic Manure

1. Introduction

Cowpea *Vigna unguiculata* L. Walp is an important tropical food grain legume for man and especially livestock in the dry savanna zone of the tropics. The annual global output of cowpea is about 3.3 million metric tonnes [1] and Nigeria is the world's largest producer with 2.15 million MT between 2006 to 2008 [2] Cowpea is an important crop because it is a cheap source of protein for human and livestock nutrition. Cowpea forage (vines and leaves), fresh or as hay or silage is often used for fodder while attempts have been made at using cowpea leaf meal in feeding pig. The haulms, residues from seed production, contain about 45-65% stems and 35-50% leaves and sometimes roots [3] are an important by-product in sub-Saharan Africa [4] Cowpea pod husks obtained after threshing are also used to feed livestock [5]. Despite the increase in cowpea production in sub-Saharan Africa, cowpea yields remain one of the lowest among all food legume crops with an average of 450 kg ha^{-1} in 2006 to 2008, which is half of the estimated yields in all other developing regions. This low grain yield notwithstanding, cowpea has continued to be a popular crop among farmers because it does not require high fertilization. However, inherently poor soil fertility, depletion of soil nutrients as a result of continuous cropping and crop removal have been reported to reduce soil fertility which limit its yield and productivity in Nigeria [6].

Maintenance of soil productivity has been one of the constraints to tropical agriculture such that crop production is usually moved between fields in order to utilize only fertile soils for some years without the use of fertilizers. However, this cannot be sustained to meet increasing demand of a rapidly growing population [7]. Most of the soils in Nigeria are strongly weathered and dominated by low-activity clay minerals with low nutrient status. Thus, they are adversely affected by sub-optimal soil fertility even as erosion causes deterioration of nutrient status and changes in soil organism's population [8]. Therefore, the soils cannot supply the quantities of nutrients required such that crop yield levels decline rapidly once cropping commences. Soil degradation and nutrient depletion have become serious threats to agricultural productivity. Adepetu, [9] noted that the downward trend in food production should prompt farmers to amend the soils with different materials in order to supply the nutrients needed to enhance growth and yield of crops. Several organic materials recommended to subsistence farmers in West Africa as soil amendments for increasing crop yield include cow dung, poultry dropping, pig dung and refuse composts [10] and [11].

The improved management practice is to use these external inputs from organic and inorganic sources to supply nutrients. The combined use of organic and mineral fertilizer has proved a sound soil fertility management strategy for high yields of cowpea but the efficiency of the applied fertilizer(s) is rarely indicated. Previous researches have shown that there is a positive interaction between the organic manures and urea as nitrogen source [12]. Makinde et al., [13] have reported that maize (Zea mays L) yields obtained from application of a combination of inorganic fertilizer and manure improved yield over manure alone. Akanbi et al., [14] noted that the combined application of 4t/ha of maize straw compost and N mineral fertilizer at 30kg/ha improved yield than other combinations. Adeniyan and Ojeniyi, [15] found that integrated application of poultry manure and NPK fertilizer increased maize yield compared with poultry manure or inorganic fertilizer applications alone. This practice has shown the superiority effect of integrated nutrient supply over sole use of inorganic or organic source in terms of balanced nutrient supply [16], control of soil acidity, extended residual effect [17], improvement on soil physical and chemical properties than can be derived from the use of either inorganic or organic manure and crop yield [18, 19]. The benefits of using organic materials have not been fully realized in Nigeria agriculture due to the large quantities required to satisfy the nutritional needs of crops, transportation and handling costs [20]. Besides, the integrated application of organic manure and chemical fertilizer in cowpea is rarely reported. Therefore, the aim of this experiment was to study the effects of NPK fertilizer and Pig manure on nutrient dynamics and cowpea production in an Alfisols at Ado-Ekiti.

2. Materials and Methods

2.1. Description of Study Area

This study was conducted at the Teaching and Research Farm, Ekiti State University, Ado-Ekiti (long. 7°47'N and lat. 5°13'E). Ado-Ekiti is located in the dry forest zone and experiences a warm sub-humid tropical climate with long term mean annual rainfall of 1,367 mm received in 112 days

between March and November [21]. The two-year field experiments evaluate influence of NPK fertilizer and Pig manure on nutrient dynamics and cowpea production at Ado-Ekiti, South-West, Nigeria. The soil in the study site is an Alfisols [22] of the basement complex, highly leached and with low to medium organic matter content. The site had been previously cultivated to some arable crops such as maize, melon, cassava, cocoyam and legumes with little or no fertilizer and intermittent fallows of short periods. At the commencement of the study, the site was covered by a weed spectrum of guinea grass (Panicum maximum), mucuna (Mucuna mucunoides), milk weed (Euphorbia heterophylla) and siam weed Chromolaena odorata. The vegetation was ploughed and harrowed.

2.2. Soil Sampling and Sample Analysis

Surface (0-15cm) soil samples were randomly collected and bulked to form a composite before planting and at the end of each cropping season another soil samples were collected. The soil samples and pig manure were air dried, crushed and allowed to pass through a 2mm sieve. Particle size distribution of the soil was carried out by the Hydrometer method, while the pH of soil and pig manure was determined using Pye unicam model 290 MK2 pH meter in a 1:2.5 soil/water suspension. Organic carbon was determined by the [23] dichromate oxidation method [24]. Total nitrogen was determined by the micro-kjeldahl digestion method as described by [25]. Available P was determined by [26] No 1 extraction method as described in [27] laboratory manuals; Exchangeable bases were extracted with neutral 1M NH_4OAC at a soil solution ratio of 1:10 and measured by flame photometry. Magnesium was determined with an atomic absorption spectrophotometry. Exchange acidity was determined by titration of 1M KCL extract against 0.05M NaOH to a pink end point using Phenolphthalein as indicator [28].

2.3. Experimental Setup: Design and Treatments

The experiment was laid out in a randomized complete block design with three replications. The treatments consisted of: (i) No NPK, No pig manure (control) (ii) 60kg NPK 15-15-15, (iii) 4t/ha Pig manure, (iv) 8t/ha Pig manure, (v) 4t/ha Pig manure + 60kg NPK 15-15-15 (T5) and (vi) 8t/ha Pig manure + 60kg NPK 15-15-15. The manure was well rooted Pig manure that had stabilized for about 100 days. Each plot size was 4 x 2m and separated by 0.5m paths while treatment blocks were 1m apart. Pig manure was applied 2 week before planting and the inorganic fertilizer NPK was applied 2 weeks after planting both for the single and combined applications. Seeds of improved cowpea variety 1190K-568.18 (Ibadan Brown) obtained from IITA Ibadan were sown at 2 seeds per hill at a spacing of 60x30cm and later thinned to one seedling to attain a population of 66,666 plants.ha^{-1}. Plots were weeded manually with hoe at frequency required. Cowpea insect pests were controlled by application of Nuvacron (Monocrotophos; Norvartis,

Switzerland) at 40ml in 15L of water. Spraying commenced at 5 weeks after planting and at 1 week interval until full pod formation. The experiment was repeated in the second year on the same plots without fertilizer application in order to assess the residual nutrients.

2.4. Data Collection

Plant data collected were plant height, no of branches, no of leaves, no of nodules per plant, dry matter yield and root length. At maturity, dry pods of cowpea were picked as from the 10th week of planting to avoid shattering. Yield components taken were: no of pods per plant, no of seeds per pod and 100 seed weight and grain yield. Six plants were selected per plot at random and were tagged for determination of plant growth and yield components parameters. Plant height was measured using a meter rule. Root length was measured using a ruler. Dry matter yields were determined by manually harvesting the six tagged cowpea plants per plot at 50% flowering. The plants were washed and cleaned to remove traces of soil and placed in a bag before oven drying at 70°C for 48 hours. The numbers of pods per plant and seeds per pod were counted at harvest. The 100 seed-weight was measured using an electronic weighing balance.

2.5. Data Analysis

Data were subjected to analysis of variance [29]. Means were separated with Duncan's Multiple Range Test at 5% level of probability.

3. Results and Discussion

3.1. Physical and Chemical Properties of the Soil Used

Table 1. The physical and chemical properties of soil and pig manure used for the study.

Parameters	Soil	Pig Manure
pH (water)	5.8	6.6
Organic carbon (g kg^{-1})	2.41	2.24
Total nitrogen (g kg^{-1})	0.09	0.26
Available phosphorus (mg kg^{-1})	5.30	4.60
Exchangeable Bases (cmol kg^{-1})		
K$^+$	0.17	0.20
Ca^{2+}	2.28	2.42
Mg^{2+}	2.92	0.45
Na$^+$	0.08	0.60
Exch. Acidity	0.17	0.16
Effective cation exchange capacity	5.62	3.83
Particle size analysis (g kg^{-1})		
Sand	746	
Silt	167	
Clay	87	
Textural class	Sandy Loam	

The results of the analysis of soil and pig manure used are presented in Table 1. The soil was a slightly acidic sandy

loam with low organic carbon content, exchangeable cations while total N and available P content were 0.09g kg^{-1} and 5.82mg kg^{-1}. The total N and available P were very low compared with the critical levels of 0.1% for N and 10-12mg kg^{-1} for available P [30] obtained for soil in South-west, Nigeria [31]. Using the critical levels of 0.16-0.20cmol kg^{-1} exchangeable K was also low [32].

3.2. Effects of NPK and Pig Manure on Plant Height, Number of Leaves of Cowpea

The effects of pig manure and NPK fertilizer and their combinations on plant height, number of leaves and number of branches are indicated in Tables 2. For both seasons, the tallest plants were obtained with the application of 8t/ha Pig manure + 60kg NPK which indicated 99% and 126% increase relative to the control plants in 2012 and 2013 respectively. Sole or combined fertilizer rates gave significant (p<0.05) increase in plant height in the two seasons. Kuldeep, [33] have observed maximum values of green pod yield, plant height with combined application of organic and inorganic fertilizer. However, in 2013 plants were taller in all the treatments and this could be attributed to the residual effects of the organic fertilizer in the second year. Number of leaves and number of branches per plant were also significantly (p<0.05) increased with the rate of fertilizer application. Highest number of leaves and number of branches were observed with the application of 8t/ha Pig manure + 60kg NPK which ranged from (26.8 – 32.8, 4.2 - 7.3) and (26.4 – 37.6, 5.2 – 8.7) for 2012 and 2013 respectively.

3.3. Effects of NPK and Pig Manure on Number of Nodules and Dry Matter

The number of nodules per plant was significantly reduced as poultry manure increased. (Table 3). Ofori, [34] and Olatunji et al., [35] have observed that nodulation in cowpea was significantly reduced at higher rate of N application. This implied that the increased plant height from the treatments was as a result of the fertilizer application and not from the effect of cowpea nodulation.

The highest mean root length of 26.6 and 25.8cm were observed in those plants that received the highest manure combined with mineral fertilizer while the least root length of 15.8 and 16.4 were observed form no fertilizer plots in both years.

The effects of fertilizer treatments on dry matter was significant (p<0.05) and the 8t/ha PM+ 60kg NPK gave the highest values of 36.9g. plant^{-1} and 43.6g. plant^{-1} in 2012 and 2013 respectively, while the lowest values (24.3 and 23.4g. plant^{-1}) were obtained in the control plot for both 2012 and 2013. However, dry matter yield obtained in 2012 was lower than the two years average of 40.3 g. plant^{-1}. Omotoso and Falade [36] had reported that application of 30mg Zn kg^{-1}soil and organo-mineral combination (Cow dung+ZnSO$_4$) significantly gave the highest plant shoot biomass (7.11 g pot^{-1}) base on dry shoot weight.

3.4. Effects of NPK and Pig Manure on Yield and Yield Characters of Cowpea

Pig manure and NPK fertilizer and their combinations significantly (p<0.05) increased the number of pods per plant, number of seeds per pod, seed weight and grain yield per plant (Table 4). The highest no of pod (24.6) was obtained at 8t/ha PM + 60kg NPK in 2012 while in 2013 no of pod were similar in 4t/ha PM + 60kg and 8t/ha PM + 60kg NPK application rates. Highest number of seed/pod (11.7 and 12.8), 100 seed weight (24.7 and 26.3g) and grain yield (1.39 and 1.42t/ha) for 2012 and 2013 respectively, was obtained with 8t/ha PM + 60kg NPK. The average no of pod/plant, average 100 seed weight and average grain yield produced by 8t/ha PM + 60kg NPK application rate for 2 years were 41, 40 and 29% increased relative to no fertilizer treatment. Increasing the rate of pig manure increased all the yield parameters. Ferreira et al., [37] studied the use of organic and mineral fertilizer for production of okra in soil of Rio de Janeiro, Brazil. The results indicated a significant increase in yield with increase in mineral fertilizer as well as manure. Nuruzzaman, et al., [38] revealed that yield characters of okra could be modified by the application of biofertilizer+cowdung. However, biofertilizers+cowdung treatments were comparable with treatment comprising of 60% N. Datt et al., [39] revealed that NPK fertilizer combined with farm yard manure significantly increased the green pod yield of pea and the maximum value was obtained in treatment where nutrients were applied at 30kg N + 39.3kg P +37.5kg K +10t FYM ha-1. Also, [33] observed that maximum values of green pod yield, plant height, number of green pods/plant, number of seeds/pod and 100 seed weight were obtained with combined application of organic and inorganic fertilizer at 20t FYM + 25kg N + 65kg P2O5 + 97.5kg K2O ha-1.

Table 2. Effects of NPK fertilizer and Pig manure on growth characters of cowpea

Treatments	Plant height (cm)		Number of leaves/plant		Number of branches/plant	
	2012	2013	2012	2013	2012	2013
No fertilizer (NF)	10.3f	12.6e	26.8e	26.4f	4.2c	5.2d
60kg/ha NPK	16.4c	21.9c	28.3d	30.3d	5.6b	7.3b
4ton/ha PM	13.2e	19.3d	28.4d	27.8e	5.3b	6.4c
8ton/ha PM	15.8cd	22.4c	31.3b	33.3b	5.4b	7.3b
4t/haPM+60kg NPK	17.3b	24.5b	30.4c	31.5c	5.5b	6.7c
8t/haPM+60kg NPK	20.5a	28.6a	32.8a	37.6a	7.3a	8.7a
Mean	15.6	21.6	29.7	31.2	5.6	6.9
SE±	0.57	0.59	1.26	1.27	0.35	0.32
CV (%)	39.12	39.10	24.67	24.96	2.13	2.26

Mean followed by the same letter(s) are not significantly different at P<0..05 using DMRT. Key: NF= no fertilizer, PM=Pig manure, NPK= 15:15:15

Table 3. Effects of NPK fertilizer and Pig manure on number of nodule, dry matter and root length of cowpea

Treatments	Number of nodules		Root length (cm)		Dry matter (g)	
	2012	2013	2012	2013	2012	2013
No fertilizer (NF)	10.3b	10.1d	15.8f	16.4e	24.3f	23.4f
60kg/ha NPK	10.5b	12.3c	17.6e	18.3d	30.8d	34.7d
4ton/ha PM	9.4bc	15.2b	20.0d	22.4c	29.3e	33.2e
8ton/ha PM	8.8d	11.3d	24.3b	24.7b	32.7c	42.1b
4t/haPM+60kg NPK	11.1a	16.4a	21.2c	24.8b	33.4b	39.2c
8t/haPM+60kg NPK	10.4b	15.2b	26.6a	25.8a	36.9a	43.6a
Mean	9.9	10.9	20.9	22.1	31.2	36.0
SE±	1.30	1.37	1.72	0.59	1.87	0.50
CV (%)	39.14	38.24	22.21	23.46	24.95	26.72

Mean follow the same letter(s) are not significantly different at P=<0.05 using DMRT. Key: NF= no fertilizer, PM=Pig manure, NPK =15:15:15

Table 4. Effects of NPK fertilizer and Pig manure on grain yield and yield components of cowpea

Treatment	No of pod/plant		Number of seed/pod		100 seed weight (g)		Yield t/ha	
	2012	2013	2012	2013	2012	2013	2012	2013
No fertilizer (NF)	16.2f	17.3e	7.4e	8.1e	14.3f	16.2f	0.89e	1.10e
60kg/ha NPK	18.4d	19.1c	8.3d	8.4e	19.4d	20.6c	1.14d	1.16d
4ton/ha PM	17.3e	18.2d	9.2c	10.3c	17.1e	18.5e	1.18c	1.20c
8ton/ha PM	19.3c	21.4b	11.6a	11.3b	20.5c	19.9d	1.19c	1.36b
4t/haPM+60kg NPK	21.4b	22.3a	10.4b	9.7d	22.4b	23.7b	1.23b	1.21c
8t/haPM+60kg NPK	24.6a	22.7a	11.7a	12.8a	24.7a	26.3a	1.39a	1.42a
Mean	19.5	20.2	9.8	10.1	19.7	20.9	1.17	1.24
SE±	0.58	0.57	0.52	0.54	0.69	0.67	0.045	0.043
CV (%)	24.6	25.9	3.84	2.13	23.9	25.4	8.48	8.54

Mean follow the same letter(s) are not significantly different at p<0.05 using DMRT. Key: NF= No fertilizer, PM=Pig manure, NPK

3.5. Effects of NPK Fertilizer and Pig Manure on Soil Nutrient Status

The nutrient status as affected by PM and NPK fertilizer after cropping season in 2012 and 2013 are presented in Table 5. Inorganic NPK fertilizer significantly reduced soil pH in both seasons and increased N and P content of the soil. This is in agreement with Olatunji et al., 2012. Plots that received higher rates of sole manure and their combination with NPK fertilizer had higher pH than the control plots but the trend was different in plots that received inorganic fertilizer where the pH of the control plot was higher than the pH of plots that received 60kg/ha NPK fertilizer alone. This is consistent with Nnaji et al., [41] who reported that soil pH was increased by combined application of mineral fertilizer NPK with cow dung and attributed this to the release of some cations from decayed organic amendments. The decline in pH of plots treated with inorganic fertilizer in this study could be attributed to their rapid rates of release of nutrient, which are immediately used up by plants, leading to poor accumulation of exchangeable bases that neutralizes soil acidity. While the sole pig manure at 8t/ha gave values of (6.4 and 6.6) and the combined application of 8t/ha Pig manure + 60kg/ha NPK gave significantly (p<0.05) higher pH values of 6.6 and 6.5 in 2012 and 2013 respectively compared to the control. This is consistent with Ibiawuchi et al, [40] who reported that all the plots treated with poultry manure + inorganic fertilizer had high residual N, P, K, Ca and Mg while soil pH increased from 5.65 to 5.71. Therefore, in this study the increase in soil pH due to corresponding increase in the rates of pig manure could be attributed to increased microbial activity during the process of decomposition leading to cations like Ca, Mg and Na released from mineralisation and organic matter formation.

Relative to no fertilizer treatment, sole pig manure and combination of pig manure and NPK significantly (p<0.05) increase soil organic caebon (SOC) with 8t/ha Pig manure + 60kg/ha NPK producing the highest values (1.50 and 1.60g/kg) in 2012 and 2013 respectively, while sole application of 60kg/ha NPK caused a decrease in SOC. Also, using the 2 years average, application of 4t/ha, 8t/ha, 4t/ha +60kg NPK and 8t/ha + 60kg/ha NPK increased SOC by 19, 27, 22 and 29% respectively, relative to no fertilizer plot.

Inorganic fertilizer NPK 15-15-15 gave significantly (p<0.05) increase in N, P, K content but caused a decrease in Ca and Mg. The pig manure alone and combination of Pig manure + NPK 15-15-15 increased total nitrogen above the control in 2012 and 2013 with 8t/ha Pig manure + 60kg/ha NPK giving the highest average N value of 0.25k/kg for the 2 years. Available P was increased above the control by all the treatments in both seasons except the plot that received inorganic fertilizer alone in 2012. This is consistent with [42] who noted that combined application of pig manure and NPK to tomato increased N, P and K content and yield of tomato. Also, Agbede et al., [43] reported that poultry manure increased plant N, P, K, Ca and Mg status in leaf of Sorghum.

Table 5. Effects of NPK fertilizer and Pig manure on Soil nutrient status after cropping season in 2012 and 2013

| Treatments | pH(H₂O) | | Org. C (g/kg) | | Total N (g/kg) | | Avail. P (mg/kg) | | Exchangeable bases (Cmol/kg | | | | | |
| | | | | | | | | | K | | Ca | | Mg | |
	2012	2013	2012	2013	2012	2013	2012	2013	2012	2013	2012	2013	2012	2013
Initial	5.80		2.41		0.09		5.30		0.18		2.28		2.92	
NF	5.20	5.10	1.20	1.20	0.09	0.10	4.62	5.10	0.11	0.13	3.10	3.10	2.62	2.02
60kg/ha NPK	4.46	4.20	1.13	1.15	0.10	0.14	5.20	5.38	0.27	0.25	2.14	2.27	1.92	2.01
4ton/ha PM	5.80	6.41	1.43	1.44	0.20	0.21	5.43	5.40	0.21	0.28	4.27	3.80	3.01	3.52
8ton/ha PM	6.40	6.60	1.53	1.54	0.22	0.20	5.71	5.92	0.39	0.31	4.42	4.12	3.04	3.46
4t/haPM+60kg NPK	6.36	6.50	1.41	1.52	0.10	0.21	5.20	5.62	0.36	0.29	3.02	3.10	2.82	3.21
8t/haPM+60kg NPK	6.60	6.50	1.50	1.60	0.24	0.25	5.90	6.10	0.42	0.44	3.93	4.63	3.06	3.16
Mean	5.76	5.89	1.40	1.45	0.17	0.21	5.33	5.59	0.29	0.28	3.76	3.73	2.87	2.78
SE±	0.86	0.85	0.38	0.04	0.041	0.043	0.04	0.07	0.03	0.03	0.085	0.085	0.09	0.09
CV (%)	4.66	4.63	8.49	8.62	1.35	1.38	4.87	3.84	1.36	1.34	3.86	4.25	2.64	2.59

Mean follow the same letter(s) are not significantly different at p<0.05 using DMRT. Key: NF= No fertilizer, PM=Pig manure, NPK= 15:15:15, respectively

4. Conclusion

The results of this study have demonstrated that the combined applications of pig manure and inorganic fertilizer NPK thus have a profound significant influence on cowpea and enhanced plant growth and development when compared to untreated plots. Maximum yield of 1.4t/ha was obtained with application of 8t/ha Pig Manure + 60kg NPK. While application of pig manure alone and its combination with NPK significantly (p<0.05) increased soil N, P, K, Ca and Mg. It can be concluded that for maximum production of cowpea the amount of manure required can reduce the chemical fertilizer that would be needed.

Acknowledgments

The author wishes to acknowledge the support of Dr O. S Shittu, the Head, Department of Crop, Soil and Environmental Sciences for allocating the research site and Mr Agbebi Joseph, Farm Director for releasing field technical staff. Special thanks also to Dr A. E Salami for helping in statistical analysis of the data.

References

[1]   FAO. Cowpea production data base for Nigeria. Food and Agriculture Organisation http://www.faostat. Fao.org/ 2005. Date assessed 19/08/14.

[2]   FAO. FAOSTAT Database. 2010 http://www.fao.org/site 342. Date assessed 24/8/2014.

[3]   U. Y Anele, K. H Sudekum, O. M Arigbede, H. Luttgenau, A.O. Oni, O.J. Bolaji and M.L Galyean. Chemical composition, rumen degradability and crude protein fractionation of some commercial and improved cowpea (*Vigna unguiculata* L. Walp) haulm varieties. Grass Forage Sci. 67 (2): 210-218, 2012.

[4]   S. Singh, S. K Nag, S. S Kundu and S. B Maity. Relative intake, eating pattern, nutrient digestibility, nitrogen metabolism, fermentation pattern and growth performance of lambs fed organically and inorganically produced cowpea hay-barley grain diets. Trop. Grassl. 44: 55-61, 2010.

[5]   J. A. Oluokun. Intake, digestion and nitrogen balance of diets blended with urea treated and untreated cowpea husk by growing rabbit. Afr. J. Biotech. 4 (10): 1203-1208, 2005.

[6]   O. E. Omotayo and K. S. Chukwuka. Soil fertility restoration techniques in sub-Saharan Africa using organic resources. African Journal of Agricultural Research. 4 (3): 144-150, 2009

[7]   M.O. Akande, F.I. Oluwatoyinbo, E.A. Makinde, A.S. Adepoju and I.S. Adepoju. Response of okra to organic and inorganic fertilization. Nature and Science. 8(11):261-266, 2010

[8]   E. C. A. Economic Commission of Africa,. State of the Environment in Africa. Economic Commission of Africa, P.O.Box 3001, Addis Ababa, Ethiopia, ECA/FSSDD/01/06, 2001.   http://www.uneca.org/water/State_Environ_Afri.pdf Date assessed 26th October, 2014.

[9]   J. A. Adepetu. Soil and Nigeria food security. Inaugural Lecture series 119, Obafemi Awolowo University, Ile-Ife. Nigeria 19pp. 1997

[10]   A. Olayinka. Carbon mineralization from poultry manure straw sawdust amended Alfisol. Ife Journal of Agriculture 1 and 2. (18) 26-36, 1996.

[11]   A. Olayinka, A. Adentunji, and A. Adebayo. Effect of organic amendment on nodulation and nitrogen fixation of cowpea. Journal of Plant Nutrition. 21 (11): 2455-2464, 1998.

[12]   S.M. Yang, S.S. Malhi, F.M. Li, D.R. Suo, M.G. Xu, P. Wang. Long-term effects of manure and fertilization on soil organic matter and quality parameters of a calcareous soil in NW China. Journal of Plant Nutrient and Soil Science. 170:234-243, 2007

[13]   E. A. Makinde, A. A. Agboola and F. I. Oluwatoyinbo. The effect of organic and inorganic fertilizes on the growth and yield of maize/melon intercrop, Moor Journal of Agricultural Research, 2001; 2: 15-20.

[14]   W. B. Akanbi, M. O. Akande and J. A. Adediran. Suitability of composted maize straw and mineral nitrogen fertilizer for tomato production. Journal of Vegetable Science, 2005;11 (1): 57-65.

[15]   O. N. Adeniyan, Ojeniyi S. O. Effect of poultry manure, NPK 15-15-15 and combination of their reduced levels on maize growth and soil chemical properties. Nigerian Journal of Soil Science 15:34-41, 2005;

[16]   M. S. Khan, N. C. Shil and S. Noor. Integrated nutrient management for sustainable yield of major vegetable crops in Bangladesh. Bangladesh J Agric Environ 4: 81-94, 2008.

[17]   P. A Adeoye, S. E Adebayo and J. J Musa. Growth and yield response of cowpea (*Vigna unguiculata* L.) to poultry and cattle manures as amendments on sandy loam soil plot. Agric. Journal. 6(5): 218–221, 2011

[18]   L. S. Ayeni, E. Omole, M. T. Adetunji and S. O. Ojeniyi. Integrated application of NPK fertilizer cocoa pod ash and poultry manure: effect on maize performance, plant and soil nutrient content. International Journal of Pure and Applied Sciences. 2 (2):34-41, 2009.

[19]   B. S. Ewulo, O. O. Babadele and S. O. Ojeniyi. Sawdust Ash and Urea Effect on Soil and Plant Nutrient Content and Yield of Tomato: Am.-Eurasian J. Sustain. Agric. 3 (1): 88-92, 2009.

[20]   O. T. Ayoola and O. N. Adeniyan. Influence of poultry manure and NPK fertilizer on yield and yield components of crops under different cropping systems in south west Nigeria. African Journal of Biotechnology. 5 (15): 1386-1392, 2006.

[21]   O. J. Ayodele and A. O. Oso. Effects of phosphorus fertilizer sources and application time on grain yield and nutrient composition of cowpea (*Vigna unguiculata* L.,Walp). American Journal of Experimental Agriculture 4 (12): 1517-1525, 2014.

[22]   A. S. Fasina, J. O. Aruleba, F. O. Omolayo, S. O. Omotoso, O. S. Shittu and T. A. Okusami. Properties and Classification of five soil formed on granitic parent materials in Ado-Ekiti, Southwestern Nigeria. Nigerian Journal of Soil Science 15 (2): 21-29, 2005.

[23]   A. Walkley and C.A Black. An examination of the Degtjareff method for determining soil organic matter and a proposed modification of the chromic acid titration method. Soil Sci. 37:29-38, 1934.

[24]   D. W. Nelson, Sommers L. E. Organic carbon. In: Page, A.L.; Miller, R.H. and Keeney, D.R. (eds). Methods of Soil Analysis Part 2. Agron 9. Madison W.I. 1982; 538-580.

[25]   J. M. Bremner and C. S. Mulvaney. Nitrogen – Total. In Methods of soil analysis, Page, A. L. (Ed). Agronomy. Amer. Soc. of Agron, Madison, Wisc. USA. 9(2):595–624, 1982.

[26]   R. H. Bray and L. T. Kurtz. Determination of total, organic and available form of phosphorus in soils. Soil Science Society of American Journal. 59: 39 – 45, 1945.

[27]   IITA. International Institute of Tropical Agriculture. Automated and semi-automated methods for soil and plant analysis. Manual series 7, 1989.

[28]   E. O. Mclean. Soil pH and lime requirement, In: Methods of Soil Analysis. Part 2 Agron. 9 (2nd ed.) ASA, SSSA, Madison, Wisc. 199 -224, 1982.

[29]   SAS Institute Inc SAS/STAT user's guide, version 8. SAS Institute Inc. Cary, North Carolina, USA 1999

[30]   G. O Adeoye, and Agboola A. A. Critical level for soil pH, available P, K, Zn and Mn and maize earleaf content of P, Cu, Zn and Mn in sedimentary soils of Southwestern Nigeria. Nutr. Cycl. Agroecosyst. 6: 65-71, 1985

[31] FMANR. Literature review on soil fertility investigations in Nigeria (in Five Volumes). Federal Ministry of Agriculture and Natural Resources, Lagos, 32-45, 1990.

[32] A. A. Agboola, G. O. Obigbesan. The response of some improved food crop varieties to fertilizers in the forest zone of Western Nigeria. In: Report of FAO/NORAD Seminar on fertilizer use development in Nigeria, Ibadan, p. 77, 1974.

[33] S. V. Kuldeep. Standardization of potassium fertilizer application technology on seed production of pea cultivar Arkel. *Thesis* submitted Dr.Yashwant Singh Parmar University of Horticulture and Forestry Nauni, Solan. pp 1-49, 2003. shodhganga.inflibnet.ac.in/bitstream/10603/9786/11/11_literature.pdf

[34] C. S. Ofori. The importance of fertilizer nitrogen in grain legume production on soils of granitic origin in the Upper Volta region of Ghana. Proceedings of First IITA Grain legume Improvement Workshop, IITA. Ibadan 1973.

[35] O. Olatunji, S. A. Ayuba, B. C. Arijembe and S. O. Ojeniyi. Effect of NPK and poultry manure on cowpea and soil nutrient composition. Nigerian Journal of Soil Science. 22 (1):108–113, 2012.

[36] S. O. Omotoso and M. J. Falade. Zinc and Organo-Mineral Fertilization Effects on Biomass Production in Maize (*Zea mays*) Grown on Acid Sand Alfisol (Typic Paleudalf). Research Journal of Agronomy. 1 (2): 62-65, 2007.

[37] M. E. Ferreira and M. C. P Cruz. Study the effect of vermicompost on nutrient absorption and dry matter production by maize and soil properties. *Scientific Sao Paulo,* 20:217-227, 2002.

[38] M. Nuruzzaman, H. Lambers, M. D. A Bollard and E. J. Veneklaas. Phosphorus uptake by grain legumes and subsequently grown wheat at different levels of residual phosphorus fertiliser. Australian Journal of Agricultural Research. 56: 1041-1047, 2005a.

[39] N. Datt, R. P. Sharma and G. D. Sharma. Effect of supplementary use of farmyard manure along with chemical fertilizers on productivity and nutrient uptake by vegetable pea (*Pisum sativum* var. *arvense*) and build up of soil fertility in Lahaul Valley. Indian J. Agric. Sci. 73:266-268, 2003.

[40] I. I. Ibeawuchi, F. A. Opara, C. T. Tom and J. C. Obiefuna. Graded replacement of inorganic with organic manure for sustainable maize production in Owerri Imo State, Nigeria. Life Science Journal 4 (2):82-87, 2007.

[41] G. U. Nnaji, J. S. C. Mbagwu and C. L. A Asadu. Influence of organic manures on cassava yield and some chemical properties of an ultisol Nsukka area of south Eastern Nigeria. Proceedings of the 29th Annual Conference of the Soil Science Society of Nigeria 5th-9th December, Makurdi, Nigeria. 2005

[42] D. D. Giwa and S. O. Ojeniyi. Effect of integrated application of pig manure and NPK on soil nutrient content and yield of tomato (*Lycopersicon esculentum* Mill). Proceeding 29th Conference of Soil Science Society of Nigeria, UNAAB, Abeokuta. 6–10, 2004.

[43] T. M. Agbede, S. O. Ojeniyi and A. J. Adeyemo. Effect of poultry manure on soil physical and chemical properties, growth and grain yield of sorghum in Southwest, Nigeria, Am.-Eurasian J. Sustain. Agric. 2 (1): 72-77, 2008.

4

Hill torrents potentials and spate irrigation management to support agricultural strategies in Pakistan

Muhammad Asif[1], Col Islam-ul-Haque[2]

[1]Scientific Officer/AAE, CAEWRI NARC/Pakistan Agricultural Research Council (PARC) Islamabad
[2]Chairman, Ecological Sustainability through Environmental Services (Eco Steps), Islamabad, Pakistan

Email address:

asifbukhari1@gmail.com (M. Asif), islamhaq3@yahoo.com (C. Islam-ul-Haque)

Abstract: Pakistan has not only been blessed with enriched hydrological cycling phenomenon which generate abandoned quantum of water in the northern part , but also possess 18.68 MAF Water Potential in water scarce areas of Pakistan , known as Rod Kohi. Spate irrigation is in practice in Pakistan, where, flood water, during monsoon season, gets generated and channelized from down-hill countered profiles (catchments area). These hill torrential are diverted to agriculture fields, by constructing earthen embankments or related hydraulic concrete structures. Since, hill sides torrential are unpredictable, temporally and spatially matrix, which poses numerous challenges to the farmers who in return forcibly integrate / pool up individual resources to partially manage this scarce resource. Though the annual rainfall (....100...mm/year) in this area is low and uncertain yet at the same time due to terrain lay out , substantial quantum of water gets accumulated due to downhill side terrain profile each rainfall event. The agriculture activities, in these areas are totally dependent on such type of phenomenal rainfalls. Unfortunately, due to lack of scientific water resource management and modern agriculture practices, major quantum of torrential flood water is not only gets wasted , but also causes huge losses to human life and property . This catastrophic situation arises, as there are neither successive layers of check-dams on the down-hill sides of these mountain ranges, nor any kind of water storage facilities exist at the foot hill areas. PARC (Pakistan Agricultural Research Council) and Pakistan Poverty Alleviation Foundation (PPAF) did funded some localized water storage , water conveyance and water distribution system interventions, but still there is a dire need of holistically adoption of watershed management approaches , based on resource integration concept and practices. This un-managed water resource must be harnessed / converted into lucrative opportunities to oxygenize the life line for millions of people residing in these areas.

Keywords: Agricultural, Spate Irrigation, Rainfall, Water Management, Storage

1. Introduction

By enlarge Pakistan is an agro based economy which contributes 21% towards the total GDP and 70% of the total population draw their livelihood from various agricultural activities. Over the past decade, due to various reasons, the agriculture industry, being the main livelihood of a sizeable population and major economic growth indicator is on the decline. Adverse impacts of climate change, coupled with bad water governance have further aggravated the problem. Required water availability at the right time and efficient irrigation led technological interventions are the key drivers for increased productivity enhancement in agriculture sector. Efficient water management and its application results in increased crop yields, more cropped area, cropping intensity and crop diversification. Water resources in Pakistan, especially outside the Indus basin irrigation system are limited/scarce to meet the crop water requirement. The productivity and sustainability of agriculture in these areas are dependent on the management of scarce water resources i.e. efficient water use. Due to saturation of agricultural production in irrigated areas of the country, search for new areas has become inevitable in order to feed the ever-increasing population in the country.

1.1. Delineation of Spate Irrigation Region

The spate Irrigation region of Pakistan lies between longitude range of 60°50′ to 72° East and latitudes 24°42′ to 34° 3′ North (Fig 1). It comprises parts of southern NWFP, south-western Punjab, western Sindh and major part of

Baluchistan province. It is bounded in the north by KPK province, west by Afghanistan and Iran, south by Arabian Sea and east by Punjab and Sindh provinces. The region covers about 41.63 million hectares of area (about 49.9% of the country area: 83.43 Mha) in the west of the country. Maximum region lies in Baluchistan province (about 82%) followed by NWFP (9.2%) and Sindh province (5%). Punjab province contributes about 4.2% to the region (Table 2). Highest elevation in the region is about 4680 meters above sea level (masl). Maximum stretch of the area is in SW-NE direction (about 1390 km). The drainage pattern is predominantly dendritic in nature (Fig 1). The highest elevation is about 4680 meters in the northwest of Baluchistan (Fig 1).

Spate irrigated areas are one of the most important option with considerable potential for increasing agricultural production. Spate irrigation is a type of irrigation which is being practiced since centuries in some parts of the world. In Pakistan, spate irrigation covers nearly 1.5 million ha, which is about 8% of the total irrigated area, as shown below in table 2.

In Pakistan's local context, It is known as Rod Kohi in the Khyber Pakhtoon Khawan (KPK) Province, and in Punjab and Baluchistan province, called Bandit/Sailaba. Commonly,

across the country, spate irrigation is also generally referred as flood irrigation. This kind of irrigation practices primarily relies on the flood water of the hill torrents, which are diverted into a plain area, locally known as Damaan. The total water potential of Rod Kohi are shown in (Table-1)

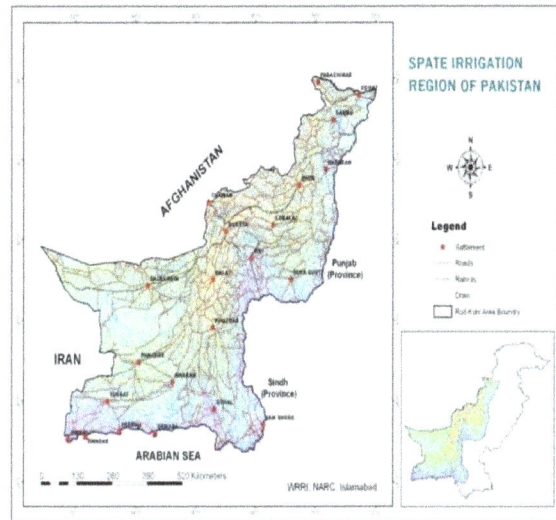

Fig 1. spate irrigation region of Pakistan

Table 1. Water Potential Rod Kohi- Pakistan

Province	Potential Area* (MA)	Potential Water* (MAF)	Ratio of Water to Area (acre feet/acre)
Federal	0.67	2.84	4.2
NWFP	2.13	4.56	2.1
Punjab	1.41	2.71	1.9
Sindh	1.36	0.72	0.5
Baluchistan	11.56	7.85	0.7
Total	17.13	18.68	1.1

Source: Feasibility Studies for Flood Management of Hill-Torrents, NESPAK, 1998.

In the indigenous systems, farmers divert the spate flow to their fields by constructing breach able earthen bunds (called Gandas) across the rivers and/or stone/gravel spurs leading towards the centre of the river (FAO,1997). Spate Irrigation farming system of Rod-Kohi areas is a unique system of farming, being practiced in Piedmont plains of Dera.Ismaiel. Khan (KPK), Dera .Ghazi. Khan (Punjab), Dadu (Sindh) and in Sulaiman ranges, Kachhi plain, Kharan and Lasbela basins of Baluchistan province. Baluchistan, the largest province in Pakistan, has about 1.2 million hector of spate or Sailaba irrigated land. These areas are often called falling flood irrigation areas and are located on extensive tracts of land along the rivers and hill streams subject to annual inundation. They utilize the moisture retained in the root zone after the flood subsides together with sub-irrigation due to the capillary rise of groundwater (Ahmed, 2000).

There are four different water supply systems in the Sailaba irrigation, namely nullah, manda, diffuse and riverine (Hamilton and Muhammad, 1995), like;

a. Nullah systems are based on a single nullahs (ephemeral stream), usually one with a mountainous catchment;

b. manda systems depend on rivers or large nullahs, which

collect water from many small ephemeral streams with quite hilly catchments.

c. Diffuse supply systems utilize large sloping areas as contributing catchments, where the runoff is collected into shallow nullahs by the time it reaches the diversion point.

d. Riverine systems are designed to divert water from perennial streams only when a sufficient flood stage is reached for the water to flow into diversion canals (Hamilton and Muhammad, 1995).

Water rights on Sailaba systems in Baluchistan are entirely controlled by the users. The government plays no role in distributing the water or maintains records of water used by the farmers. Water is distributed between the irrigation systems' participants according to the principle of 'first come, first served.' There is no formal government-sanctioned entity to manage the system. The government agencies mainly serve as Facilitators such as making available the equipment necessary for the building or reconstruction of Sailaba earthen bunds / embankments. This is unlike the spate irrigation systems in KPK and Punjab provinces, where the civil administration actively intervenes in instructing the farmers to plug breaches and to connect flood canals (Van

Steenbergen, 1997). Upper portion of Dera Ismail Khan (DI Khan), Tank and Kullachi Tehsil are the three districts in KPK province, where spate irrigation is still prevailing. The total area of the districts is about 9 million hector, out of which the cultivated land is 700,000 hectors. Spate irrigation covers nearly 250,000 hectors in KPK, minor spate flows occur in spring and the major floods come in summer as a result of monsoon rainfall on the Suleiman range and Lakai-Marwat hills during July and August (Hamilton and Muhammad, 1995)

2. Objectives

The spate irrigation system study objectives are as under;
a. To discuss existing spate irrigation infrastructure system and water application practices in Pakistan associated with flood damages, hazards & risks.
b. To compare crop yield productivity between spate irrigation / flood water & managed irrigation system, through perennial canal system & tube wells water application system
c. To analyze and discuss the role of all other related government departments for effective and prudent hill torrents watershed area management.

3. Methods and Material

Secondary data/reports/studies have been consulted, as in the past, lot of work on potentials of hill torrents has been carried out by various organizations, including " Master Feasibility Studies for Flood Management of Hill-Torrents, by NESPAK, 1998". Besides this, field visits of various have been carried out and personal interaction with the various government departments and with on-site farmers/ communities was also carried out.

4. Results and Discussion

4.1. Climate and Geo-Technical Characteristics, Soil Texture Etc

The climate of the Rod-Kohi areas in KPK, Punjab and Sindh provinces is arid to semi-arid with precipitation ranging from less than 100 mm in South West (Dadu, Sindh) to 300 mm in Northwest (D.I. Khan, KPK). In Kharif season, rain is received in the months of July and August and in 'Rabi' season in the months of March and April. June is the hottest month with a mean maximum temperature of 44°C while January is the coldest month with a minimum temperature of 4°C. In Baluchistan province, however, the climatic conditions differ from the rest of hill torrent areas in the country. Its climate is arid to semi-arid in winter and arid to hyper-arid in summer. Rainfall is erratic and is received in monsoon as well as in winter. Due to wide variability of temperature regime, the climate of the province varies from cool temperate to tropical allowing an amazing variety of crops to grow economically. Soil texture and structure, bulk

and particle densities, and porosity are the major soil physical properties that determine the extent of the water-storage capacity of the soil. A number of studies (Thomas, et al., 2004 and Randall and Sharon, 2005) have shown that estimates of many physical and chemical characteristics of soils can be done if their texture is accurately assessed.

4.2. Distribution of Spate Flow

Spate flow draining upland areas generally have little base flow and rise rapidly after rainfall on the catchments area. Rapid runoff coupled with steep stream gradients, results in high peak flows. Serious flooding results when these streams reach the lowlands where the hydraulic gradient is much flatter. Rod Kohi system of irrigation uses earthen embankments, called bund or ganda, to divert flow from hill torrents. When the upstream diversion has received adequate water, the bund is breached and the next farmer receives the water. Almost all the spate irrigated areas in Pakistan lie in the most marginalized and socially low-ranking districts. This had a negative impact on the decision-making at the national level as far as resource allocation for the irrigation sector is concerned. A review of budgetary records clearly indicates that the bulk of investment in agricultural research and physical development has gone into the perennial irrigated agriculture (Nawaz, 2003)

4.3. Flood Damages in Spate Irrigated Areas

Fig 2. Flood damages in spate irrigated areas

The hill torrent brings flashy flood of shorter duration but very high magnitude. High flows breach earthen diversion bunds and deprive the cultivators from the use of this water. The flood water thus rushes downstream and damages crops, houses and other infrastructures. During the year 2004, a flood of about 67,600 cusecs discharge was observed in Sanghar Hill Torrent at Taunsa on June 6, 2004. A serious embayment along the left bank downstream of Indus Super Highway Bridge partially damaged the residential area. Sloughing of bank in a very large chunk took away a part of the Kacha colony whereby about 13 houses were fully taken by the flood currents along with their debris. One man and a number of animals were killed; a lot of household and food grain were taken away by the flood currents. In August 2008, a flood of 80,000 to 100,000 cusec passed through different parts of the Rajanpur district and more than 160 villages were submerged due to the flooding; resulting in the migration of over 6,000 people to safe places (Fig 2).

4.4. Area under Spate Irrigation

It is difficult to give exact figures about the area under spate irrigation because the system has never had the same amount of attention as perennial irrigation from governments, non-government development institutions and the donor community, due highly unreliable and unpredictable flood water occurrences. An estimate of the land coverage of spate irrigation systems in some countries compiled from different sources (FAO, 2005; Ahmed, 2000; Al-Shaybani, 2003 and Mehari, et al., 2005a) is presented in Table 2.

Table 2. Spate irrigated versus total irrigated areas in some countries

Country	Year of data collection	Total irrigated area in (ha) (!)	Spate-irrigated area in ha (2)	% of total irrigated area covered by spate irrigation (2)/ (!)*100
Algeria	1997	560,000	70,000	13
Eritrea	2005	28,000	15,630	56
Kazakhstan₁	1993	3,556,400	1,104,600	31
Libya	1997	470,000	53,000	11
Mongolia	1993	84,300	27,000	32
Morocco	1997	258,200	165,000	13
Pakistan	2000	17,580,000	1,450,000	8
Somalia	1984	200,000	150,000	75
Sudan	1997	1,946,000	280,000	14
Tunisia	1997	481,520	98,320	20
Yemen	2003	485,000	193,000	40

FAO, 2005; 2 Mehari, et al.,2005c; 3 Ahmed, 2000; 4Al-Shaybani,2003

4.5. Crop Production

The crop production in the area is traditionally of subsistence level in spite of the fact that farmers' resources, fertility of land, large holdings and long planting season are productivity oriented. The major limiting factor is the unreliable source of water. Cereal crops are the most important crops that occupy 60% of the area. The major crops grown during Rabi are wheat, barley and oilseed while millets, sorghum, mungbeans are the summer crops. Farmers have been diverting flood flows into small channels for irrigation purposes since decades, but many have become defunctional due to physical, technical, financial and social constraints. Yields are low, averaging about 298 kgs/acre for Sorghum, 204 kgs/acre for Oilseeds and 367 kgs per acre for Wheat. Lack of optimal use of water; poor management practices, inefficient flood management and improper use of other essential inputs are main causes for low agricultural outputs in these areas. The prevailing Rod Kohi irrigation practices are traditional in nature and provide subsistence based livelihood to the majority of the farmers in the area. Production oriented agriculture is not practiced and therefore, crop yields are quite low. Consequently, the average yields per unit area under Rod Kohi irrigation system are far below the national average (Mumtaz, 1989). Pakistan agricultural research council (PARC) has developed many measures for attaining reasonable crop yield. The promising ones could be: application of optimum irrigation at the correct time; switching of crops from low value to high value, switching from high delta of water crops to low delta of water crops, use of high efficient irrigation methods, etc. But this could only be possible when there is assured availability of flood water. This can be possible only when provide some kind of irrigation structures which stored water so as to provide supplement irrigation during the dry period particularly during the critical period of crop growth.

4.6. Past Spate Flow Management

In spite of the fact that the government of Pakistan favors allocation of resources to perennial irrigation systems, around 74 permanent headworks have been constructed in Balochistan in the past decades (Van Steenbergen, 1997). The failure rate of these modern structures has, however, been very high for a number of reasons. The main ones include: sedimentation, discrepancy with the indigenous water rules and water sharing arrangements, lack of flexibility of the structures to cope with the unpredictable nature of the floods. An extensive evaluation of 47 modernized systems constructed in the past 30 years has revealed that only 34% still function satisfactorily, 32% have serious operational problems and 34% are completely non-functional (Van Steenbergen, 1997) (Table 2). Fig. 3 shows the unmanaged or traditional spate irrigation system while Fig.4 shows the managed and well established system. The irrigation and Power Department in Balochistan is responsible for the operation and maintenance of the established spate irrigation systems. The annual budget of the Department for the maintenance of the structures is on the decline. The maintenance work is limited to posting of linemen and guards, and the major repair work is done on ad-hoc basis (Van Steenbergen, 1997). This has already made many of the structures listed as 'with serious operational problems' in Table 3, non-functional. If these problems are not fixed, a number of the functional structures could soon become out of use. Some modernized structures were not acceptable by the local people (upstream & downstream) e.g. In Anambar Plain in Balochistan, Pakistan, one of the introduced modern weirs significantly changed the indigenous water distribution system (Van Steenbergen, 1997 and Mehari, etal., 2005c).

Table 3. Performance of government constructed spate schemes in Balochistan, Pakistan

Date of construction	Total head works constructed	Functional		With serious Operational problems		Non-functional	
		Nos.	%	Nos.	%	Nos.	%
Prior to 1973	20	7	35	6	30	7	35
1974-1984	14	4	29	2	14	8	57
After 1984	13	5	38	7	54	1	8
Total	47	16	34	15	32	16	34

Source: Groundwater Consult 1991

The weir was constructed to divert spate flows to upstream fields. It performed this function, but it also considerably reduced the base flow to the downstream fields' which caused many tensions and conflicts. As conflicts became unbearable, the two communities (upstream & downstream farmers) reached a mutual agreement: they purposely blew up the weir (Fig. 5) and returned to their indigenous structures and water-sharing arrangement.

Fig. 3. Unmanaged Non perennial spate irrigation

Fig. 4. Managed Non perennial spate irrigation

Fig. 5. Deliberately destroyed weir in Anambar Plain

4.7. Present & Future Spate Flow Management

As mentioned above that the upstream & downstream farmers don't compromise on the water distribution. So the need was to make such structures which are acceptable by both communities. Considering these facts PARC emphasized on the improvement of water conveyance system, distribution, diversion and water application structure. The local communities adopted these technologies successfully.

4.8. Remodeling of Rod-Kohi Conveyance Irrigation

Remodeling of the system improved control over water, reduced operational losses and increased the quantity of flood water. Silted channel not only reduces the carrying capacity of water channel but also causes over topping of flood water which not only damage crops but also results devastation to property. The cost effective remodeling conveyance system not only provided full supply level of floodwater but also provided opportunity to increase the irrigated area with enhanced water conveyance efficiency. Improved/enhanced water conveyance system has increased the reliability of flood water flow which ultimately furnished optimum moisture level corresponding to each diversion point. Therefore results showed in Fig-6 that after refinement of water channel the grain wheat yield has been increased from 1520 to 1870(19%), 1300 to 1780(27%) kg/ha at head, 1020 to 1790(43%), 700 to 1830(62%)kg/ha at middle and from 630 to 1465(57%) and from 410 to 1280(68%) kg/ha at tail reaches (Asif at el, 2014);

The comparison of discharge & crop yield production, before and after remodeling is shown below in Fig-6;

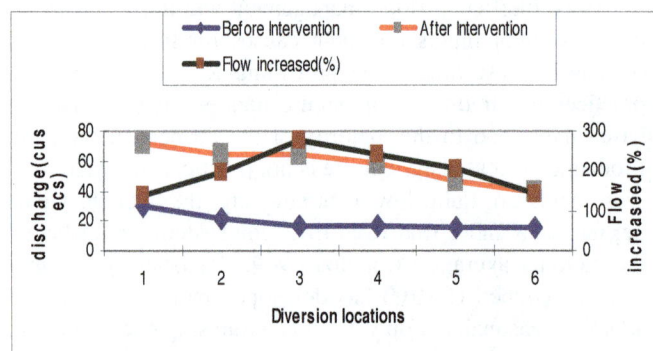

Fig 6. Frequency of receiving flood pre and post intervention

4.9. Water Distribution Structures

Most critical element in managing Rod-Kohi System is diversion and distribution structures, which needs to perform both under low-and high flows and should provide equitable water according to their water rights. Such structures have been constructed in spate irrigated areas of Pakistan using different construction material with the participation of local communities. These structures helped farmers to regulate floodwater at desired level without any extra labor and time. Both communities (upstream & downstream) has consensus on these structures which also helped to remove the conflicts among them (Fig 7).

4.10. Water Application Structures

The cultivated fields in Rod-Kohi irrigation System are large to the extent of 8 - 10 acres or more. The water application to the fields is crucial to control water in the flood season. Cost effective water application structures of

different designs capacities and construction materials (pre-casted pipe nacca structure) have been developed. The structures were constructed on the basis of cost effectively, feasibility, adaptability and farmers perception.

4.11. Water Diversion Structures

Water regulating and diversion structures are used to regulate the flow by raising the water level for diversion of water to command area. Initially, the activities were initiated at the farm level to improve water conveyance, distribution and application systems. Cost-effective technologies for water distribution and application have been tested and introduced successfully. Local community is adopting these technologies due to cost effectiveness and easy in operation.

(a)   (b)   (c)   (d)

Fig. 7. Development of hill torrent structures made by PARC

4.12. Role Sister Departments for Rod Kohi Watershed Management

At present, no other provincial departments or federal ministries, like forest, agriculture, fisheries, soil conservation and environment have not undertaken any kind of tangible interventions to manage the vast Rod Kohi watershed area. As a matter of fact, an integrated approach should have been adopted for downhill tree plantations, river training works in the various drains flowing from hill tops to down wards and construction of series of check dams on downhill slopes

5. Conclusion

Rod Kohi system of irrigation is the least known and the most unattended among irrigation systems in Pakistan, and therefore, remained undeveloped. The major reasons include poor resources of Rod Kohi farmers, ignorance of farmers to advanced irrigation practices, excessively high flows, non-existence of control structures, lack of scientific investigations about the farmer's irrigation practices and performance evaluation Agriculture in these areas totally depends on hill torrent flows that are un-predictable in terms of timing and magnitude making scheduled irrigations impossible. Although, the production level of these areas cannot be brought at par with those in irrigated areas, however, it can certainly be increased if suitable cultivars, appropriate technologies of soil and water conservation, best suitable with the agro-climatic conditions, are evolved and

developed for adoption by the farmers. Each structure should be designed based on careful research of site-specific conditions and field experiences in construction, operation and maintenance. The flood control structures have to work during high floods and thus are to be durable against extreme conditions.

6. Recommendations

To attain Agricultural enhancement, employment, income and thereby to alleviate poverty and hunger in spate irrigated areas, we need to carry out proper water management practices so that water is made available as per requirement. For this we have to design/develop two kinds of water management practices such as

i. Development of permanent diversion and distribution structures by involving the local communities so that the water rights of the people at up/down streams could be protected. PARC is providing great services in spate irrigated areas with promising results but it needs to be enhanced on large scale because we have land potential to 17 M Acre and average water potential of 19MAF for development of spate irrigation system (NESPAK-1998). If this potential area is bring under irrigation through proper water management practices along with the use of productive oriented technology then no doubt agricultural enhancement will takes place in these areas and food security problems will be solved. Furthermore the impact of these interventions will create job opportunities which will ultimately stop out-migration to urban areas and more importantly the since of deprivation among the people of these areas will be somehow tone down.

ii. As the rushing flood is unpredictable in term of timing and magnitude making scheduled irrigation impossible so therefore storage facility needs to be provided at suitable location so that precious flood water could be stored and then can be used as subsequent irrigation during dry periods particularly at critical period of crop growth.

References

[1]   Ahmed, S., 2000. Indigenous water harvesting systems in Pakistan. Water Resources Research Institute (WRRI), National Agricultural Research Centre (NARC), Islamabad, Pakistan.

[2]   Al-Shaybani, S.R., 2003. Overview of non-modernized spate irrigation systems in Yemen.

[3]   FAO. 1997. Modernization of irrigation schemes. In: Past Experiences and Future Options; Proceedings of the Expert Consultation, Bangkok, November 1996

[4]   FAO, 2005. Crop water management. FAO Land and Water Development Division.

[5]   GROUNDWATER CONSULT (1991).Baluchistan flood water irrigation systems. Islamabad: Royal Netherlands Embassy

[6]   Hamilton, R., Muhammad, A.K., 1995. Sailaba irrigation practices and practices. Arid Soil Research and Rehabilitation. University of Idaho, ID 83844-2334, USA.

[7]   Mehari, A., Schultz, B., Depeweg, H., 2005a. Hydraulic performance evaluation of the spate irrigation systems in Eritrea. Irrigation and Drainage 54.4: 1-18.

[8]   Mehari, A., Van Steenbergen, F., Schultz, B., 2005b. Water rights and rules and management in spate irrigation systems in

Eritrea, Yemen and Pakistan. In: African Water Laws: Plural Legislative Frameworks for Rural Water Management in Africa;

[9]   Proceedings of the International Water Management Institute Workshop, Johannesburg, South Africa, 26 - 28 January, 2005 Mehari, A., Schultz, B., Depeweg, H., 2005c. Where indigenous water management practices overcome failures of structures. Irrigation and Drainage 54.1: 1-14.

[10] Mumtaz, A (1989). Keynote Address, In: BARD Rod Kohi Agricultural Problems and Prospects Symposium, Nov. 27-29 1989, Islamabad, Pakistan Agriculture Research Council, 2-9.

[11] Nawaz, K., 2003. Spate irrigation in Dera Khazi Khan, Pakistan [online]. Available at http://www spate irrigation.org

[12] NESPAK. (1998). Master Feasibility Studies for Flood management of Hill Torrents of Pakistan. Supporting Vol-V, Balochistan,

[13] Randall, J.S., Sharon, A., 2005. Soils: Genesis and geomorphology. Cambridge University Press, UK, ISBN: 0521812011, 832 pp.

[14] Sithole, B., 2000. Telling it like it is: Devolution in the water reform process in Zimbabwe. In: Constituting the Common: Crafting Sustainable Commons in the New Millennium; Proceedings of the 8th Biennial Conference of the International Association for the Study of Common Property (IASCP), Bloomington, Indiana, USA, 31 May - 4 June 2000.

[15] Thomas, W.L., Robert, G.S., Richard, R.T., Howard, W.N., 2004. Soil water monitoring and measurement. A Pacific Northwest Publication, Washington State University, Oregon, Idaho   [online].   Available   at http://cru.cahe.wsu.edu/CEPublications

[16] Van Der Zaag, P., 2006. Water's vulnerable value in Africa. Value of Water Research Report Series No. 22, UNESCO-IHE Institute for Water Education, Delft, the Netherlands; University of Twente, Enschede, the Netherlands; Delft University of Technology, Delft the Netherlands.

[17] Van Steenbergen, F., 1997. Understanding the sociology of spate irrigation: Cases from Balochistan. Journal of Arid Environments 35: 349-36

5

Effects of phosphorus availability on plant growth and soil nutrient status in the rice/soybean rotation system on newly cultivated acidic soils

Laye Djouba Conde[1,2], Zhijian Chen[1], Hongkao Chen[1], Hong Liao[1]

[1]State Key Laboratory for Conservation and Utilization of Subtropical Agro-Bioresources, Root Biology Center, South China Agricultural University, Guangzhou, 510642, People's Republic of China
[2]National Department of Agriculture, Ministry of Agriculture and Livestock, Conakry BP: 576, Republic of Guinea

Email address:
hliao@scau.edu.cn (Hong Liao), ladjoc@hotmail.com (L. D. Conde)

Abstract: Acid soils are worldwide spread, where low phosphorus (P) availability is considered as the major limiting constraint for crop growth, particularly on the newly cultivated acidic soils. Traditionally, the rotation system of rice with leguminous crops has been often used on acid soils. However, little is known about how P availability affects this traditional rotation system on acid soils. In the present study, two years of soil pot experiments had been done using rice (*Oryza sativa* L.) as the first crop and soybean (*Glycine Max* L.) as the second crop. The results showed that rice growth were significantly affected by P fertilization on acid soils. Sufficient P application increased plant height, shoot biomass, tiller number, and panicle dry weight compared to that of no P fertilization in both two years' studies. The growth of following crop soybean was also influenced by P supply, and the P efficient genotype HX1 exhibited more adaptive to low P than the P inefficiency genotype BD2, as reflected by better growth of HX1 than BD2. Rhizosphere pH and soil nutrient status was significantly influenced by the rotation system. An increased tendency of rhizosphere pH was observed after the growth of rice and soybean. Soil N concentration was significantly increased after planting HX1 but not BD2. Furthermore, rice rotated with HX1 resulted in higher P fertilizer use efficiency (PFUE). Taken together, we conclude that the rice-soybean rotation with optimal P supply is a suitable agricultural mode on acid soils, and rotating with the P efficient soybean genotype could benefit more in soil nutrient status, which might increase the agriculture sustainability on acid soils.

Keywords: P Availability, Crop Rotation, P Fertilization, Nutrient Status, Acid Soils

1. Introduction

Acid soils (pH<5) are widely distributed in the world, and over 40% of the world's arable lands are acidic [1]. In China, there are about 2.1 million ha of acid soils, which mainly existed in South China [2]. Although the abundant rainfall and high temperature make this region beneficial for agricultural production, there are still a lot of limiting factors for crop growth, such as low pH, low cation exchange capacity (CEC) and organic matter, which easily lead to nutrient deficiency and element toxicity, such as phosphorus (P) deficiency and aluminum (Al) toxicity, especially on the newly cultivated acid soils [3]. The newly cultivated acid soils are much severely affected by all above characters and required great amount of fertilizers to maintain or improve

crop yield, and thus believed to have significant negative effects on the quality and production of crops on these soils [4]. Therefore, the crop productivity of these soils is generally low, and proper utilization and amelioration for the soil constraints, especially low P bioavailability, remain a great challenge for agricultural production [5].

Phosphorus is an essential macronutrient for plant growth [6]. Although the total amount of P is high in soils, available P for plant growth is often limited, especially in acid soils, where P is easily fixed by soil components into unavailable forms [7]. Low P availability is a major limiting constraint for crop production on acid soils [8]. In spite of application of P fertilizer is essential to maintain crop yield, applied P fertilization efficiency is usually low (only 20%) and readily leading to P accumulation in soils and resulting in potentially environmental pollution, and thus not economic for agricultural production at developing countries or even at developed countries [9]. Furthermore, P fertilizers are produced from phosphate rock which is a non-renewable resource, and will be fully consumed within next few decades [10]. Therefore, to improve P fertilizer management as well as to enhance P efficiency in crops is absolutely necessary for environment-friendly agriculture.

Appropriate crop cultivation is absolutely necessary to increase crop biomass and yield, such as rotation and intercropping among various crop species [11]. Crop rotation has been used for thousands of years. Improvements in soil physical properties and soil organic matter including multiple years of sod, pasture, or hay, probably play a beneficial role in rotations [12]. In cereals and legumes rotation system, legumes are considered as the resource of nitrogen (N), since legume nodules have the strongly ability of N_2 fixation, and thus increased N availability to the following crops, which also contributes to increase soil fertility and avoid some diseases proliferation in crops [13,14]. Furthermore, it has been reported that some pre-culture crops, such as rice and wheat, have a potential ability to mobilize insoluble P from soils and thus make the P available to the following crop species [15]. Therefore, rice rotated with legumes has been traditionally used for increasing crop yield as well as improving soil fertility.

Rice (*Oryza sativa* L.) is the world's most important crop and a primary food for half of the world's population [16]. Soybean (*Glycine max* L.) is the world's foremost provider of protein and oil, which could form symbiosis with rhizobia to form nodule and thus fix atmospheric N [17]. In this study, pot experiments on rice-soybean rotation were conducted for two years (2009 and 2010) using newly cultivated acid soils at the Ningxi experimental site of South China Agricultural University in Guangdong Province. Rice (*Oryza sativa* L., var. Hua han xi) was grown as the first crop during the first season (April-August) and then followed by two soybean genotypes (*Glycine max* L., var. HX1 and BD2) contrasting in P efficiency in the second season (August-November). Effects of P availability on the growth of rice and soybean plants as well as soil nutrient status were investigated under greenhouse conditions. The beneficial effects of rice rotating with the P efficient soybean genotype on soil nutrient status were further discussed.

2. Materials and Methods

2.1. Plants Materials and Growth Conditions

The rice-soybean rotation experiment had been conducted in 2009 and 2010 in a greenhouse located at the Ningxi experimental site of South China Agricultural University in Guangdong Province, China (22°56' N, 1132°57' E), where the climate is subtropical and the average annual rainfall is about 1200 mm, out of which 70-80% occurs during May-

October. The soil has developed on a basaltic alluvium of very fine montmorillonite. The textural class of soil is Ultisols, often known as red clay soils (USDA soil taxonomy). Basic soil chemical characteristics as follows: pH, 4.38; organic matter, 0.88%; available P Bray II method [18], 8.96 mg P kg^{-1}; available nitrogen (N), 19.58mg N kg^{-1}; available potassium (K), 64.25 mg K kg^{-1}.

One rice cultivar (*Oryza sativa* L., var. Hua Han xi) and two soybean genotypes (*Glycine Max* L., var. HX1 and BD2) were used in these two years' experiments. Among the two soybean genotypes, HX1 had been previously proved as the P efficient genotype and BD2 as the P inefficient genotype through field screening studies for P efficiency on acid soils [19]. Thirty days old rice seedlings stemming from nurses were transplanted into soil pots as the first crop from April to August. After rice plants harvested, the soil of each pot was separately sifted to eliminate the left roots, and then two soybean genotypes were grown in the same pot with the same soil materials without any fertilization from August to November at the same year.

There were three P treatments in the experiments, including high P (200 mg P_2O_5 kg^{-1} soil added as calcium super phosphate, HP), low P (100 mg P_2O_5 kg^{-1} soil added as calcium super phosphate, LP), and non P (none P fertilizer added, NP). The P fertilizer was applied to the soil and fully mixed, and then kept under submerged conditions for 1 d before transplanting. 200 mg kg^{-1} of N from urea (CO (NH$_2$)$_2$) and 200 mg kg^{-1} of K from potassium chloride (KCl) were applied for each treatment. All the fertilizers were applied one time as basal fertilizer before rice growth. The same soils with P fertilizers (NP, LP and HP) used for rice plants were maintained for soybean growth without any other fertilization. Each treatment had eight replicates for rice and four replicates for each soybean genotype, and thus there were 24 pots in total in each year. Plastic boxes with 25 cm × 19 cm and 23 cm × 23 cm (diameter × depth) were employed in the pot experiments in 2009 and 2010, respectively. The rice and soybean plants were harvested at 105 d and 92 d after P treatments, respectively. After washing soils away with water, shoot and root samples were collected for further measurements.

2.2. Determination of Root Growth Parameters and P Content

Total root length and root surface area were measured using an Epson 1640XL scanner (Epson Corp., Nagano, Japan) and further quantified with the WinRhizo software (Regent Instruments Inc., Sainte-Foy, Canada). After that, shoot and root samples were oven dried at 105°C for 30 min and further dried at 75°C for dry weight determination. Then, shoots were ground into powder, and 0.2 g sample was digested with a mixture of H_2SO_4-$HClO_4$ and total P content was measured via the phosphorus-molybdate blue color reaction [20]. The root/shoot ratio was determined as dry weight of roots divided by those of shoots. The P fertilizer use efficiency (PFUE) was calculated according to [21] as follows:

$$PFUE = \frac{\text{P uptake in the fertilized treatment} - \text{P uptake in the control treatment}}{\text{Amount of P applied for the fertilized treatment}} \times 100$$

Namely P uptake by plants with LP or HP treatment minus those of with NP treatment, and then divided by the amount of P fertilized in soils (LP or HP).

2.3. Analysis of Soil Nutrient Status

For determination of N, P and K content in soils, the soil samples were collected after rice or soybean plants harvested. The soil available P was extracted by the Bray II method digesting with 25 mL mixture of $HClO_4$-H_2SO_4 and measured as described above. N and K in soils were measured according to [20].

2.4. Data Analysis

All the statistical analysis was performed in Microsoft Excel 2003 (Microsoft Company, Redmond, WA, USA) for calculating means and standard errors. Statistical Package for Social Sciences (SPSS 15.0, 2001) was used for ANOVA and multiple comparison analysis.

3. Results

3.1. Rice Growth as Affected by P Availability

Table 1. Rice growth was affected by P availability in 2009 and 2010.

Year	Treatment	Plant height / cm plant^{-1}	Shoot biomass / g (3 plants)$^{-1}$	Tiller number / # (3 plants)$^{-1}$	Panicle dry weight / g (3 plants)$^{-1}$
2009	NP	50.87±13.21b	7.75±3.44b	7.37±2.32c	3.18±2.55b
	LP	78.75±6.40a	56.53±17.82a	33.87±5.56b	40.62±16.24a
	HP	82.12±7.33a	63.54±23.17a	40.37±7.76a	40.25±13.47a
	F-value	26.18***	25.57***	75.87***	24.58***
2010	NP	83.19±2.64c	51.62±20.94b	22.13±3.314c	25.02±11.47b
	LP	99.75±3.61b	61.08±6.409b	40.25±4.13b	62.17±14.38a
	HP	105.60±5.04a	80.73±8.05a	47.38±7.19a	85.60±18.84a
	F-value	71.44***	9.72**	51.00***	16.12***

HP (200 P_2O_5 mg kg^{-1}) and LP (100 P_2O_5 mg kg^{-1}) added as calcium super phosphate. NP means none P fertilizer added. All the data are the means of eight replicates with standard error. The same letter in the same column indicated no significant difference between P treatments at the $P<0.05$ level. F-value was calculated by two-way ANOVA. Asterisk indicated significant difference. **: $0.001<P<0.01$; ***: $P<0.001$.

As the first crop in the rice-soybean rotation during April to August in 2009 and 2010, rice growth was significantly affected by P fertilization ($P<0.05$) (Table 1). Although similar effects on rice growth were observed between LP and HP treatments, rice growth was enhanced with increasing P application compared to that of no P fertilization (NP). For example, with sufficient P supply (HP) in soils in 2009, plant height, shoot biomass, tiller number, and panicle dry weight were significantly increased by 61.43%, 719.87%, 447.76% and 165.72% compared with that in NP conditions, respectively (Table 1). Similarly, increased P supply in 2010 also resulted in dramatically increase in plant growth. Plant height, tiller number, shoot biomass and panicle dry weight were 26.93%, 56.39%, 114.09% and 242.12% higher in HP than that in NP conditions, respectively (Table 1). Furthermore, we also observed that plant growth of all the above parameters in 2010 were better than that in 2009, which probably due to soil nutrient status in 2010 was more fertile than that in 2009.

3.2. Soybean Growth as Affected by P Availability

After rice plants harvested, two soybean genotypes contrasting in P efficiency were separately cultured in the same soils from August to November in the same year without any fertilization. For the P inefficient soybean genotype BD2 in 2009, significantly higher plant height (37.75 cm plant^{-1}), plant biomass (9.65 g (3 plants)$^{-1}$) and

root/shoot ratio (0.95 g (3 plants)$^{-1}$) were observed at HP treatment, but the above parameters were not different between LP and NP treatments (Table 2). Furthermore, P supply significantly affected BD2 root growth, including total root length and root surface area. The longest root length was observed at HP treatment followed by LP treatment, which was 326.27% and 204.71% higher than those at NP treatment. Similar response was also observed in root surface area, HP had the highest value of root surface area, followed by LP and NP (Table 2).

For the P efficient soybean genotype HX1 in 2009, the growth of HX1 was also affected by P fertilization. As shown in Table 3, increasing P supply (LP and HP) enhanced the growth of HX1, especially under HP conditions, which plant height and plant biomass was 51.06% and 231.63% higher than that in NP treatment, respectively. Furthermore, growth of HX1 roots was also influenced by P supply. The root/shoot ratio, total root length and root surface area under HP were 526%, 132.19% and 409.04% higher than that in NP conditions, respectively (Table 3).

Consistent to the results from 2009, most of the growth parameters of BD2 were increased with increasing P fertilization in 2010, especially in HP conditions (Table 2). For HX1, plant height, plant biomass and root surface area were increased with increasing P fertilization, whereas root/shoot ratio, total root length were similar within P treatments in 2010 (Table 3). Furthermore, we also found that most of the growth parameters in HX1 were higher than those in BD2 among different P treatments, especially in 2010 (Tables 2 and 3), confirmed that the P efficient soybean genotype HX1 have a higher P acquisition and/or utilization efficiency than BD2.

Table 2. Soybean (BD2) growth was affected by P availability in 2009 and 2010.

Year	Treatment	Plant height / cm plant^{-1}	Plant biomass / g (3 plants)$^{-1}$	Root/Shoot ratio	RL / cm (3 plants)$^{-1}$	RSA / cm^2 (3 plants^{-1})
	NP	26.00±1.63b	2.01±0.43b	0.72±0.61b	61.55±8.72c	3.22±0.53c
2009	LP	30.00±2.94b	4.58±1.25b	0.83±0.51b	126.00±47.97b	5.92±1.98b
	HP	37.75±6.55a	9.65±5.21a	0.95±0.23a	200.82±38.10a	9.27±1.50a
	F-value	7.89*	7.28*	6.34*	15.22**	17.02**
	NP	21.00±2.16c	2.66±1.18c	0.59±0.32b	487.25±207.91b	64.50±14.05b
2010	LP	24.00±1.63b	6.97±2.65b	0.62±0.51b	695.50±111.21b	77.95±13.56b
	HP	28.25±1.70a	10.53±6.43a	0.96±0.80a	956.75±328.28a	141.47±35.29a
	F-value	15.53**	13.62**	7.59*	4.06*	12.46**

HP (200 P_2O_5 mg kg^{-1}) and LP (100 P_2O_5 mg kg^{-1}) added as calcium super phosphate. NP means none P fertilizer added. RL: Root length. RSA: Root surface area. All the data are the means of four replicates with standard error. The same letter in the same column indicated no significant difference between P treatments at the $P<0.05$ level. F-value was calculated by two-way ANOVA. Asterisk indicated significant difference. *: $0.01<P<0.05$; **: $0.001<P<0.01$.

Table 3. Soybean (HX1) growth was affected by P availability in 2009 and 2010.

Year	Treatment	Plant height / cm plant^{-1}	Plant biomass / g (3 plants)$^{-1}$	Root/Shoot ratio	RL / cm (3 plants)$^{-1}$	RSA / cm^2 (3 plants)$^{-1}$
	NP	25.32±5.06b	3.78±1.52b	0.13±0.08a	143.20±65.20b	5.75±2.04c
2009	LP	34.00±3.16a	5.83±02.18b	0.25±0.12a	223.05±84.5ba	15.37±2.13b
	HP	38.25±2.21a	11.38±4.21a	0.94±0.23b	332.50±62.34a	29.27±2.51a
	F-value	12.82**	18.49**	5.20*	7.09*	111.29***
	NP	30.50±1.29b	8.92±3.14c	0.39±0.17a	521.00±169.97a	71.30±29.85b
2010	LP	40.50±3.69a	13.66±5.25b	0.33±0.11b	988.42±568.90a	140.32±70.58a
	HP	42.25±3.30a	23.97±4.37a	0.30±0.09b	1168.83±349.49a	241.02±86.34a
	F-value	18.37**	26.22***	5.20*	2.82ns	6.55**

HP (200 P_2O_5 mg kg^{-1}) and LP (100 P_2O_5 mg kg^{-1}) added as calcium super phosphate. NP means none P fertilizer added. RL: Root length. RSA: Root surface area. All the data are the means of four replicates with standard error. The same letters in the same column indicate no significant difference between P treatments at the $P<0.05$ level. F-value was calculated by two-way ANOVA. Asterisk indicated significant difference. *: $0.01<P<0.05$; **: $0.001<P<0.01$; ***: $P<0.001$. ns, indicated no significant difference at the $P<0.05$ level.

3.3. Effects of P Availability on Soil pH and Nutrient Status in the Rice-Soybean Rotation System

We further investigated the changes of pH and nutrient status in rhizosphere after rice and soybean growth at three P levels. As shown in Table 4, the initial pH of the newly cultivated acid soils was 4.38 and 4.50 in 2009 and 2010, respectively. An increased tendency of rhizosphere pH was observed after the growth of rice and soybean at all P levels, especially in 2010. Furthermore, rhizosphere pH was higher after growth of HX1 as compared to BD2 (Table 4). The nutrient status in rhizosphere showed that N, P and K concentrations in soils were decreased after rice growth at three P levels, and the three tested nutrient concentrations were decreased when the P level was increased both in 2009 and 2010 (Table 5). Similarly, P and K concentration in soils were also decreased after BD2 and HX1 growth in both two years (Table 5). Interestingly, N concentration in soils was increased after soybean (both BD2 and HX1) growth both in 2009 and 2010. For example, after the growth of the P efficient soybean genotype HX1, N concentration in soils were 121.16%, 173.48% and 101.29% increasing at NP, LP and HP treatments, respectively (Table 5), which was probably attributed to the strong symbiotic N fixation ability of soybean plants to increase N concentration in soils.

Table 4. Changes of pH in rhizosphere of rice and soybean plants in 2009 and 2010.

Year	Treatment	Before planting	After rice	After soybean BD2	HX1
	NP		4.45±0.09a	5.19±0.04a	5.30±0.06a
2009	LP	4.38	4.56±0.10a	5.20±0.05a	5.20±0.05a
	HP		4.62±0.10a	5.21±0.05a	5.25±0.03a
	F-value		2.01ns	1.21ns	3.11ns
	NP		5.29±0.17a	5.84±0.07a	5.95±0.08a
2010	LP	4.50	5.30±0.22a	6.12±0.15a	6.17±0.16a
	HP		5.31±0.24a	6.25±0.16a	6.33±0.19a
	F-value		1.51ns	2.81ns	2.91ns

HP (200 P_2O_5 mg kg^{-1}) and LP (100 P_2O_5 mg kg^{-1}) added as calcium super phosphate. NP means none P fertilizer added. All the data are the means of four replicates with standard error. The same letters in the same column indicate no significant difference between P treatments at the $P<0.05$ level. F-value was calculated by two-way ANOVA. ns, indicated no significant difference at the $P<0.05$ level.

Table 5. Changes of nutrient status in rhizosphere of rice and soybean plants in 2009 and 2010.

Year	Nutrient	Rice			BD2			HX1		
		NP	LP	HP	NP	LP	HP	NP	LP	HP
2009	N	-68%	-82.89%	-80.18%	+121.16%	+173.48%	+101.29%	+25.84%	+42.45%	+7.3%
	P	-5.58%	-72.29%	-64.82%	-7.03%	-16.95%	-40.28%	-42.41%	-14.40%	-21.32%
	K	-38.66%	-47.76%	-61.33%	-47.76%	-61.33%	-39.42%	-47.32%	-59.23%	-68.20%
2010	N	-62.21%	-83.55%	-83.09%	+78.72%	+172.91%	+140.54%	+84.47%	+180.90%	+150.61%
	P	-52.00%	-87.79%	-88.56%	-27.80%	-44.61%	-44.33%	-18.25%	-59.95%	-53.44%
	K	-30.51%	-45.40%	-65.10%	-16.95%	-9.98%	-4.09%	-33.94%	-25.94%	-3.74%

HP (200 P_2O_5 mg kg^{-1}) and LP (100 P_2O_5 mg kg^{-1}) added as calcium super phosphate. NP means none P fertilizer added. '+' means increasing rate between soil nutrient concentration before rice or soybean and after rice/or soybean; '-' means decreasing rate between soil nutrient concentration before rice/or soybean and after rice/or soybean.

3.4. P Fertilizer Use Efficiency (PFUE) in Rice-Soybean Rotation on Acid Soils

The results showed that the PFUE for rice, HX1 and BD2 under HP conditions were higher than that in LP conditions in 2009 (Figure 1A), and the highest PFUE was observed in rice (42.31%) followed by HX1 (4.89%) and BD2 (3.94%) under both LP and HP conditions in 2009 (Figure 1A). Moreover, the total PFUE for rice/HX1 rotation were 46.53% and 67.42%, which were higher than that for rice/BD2 rotation (37.78% and 46.68%) in LP and HP treatments in 2009, respectively. However, in contrast to 2009, PFUE for rice under LP conditions was significantly higher than that in HP conditions in 2010, which resulted in lower total PFUE for both rice/HX1 and rice/BD2 rotation under HP conditions compared to LP treatment in 2010 (Figure 1B). These results indicated that rice rotated with the P efficient soybean genotype HX1 had higher PFUE, which is better for development of sustainable agriculture through reduced P fertilization.

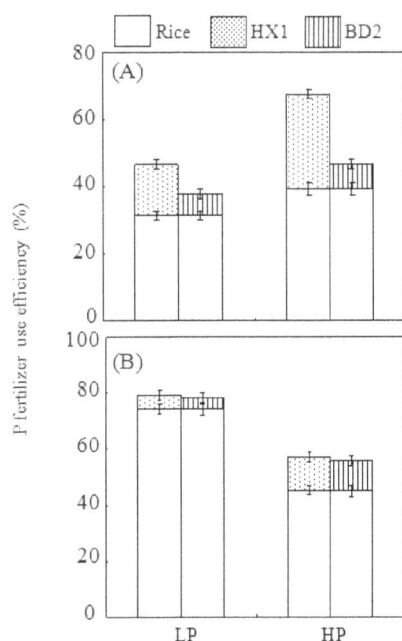

Figure 1. Phosphorus fertilizer use efficiency (PFUE) at rice-soybean rotation. (A) PFUE in 2009. (B) PFUE in 2010. HP (200 P_2O_5 mg kg^{-1}) and LP (100 P_2O_5 mg kg^{-1}) added as calcium super phosphate. Each bar represents the means of eight replicates for rice and four replicates for soybean with standard error.

4. Discussion

The newly cultivated acidic soils are not suitable for crop growth because of many poor soil properties, which make it unproductive for agricultural development [20]. Low pH and low P availability are consider as the two coexisted important limiting factors for crop growth and productivity on acid soils. It has been reported that rice growth in newly cultivated acidic soils was seriously influenced and thus decreased rice productivity [22, 23]. The growth of other crop species, such as cotton and maize, was also influenced attributed to adverse soil fertility and the proliferation of plant disease [24]. Until now, diverse appropriate crop systems were improved by farmers to adapt to the infertility of newly cultivated acid soils, such as intercropping and crop rotation [25].

Crop rotation system is an effective approach to improve soil fertility and further support crop growth [26]. Rotations of cereals with legumes or waterlogged crop with legumes have positive effects on improvement in the soil physical properties [27]. Increased organic matter or plant residues returned to soils after the first cereal crop were considered as a great advantage for the soil fertility that contributed more benefits to the following crop under crop rotation system [28]

It has been demonstrated that rice and soybean rotation could increase rhizosphere pH or organic matter, which subsequently affect the growth and activity of microorganisms, and thus benefit for nutrient cycling processes in soils [29]. Presently, our results showed that after both rice and soybean growth, pH of the newly cultivated acidic soils was increased about 1 unit compared to that before crop planting, especially in 2010 (Table 4), which might benefit in improvement of soil quality and growth of the following crop. Consistently, it has been also revealed that soil pH was also increased in wheat/common bean, and maize/faba bean rotation system [30, 31]. Based on the pH changes in the 10 years' soybean and mungbean-based rotation systems, researchers found that rotation increased up to 1 pH unit in these crop systems [32]. The near neutral pH probably resulted in maximum dissolution of P from Fe and Al complexes [33]. Additionally, in the cereal legume rotation system, legumes are prone to improve soil microbial activity, and fertility properties through increased water holding capacity, soil organic C and N content, which benefit for the next crop growth [34]. Therefore, rice and soybean rotation could be one of the important approaches to improve crop production and maintain soil fertility.

In this study, after rice harvested, the concentration of N, P and K in the soil was decreased significantly as compared to soil nutrient status before rice planting (Table 5). This showed that rice plants took up significant amount of N, P and K from the soil for its proper and vigorous growth. Similar effects were also found in maize and wheat that cereal crops have to absorb adequate nutrients from soils in order to meet the demand of shoot and root growth, and thus only a few contribution to improve soil fertility [35, 36]. In contrast, N concentration in the soils was significantly increased after soybean growth, especially under LP application, as compared to soils nutrient conditions before soybean planting, while the concentration of P and K decreased significantly as compared in the soils before soybean planting (Table 5). Similar results have been reported that legume crops have positive effects on nutrient status, especially on N and P status [37]. For example, it have been reported that legume corps with the ability of symbiotic N_2 fixation in nodules could lead to significantly increase in the available soil N, and thus increased N availability for the cereal crops in the legume/cereal rotation system [38]. Thus, in our rice and soybean rotation, soybean plants play an important role in increasing the N status in soils, which may be further increased the fertility of soils.

Phosphorus fertilizer use efficiency (PFUE) is a parameter that reflected the crop capacity to take up available P from rhizosphere [39]. In this study, we found that PFUE of the P efficient soybean genotype HX1 was higher than that of the P inefficient genotype BD2, especially in 2009 (Figure 1A). It has been demonstrated that crops with higher P efficiency exhibited higher PFUE in soils, such as wheat, fafa bean, chickpea and rapeseed [40]. Root architecture was closely related to P efficiency in soybean [17]. Root architecture determines the distribution of the whole root system in the different soil layers hence may affect soil exploration and exploitation of nutrients (especially for nutrients with low mobility, such as P [41, 42]. Additionally, crops with suitable root morphology probably modified rhizosphere pH and exudates organic acids into rhizosphere, and thereby increased soil P availability [43]. In this study, HX1 exhibited appropriate root/shoot ratio, root length and root surface area in all P supply conditions (Table 3), which probably helps its root system to acquire more P than BD2. Similar response has been previously reported in wheat [44] and rice [45], where P efficient genotypes highly facilitated biomass accumulation, root growth and plant P content. Therefore, P efficient genotype HX1 with appropriate root morphology was probably able to mobilize P from poorly soluble sources or to take up the available P in soil solution, and thus increased the PFUE with rice rotation on acid soils.

5. Conclusions

Rice/legume rotation was significantly influenced by P availability on the newly cultivated acidic soils. Sufficient P application promoted rice growth. The growth of following crop soybean was also influenced by P supply, and the P efficient genotype HX1 exhibited more adaptive to low P than the P inefficiency genotype BD2. Furthermore, rice rotated with HX1 resulted in higher PFUE probably attributed to fine root system to acquire more P. Taken together, rice-soybean rotation with optimal P supply is a suitable agricultural mode on acid soils, and rotating with the P efficient soybean genotype could benefit more in soil nutrient status, and thereby increase the agriculture sustainability on acid soils.

Acknowledgments

This work was funded by the grant of National Key Basic Research Special Funds of China (2011CB100301) and National Natural Science Foundation of China (No. 31025022).

References

[1] L.V. Kochian, O.A. Hoekenga, and M.A. Piñeros. How do crop plants tolerate acid soils? mechanisms of aluminum tolerance and phosphorus efficiency. Ann. Rev. Plant Biol. 55: 459-493, 2004.

[2] D.Q. Feng, Z.D. Chen, X.S. Huang, and L.H. Tang. Utilization of quality forage as fish fodder in the mountainous red soil area of south China. Pratacultural Sci. 23:57-60, 2006.

[3] H.M. Zhang, B.R. Wang, M.G. Xu, and T.L. Fan. Crop yield and soil responses to long-term fertilization on a red soil in Southern China. Pedosphere 19:199-207, 2009.

[4] N.K. Fageria, and V.C. Baligar. Fertility management of tropical acid soil for sustainable crop production. In: Rengel, Z., ed. Handbook of soil acidity. New York, Marcel Dekker, p. 359-385, 2003.

[5] Y. He, H. Lian, and X. Yan. Localized supply of phosphorus induces root morphological and architectural changes of rice in split and stratified soil cultures. Plant Soil 248:247-256, 2003.

[6] K.G. Raghothama and A.S. Karthikeyan. Phosphate acquisition. Plant Soil 274:37-49, 2005.

[7] F. Aref. Influence of zinc and boron nutrition on copper, manganese and iron concentrations in maize leaf. Aust. J. Basic Appl. Sci. 5:52-62, 2011.

[8] X.R. Wang, J.B. Shen, and H. Liao. Acquisition or utilization, which is more critical for enhancing phosphorus efficiency in modern crops? Plant Sci. 179:302-306, 2010.

[9] X.T. Ju, C.L. Kou, P. Christie, Z.X. Dou, and F.S. Zhang. Changes in the soil environment from excessive application of fertilizers and manures to two contrasting intensive cropping systems on the North China Plain. Environ. Pollut. 145:497-506, 2007.

[10] W.M. Stewart, L.L. Hammond, and S.J.V. Kauwenbergh. Phosphorus as a natural resource. Phosphorus: agriculture and the environment. P 3-22, 2005.

[11] C.L.A. Asadu, S.C. Obasi, A.G.O. Dixon, N. Ugele, and G.U. Chibuike. Soil fertility recovery in a cleared forestland cultivated and fallowed for seven years. J. Agri. Biodivers. Res. 25: 110-116, 2013.

[12] D.G. Bullock. Crop rotation. Crit. Rev. Plant Sci. 11:309-326, 1992.

[13] C.E. Ariel, A.E. Ostolaza, E.B. Giardina, and L. Giuffré. Effects of two plant arrangements in corn (Zea Mays L.) and soybean (Glycine Max L. Merrill) intercropping on soil nitrogen and phosphorus status and growth of component crops at an Argentinean argiudoll. Am. J. Agr. Forest. 1:22-31, 2013.

[14] J. Ehrmann, and R. Karl. Plant: soil interactions in temperate multi-cropping production systems. Plant Soil 376:1-29, 2013.

[15] H.G. Li, J.B. Shen, F.S. Zhang, P. Marschner, G. Cawthray, and Z. Rengel. Phosphorus uptake and rhizosphere properties of intercropped and monocropped maize, faba bean, and white lupin in acidic soil. Biol. Fert. Soils 46:79-91, 2010.

[16] N.G. Dowling, S.M. Greenfield, and K.S. Fischerf. Pacific basin study center sustainability of rice in the global food system. Pacific Basin Study Center, Los Banos, Philippines, International Rice Research Institute, 1998.

[17] H. Fujikake, H. Yashima, T. Sato, N. Ohtake, K. Sueyoshi, and T. Ohyama,. Rapid and reversible nitrate inhibition of nodule growth and N fixation activity in soybean (Glycine Max L. Merr). Soil Sci. Plant Nutr. 48:211-217, 2002.

[18] Bray, R. H., and L. T. Kurtz. 1945. Determination of total organic and available forms of phosphorus in soils. Soil Sci. 59: 39-45.

[19] J. Zhao, J.B Fu, H. Liao, Y. He, H. Nian, Y.M. Hu, L.J. Qiu, Y.S. Dong, and X.L. Yan. Characterization of root architecture in an applied core collection for phosphorus efficiency of soybean germplasm. Chinese Sci. Bull. 49:1611-1620, 2004.

[20] X. Gao, X. Lu, M. Wu, H. Zhang, R. Pan, T. Jian, L. Xian, and H. Liao. Co-Inoculation with rhizobia and AMF inhibited soybean red crown rot: From field study to plant defense-related gene expression analysis. PLoS ONE 7(3):e33977, 2012.

[21] J.M. Mohammad, A. Hammouri, and A.E Ferdows. Phosphorus fertigation and preplant conventional soil application of drip irrigated summer squash. J. Agron. 3:162-169, 2004.

[22] P. Gruhn, F. Goletti, and M. Yudelman. Integrated nutrient management, soil fertility, and sustainable agriculture: current issues and future challenges. International Food Policy Research Institute, Washington, D.C., p. 31, 2000.

[23] T.D. Khanh, M.I. Chung, T.D. Xuan, and S. Tawata. The exploitation of crop allelopathy in sustainable agricultural production. J. Agron. Crop Sci. 191:172-184, 2005.

[24] A.S. Lithourgidis, C.A. Damalas, and A.A. Gagianas. Long-term yield patterns for continuous winter wheat cropping in northern Greece. Eur. J. Agron. 25:208-214, 2006.

[25] N.K. Fageria, O.P. Morais, and V.C. Baligar. Response of rice cultivars to phosphorus supply on oxisol. Fert. Res. 16:195-206, 1988.

[26] D. Tilman, K.G. Cassman, P.A. Matson, R. Naylor, and S. Polasky. Agricultural sustainability and intensive production practices. Nature 418:671-677, 2002.

[27] A.G. Good, and H.B. Perrin. Fertilizing nature: a tragedy of excess in the common. PLoS Biol. 9:e1001124, 2011.

[28] J. Peigné, B.C. Ball, J. Roger-Estrade, and C. David1. Is conservation tillage suitable for organic farming? A review. Soil Use Manage. 23:129-144, 2007.

[29] D.L. Karlen, G.E. Varvel, D.G. Bullock, and R.M. Cruse. Crop rotations for the 21st century. Adv. Agron. 53:1-45, 1994.

[30] M.J. Brimecombe, F.A. DeLel, and J.M. Lynch. The effect of root exudates on rhizosphere microbil populations. In: Pinton R, Varanini Z, and Nannipieri P. Ed. The rhizosphere biochemistry and organic substances at the soil-plant Interface. Marcel Dekker, New York, p. 95-140, 2001.

[31] E. Zagal, C. Munoz, M. Quiroz, and C. Cordova. Sensitivity of early indicators for evaluating quality changes in soil organic matter. Geoderma 151:191-198, 2009.

[32] R. Chintala, M.M. Louis, and B.B. William. Effect of soil water and nutrients on productivity of Kentucky bluegrass system in acidic soils. J. Plant Nutr. 35:288-303, 2012.

[33] Y. Imai, S. Miyake, D.A. Hughes, and M. Yamamoto. Identification of a GTPase-activating protein homolog in Schizosaccharomyces pombe. Mol. Cell Biol. 11:3088-3094, 1991.

[34] G.V.N. Powell, and RD Bjork. Implications of altitudinal migration for conservation strategies to protect tropical biodiversity: a case study of the Quetzal Pharomachrus mocinno at Monteverde, Costa Rica. Bird Conserv. Int. 4:243-255, 1994.

[35] V.O. Biederbeck, H.H. Janzen, C.A. Campbell, and R.P. Zentner. Labile soil organic matter as influenced by cropping practices in an arid environment. Soil Biol. Biochem. 26:1647-1656, 1994.

[36] C.A. Grant, D.N. Flaten, D.J. Tomasiewicz, and S.C. Sheppard. The importance of early season phosphorus nutrition. Can. J. Plant Sci. 81:211-224, 2001.

[37] Z.H. Wang, S.X. Li, and S Malhi. Effects of fertilization and other agronomic measures on nutritional quality of crops. J. Sci. Food Agr. 88:7-23, 2008.

[38] P. Jeffries, S. Gianinazzi, S. Perotto, K. Turnau, and J.M. Barea. The contribution of arbuscular mycorrhizal fungi in sustainable maintenance of plant health and soil fertility. Biol. Fert. Soils 37:1-16, 2003.

[39] M.J. Unkovich, J.S. Pate, and P. Sanford. Nitrogen fixation by annual legumes in Australian Mediterranean agriculture. Aust. J. Agric. Res. 48:267-93, 1997.

[40] V.V. Shenoy, and G.M. Kalagudi. Enhancing plant phosphorus use efficiency for sustainable cropping. Biotechnol. Adv. 23:501-513, 2005.

[41] E.S. Jensen, B.P. Mark, and H.N. Henrik. Faba bean in cropping systems. Field Crops Res. 115:203-216, 2010.

[42] J. P. Lynch. Root architecture and plant productivity. Plant Physiol. 109:7-13, 1995.

[43] X.L. Yan, H. Liao, and Z.Y. Ge. Root architectural characteristics and phosphorus acquisition efficiency in plants. Chinese Bull. Bot. 17:511-519, 2001.

[44] W.J. Horst, M. Abdou, and F. Wiesler. Genotypic differences in phosphorus efficiency of wheat. Plant Soil 155:293-296, 1993.

[45] Y.Y. Sui, X.G. Jiao, and X.Y. Zhang. Research on the integrated assessment of black soil fertility in farmland. Soils Fert. 5:46-48, 2005.

6

Adaptability study of banana (*Musa paradisiacal var. sapiertum*) varieties at Jinka, southern Ethiopia

Tekle Yoseph[1, *], Wondewosen Shiferaw[1], Zemach Sorsa[2], Tibebu Simon[2], Abraham Shumbullo[2], Woineshet Solomon[3]

[1]Southern Agricultural Research Institute, Jinka Agricultural Research Center, Department of Crop Science Research Process, Jinka, Ethiopia
[2]Department of Plant Sciences and Horticulture, Wolaita Sodo University, Wolaita Sodo, Ethiopia
[3]Southern Agricultural Research Institute, Hawassa Agricultural Research Center, Department of Crop Science Research Process, Hawassa, Ethiopia

Email address:
tganta@yahoo.com (T. Yoseph)

Abstract: A field experiment involving eleven improved banana (Musa paradisiacal var. sapiertum) varieties and one local check was carried out at Jinka Agricultural Research Center during the 2006 to 2009 cropping seasons under rain fed conditions to identify the best performing variety to the target areas of South Omo Zone. The banana varieties included in the field experiment were eleven improved (Kampala, Pisang, Lacatan, Poyo, Dwarf Cavendish, Giant Cavendish, Butuzua, Grand Naine, Robusta, Williams-1, Williams-2) and a local check. The experimental design was a randomized complete block design (RCBD) with three replications. Phenological and growth parameters, bunch yield and yield components were studied. The result showed that days to flowering were significantly affected by variety while days to maturity were not significantly influenced by variety. Psedostem height was significantly affected by variety; whereas, variety had brought no significant effect on psedostem circumference. All the yield and yield components studied were significantly affected by variety except finger diameter. Bunch yield advantages of 59.11%, 55.87% and 47.55%, were obtained from the improved banana varieties Dwarf Cavandish, Giant Cavandish and Poyo, respectively over the local check. The highest bunch yields of (45.333 t ha^{-1}) and (42.000 t ha^{-1}) were recorded for the varieties Dwarf Cavendish and Giant Cavendish, respectively. Therefore, it can be concluded that use of the improved banana varieties such as Dwarf Cavendish or Giant Cavendish is advisable and could be appropriate for banana production in the test area even though further testing is required to put the recommendation on a strong basis.

Keywords: Banana Variety, Bunch Yield, Growth Parameters, Phenological Parameters, Yield Components

1. Introduction

Banana (*Musa paradisiacal var. sapiertum*) is one of the most important tropical fruits and evolved in the humid tropical regions of South East Asia with India as one of its centers of origin. Banana represents the world's second largest fruit crop with an annual production of 129,906,098 metric tons [1]. It ranks as the fourth most important global food commodity after rice, wheat and maize in terms of gross value of production [2]. About 70 million people are estimated to depend on banana fruit for a large proportion of their daily carbohydrate intake [3]. Banana is the major staple food in developing countries. The fact that it produces fruit throughout the year adds to its importance as a food security crop in Africa. It is a primary food and cash crop for over 30 million people in East Africa. Banana is now a major food crop in Africa estimated to meet more than a quarter of the food energy requirements in the continent [4]. It is a staple food and good source of income for a number of African countries especially East and Central Africa [5]. Banana is a source of potassium, magnesium, copper, manganese and vitamin C, but is low in iron and vitamin A [6].

Uganda is Africa's largest producer while Rwanda and Burundi are the second and third largest producers in East Africa, respectively [7]. Banana has been cultivated for several years in Ethiopia as a garden plant. In Ethiopia, the major banana producing regions are Southern, Oromia and Amhara regions [8]. During the 2010/2011 production season about 31, 885.86 hectares of land has been covered with banana and the estimated annual production was about 270571.516 tones [9]. The actual yields are less than 40 t ha^{-1} year^{-1}[10]; whereas, the potential yield of banana is greater than 70 t ha^{-1} year^{-1} [11]. The poor productivity of banana has been attributed to a number of biophysical factors [12].

Banana is the most important crop in Ethiopia, but over the years a number of problems tend to faced against the production of this crop in the country. Out of these, lack of improved varieties is the critical problem to banana. It is the most important cash crop in some parts of Southern Ethiopia, especially Gamo Gofa Zone. But, banana production is also familiar in South Omo Zone of Southern Ethiopia. Though, the crop is important in the target area, a number of factors constrained productivity of the crop in the target areas. This is associated with the lack of improved varieties has been appreciated as one of the primary sources of lower banana production in the target areas. There had no trend of using improved of banana varieties in the existing production system, so that it was the number one problem in the study areas. Hence; there is need to introduce improved banana varieties to the target area is crucial for banana production and productivity. Therefore, this study is aimed at and

initiated with the objective of selecting the best performing banana varieties to the target area.

2. Materials and Methods

2.1. Description of the Study Area

The experiment was conducted at research farm of Jinka Agricultural Research Center located 729 kms South West of the capital Addis Ababa at E 36^0 33' 02.7" Longitude and N 05^0 46' 52.0" Latitude and at an altitude of 1383 meters above sea level. The long term weather data for the center revealed that the maximum and minimum monthly average temperature of the center is 27.55^{0}C and 16.55^{0}C, respectively; whereas, the maximum and minimum monthly average temperature of the growing periods was 27.576^{0}C and 16.622^{0}C, respectively. The long term rainfall data for the area showed that the mean annual rainfall of the area is 1274.67 mm; while the mean monthly rainfall of the area for the growing seasons was 121.7188 mm. Rainfall pattern of the area over the years have been bi-modal with peaks around September and October and spans from February to November. The experiment was conducted during the 2006 to 2009 cropping seasons under rain fed conditions.

Table 1. *The Weather Data for Jinka, During the Years 2006 to 2009.*

Month	MaximumTemp. (°c)	Minimum Temp. (°c)	Rainfall (mm)
January	31.22	15.71	56.65
February	30.71	17.22	67.05
March	28.35	17.25	115.28
April	26.51	17.23	190.98
May	26.43	17.61	150.85
June	26.66	16.66	172.70
July	26.23	16.55	59.00
August	25.92	16.76	105.23
September	27.18	17.38	130.35
October	26.66	17.24	188.73
November	25.71	16.17	128.63
December	25.32	16.09	95.20

2.2. Treatments and Experimental Design

The experiment was executed by using eleven improved banana varieties and one local check. The field experiment was laid out in a randomized complete block design (RCBD) with three replications. Four banana plants were used in a single plot basis by using square planting method to make a unit plot area in spacing of 2.5 m between rows and 2.5 m between plants within a row making a gross plot area of 25 m^2.

2.3. Data Collection

2.3.1. Phenological Parameters and Growth Parameters

Phenological parameters such as days to flowering and days to maturity were recorded. Days to flowering was recorded by counting the number of days after establishment when 50% of the plants per plot had the first open flower. Days to maturity were recorded when 90% of flowers per

plot was matured. At mid flowering stages crop growth parameters such as psedostem height and Psedostem Circumference were measured.

2.3.2. Bunch Yield and Yield Components

The matured bunch was harvested for determination of bunch yield. Number of hands per bunch, number of fingers per hand, bunch weight, finger weight per hand and finger diameter was measured. All the phenological, growth, yield and yield components were recorded at every harvest of the growing period. All the data recorded throughout the growing periods were averaged over every harvest in the growing seasons for data analysis and computation. The weight of a bunch is determined by the total number of hands per bunch and fingers produced per hand, therefore, the weight of bunch is a function of the total number of hands and fingers obtained from the entire bunch.

2.4. Statistical Analysis

Analysis of variance was performed using the GLM procedure of SAS Statistical Software Version 9.1 [13]. Effects were considered significant in all statistical calculations if the P-values were ≤ 0.05. Means were separated using Fisher's Least Significant Difference (LSD) test.

3. Results and Discussion

The analysis of variance results for mean squares revealed that days to flowering and days to maturity were significantly (P< 0.01) influenced by varieties (Table 2). The analysis of variance result for mean squares also depicted that psedostem height was significantly (P< 0.001) affected by varieties while; psedostem circumference was not significantly affected by varieties (Table 2).

Table 2. Mean Square Values for Crop Phenology and Growth Parameters of Banana at Jinka, in 2006 to 2009.

Source	DF	Days to Flowering	Days to Maturity	Psedostem Height (m)	Psedostem Circumference (cm)
Replication (R)	2	420.1512ns	2975.076ns	0.3673ns	. 1.98ns
Variety (VAR.)	11	2141.315**	15461.48**	1.0103***	5.98ns
Error	22	548.15767	4303.0304	0.1537	9.62

*, ** and *** indicate significance at P< 0.05, P< 0.01 and P< 0.001, respectively and 'ns' indicate non significant

Table 3. Crop Phenology and Growth Parameters of Banana as Affected By Variety at Jinka, in 2006 to 2009.

Treatments	Days toFlowering	Days to Maturity	Psedostem Height (m)	Psedostem Circumference (cm)
Variety(Var.) Kampala	221.06ab	451.25bc	1.9533bc	44.00bcd
Pisang	202.22ab	380.19bcd	1.3233cde	40.333cd
Lacatan	212.86ab	451.48bc	1.9567bc	41.667cd
Poyo	187.89b	353.56cd	2.1433b	45.000bc
Dwarf Cavendish	227.97ab	311.67d	1.0333e	44.000bcd
Giant Cavendish	234.84a	483.00ab	1.9000bcd	50.667a
Butuzua	187.96b	399.67bcd	1.8000bcd	41.333cd
Grand Nain	206.89ab	415.93bcd	1.6567bcde	49.667ab
Robusta	225.37ab	357.04bcd	1.6933bcde	37.000d
Williams-1	212.96ab	419.78bcd	1.1833de	40.667cd
Williams-2	134.63c	363.81bcd	1.1933de	41.667cd
Local Check	214.81ab	580.93a	3.2000a	37.667cd
LSD 0.05	39.645	111.08	0.66	4.46
CV (%)	11.37	15.84	22.36	6.15

Note: Means with the same letters within the columns are not significantly different at P < 0.05.

The result of analysis of variance for mean squares depicted that bunch weight was significantly (P< 0.001) affected by varieties, finger weight was significantly affected (P< 0.01) by varieties (Table 4). This finding has confirmed the previous report [14]. According to the result of analysis of variance for mean squares; number of hands per bunch was significantly (P< 0.05) affected by varieties, number of fingers per hand was significantly (P< 0.001) influenced by varieties whereas; varieties had not brought a significant effect on finger diameter (Table 4). The maximum number of hands per bunch of (7.3333) was recorded for the improved banana variety Pisang and the minimum number of hands per bunch of (4.3333) was recorded for the local check (Table 5). The maximum number of fingers per hand of (80.000), (79.000) and (77.333) were recorded for the improved banana varieties Dwarf Cavendish, Giant Cavendish and Poyo, respectively and the minimum number of fingers per hand of (27.000) was noted for the local check (Table 5). The highest finger weights of (10.000 kg

hand⁻¹), (9.667 kg hand⁻¹) and (9.000 kg hand⁻¹) were noted from the improved banana varieties Dwarf Cavendish, Robusta and Giant Cavendish, respectively and the least finger weight of (3.167 kg hand⁻¹) was recorded from the local check (Table 5). The maximum bunch yields of (45.333 t ha⁻¹), (42.000 t ha⁻¹) and (35.333 t ha⁻¹) were recorded from the improved banana varieties Dwarf Cavendish, Giant Cavendish and Poyo, respectively and the minimum bunch yield of (18.533 t ha⁻¹) was noted from the local check (Table 5). The bunch yield advantages of 59.11%, 55.87% and 47.55% were obtained from the improved banana varieties Dwarf Cavendish, Giant Cavendish and Poyo, respectively over the local check in this study. The bunch yield advantage obtained from the improved banana varieties is related with the increased number yield attributing parameters such as number of fingers per hand in improved banana varieties than the local check.

According to the above findings, the improved banana varieties had resulted in greater bunch yield than the local check. This finding has confirmed the previous reports that indicate the potential of improved banana varieties over the local check [14, 15, and 16]. From the above findings it could be suggested that use of the improved banana varieties had brought a proportional yield increment than the local check.

Table 4. *Mean Square Values for Yield and Yield Components in Banana at Jinka, in 2006 to 2009.*

Source	DF	Bunch Yield (t ha⁻¹)	Finger Weight (kg hand⁻¹)	Number of Hands (bunch⁻¹)	Number of Fingers (hand⁻¹)	Finger Diameter (cm)
Replication (R)	2	29.204ns	3.0044^{ns}	3.527ns	271.194ns	0.3027^{ns}
Variety (Var.)	11	229.792***	13.109**	2.656*	647.868***	0.1563ns
Error a	22	33.907	3.387	1.1324	111.861	0.1008

*, ** and *** indicate significance at P< 0.05, P< 0.01 and P< 0.001, respectively and 'ns' indicate non significant

Table 5. *Yield and Yield Components of Banana as Affected By Variety at Jinka, in 2006 to 2009.*

Treatments	Bunch Yield (t ha⁻¹)	Finger Weight (kg hand⁻¹)	Number of Hands (bunch⁻¹)	Number of Fingers (hand⁻¹)	Finger Diameter (cm)
Variety (Var.)					
Kampala	24.000defg	3.667cd	5.3333abc	64.333ab	3.8067a
Pisang	34.000bcd	7.167ab	7.3333a	60.000b	3.0833a
Lacatan	33.333bcde	6.333bcd	6.6667ab	53.000b	3.0900a
Poyo	35.333abc	6.667abc	4.6667bc	77.333a	3.5067a
Dwarf Cavendish	45.333a	10.000a	6.6667ab	80.000a	3.4967a
Giant Cavendish	42.000ab	9.000ab	5.3333abc	79.000a	3.5467a
Butuzua	22.667efg	7.167ab	6.6667ab	46.000b	3.2700a
Grand Nain	33.333bcde	7.633ab	6.0000abc	58.667b	3.5700a
Robusta	30.667cdef	9.667ab	5.6667abc	49.667b	3.4267a
Williams-1	21.333fg	8.167ab	5.0000bc	45.333b	3.0967a
Williams-2	20.000fg	7.000abc	5.0000bc	49.000b	3.3100a
Local Check	18.533g	3.167d	4.3333c	27.000c	3.2133a
LSD 0.05	9.86	3.12	1.80	17.91	NS
CV%	19.38	25.79	18.60	18.90	9.42

Note: Means with the same letters within the columns are not significantly different at P <0.05.

The result of the Pearson correlation coefficient depicted that, among yield and yield components and some growth and phenological traits in this study, number of hands per bunch (r = 0.24807), finger weight (r = 0.23027), finger diameter(r= 0.09211), psedostem height (r = 0.16236), psedostem circumference (r = 0.30515), days to flowering (r=0.20209) and days to maturity (r = 0.18577) were positively correlated with bunch weight (Table 6). Bunch weight was also correlated significantly positively (r = 0.577***) with the number of fingers per hand (Table 6). This result is in agreement with the previous report [14]. The number of hands per bunch was negatively correlated with finger weight, finger diameter, psedostem height, psedostem circumference and days to maturity; whereas, it was associated positively with the number of fingers per hand and days to flowering (Table 6). The number of fingers per hand was positively correlated with psedostem height and days to flowering; while, it was correlated negatively with psedostem circumference and days to maturity (Table 6). The number of fingers per hand was positively (r = 0.29422) associated with finger weight (Table 6). On the other hand, the total number of fingers per hand was correlated negatively with fruit weight [14]. Finger weight was positively correlated with finger diameter, psedostem height, days to flowering and days to maturity but it was associated negatively with

psedostem circumference (Table 6). Finger diameter was positively correlated with psedostem height, psedostem circumference, days to flowering and days to maturity (Table 6). Psedostem height was positively correlated with psedostem circumference but it was correlated negatively with days to flowering (Table 6). On the other hand, psedostem height was correlated significantly positively (r = 0.518**) with days to maturity (Table 6). Psedostem circumference was positively associated with days to flowering and days to maturity (Table 6). Days to flowering was positively correlated with days to maturity (Table 6). This result has confirmed the previous findings being reported [14].

It is observed from this result that the major variables contributing to the bunch yield were biologically related and the contributions of such correlated and related variables influence positively the performance of the other, hence, the variables that showed negative association will inhibited the performance of the other and this largely depends on their attributes to the performance of the particular traits measured. From this study, it was possible to observe that for example, the variable bunch weight was positively correlated with all the entire traits in this experiment. This study has confirmed that total number of hands per bunch, number of fingers per hand, finger weight and finger diameter are the major contributing factors to bunch yield.

Table 6. Pearson Correlation Coefficient for Nine Traits of the Improved Banana Varieties at Jinka, in 2006 to 2009.

Traits	BWT	NHD	NFG	FWT	FDM	PSHT	PSCM	DTF	DTM
BWT	1	0.24807	0.577***	0.23027	0.09211	0.16236	0.30515	0.20209	0.18577
		0.1446	0.0002	0.1767	0.5931	0.3441	0.0703	0.2372	0.278
NHD		1	0.15969	-0.26726	-0.03607	-0.01087	-0.05164	0.12068	-0.1821
			0.3522	0.1151	0.8346	0.9498	0.7649	0.4832	0.2878
NFG			1	0.29422	0.25639	0.14796	-0.06632	0.02189	-0.10999
				0.0815	0.1312	0.3891	0.7007	0.8992	0.5231
FWT				1	0.008	0.07282	-0.22296	0.03026	0.20097
					0.9631	0.673	0.1912	0.8609	0.2399
FDM					1	0.12902	0.04579	0.05074	0.00836
						0.4533	0.7909	0.7688	0.9614
PSHT						1	-0.11816	0.02899	0.518**
							0.4925	0.8667	0.0012
PSCM							1	0.22928	0.00152
								0.1786	0.993
DTF								1	0.12739
									0.4591
DTM									1

DTF = days to flowering, DTM = days to maturity, PSCM = psedostem circumference, PSHT= psedostem height, FDM = finger diameter, FWT = finger weight, NFGS = number of fingers per hand, NHD = number of hands per bunch, BWT = bunch weight

4. Summary and Conclusion

Using improved varieties of banana could make an important contribution to increase agricultural production and productivity in areas like Jinka where there is low practice of using improved technologies such as improved crop varieties. To this end, use of improved banana technologies such as improved varieties could be one of the alternatives to improve productivity by small farmers. However, the use of improved banana varieties is not yet studied in the area. Thus, this research work is initiated to investigate the impact of including improved banana varieties on the existing production system is of paramount important.

Study on banana variety was conducted at Jinka under rain fed conditions in 2006 to 2009. The objective of the study was to determine the best performing banana varieties that will improve banana production and productivity in the target area. The experiment was carried out using the randomized complete block design (RCBD) with three replications at Jinka in 2006 to 2009. During the field implementation, eleven improved banana varieties and one local check were used. According to the results of analysis of variance, all the phenological and growth parameters were significantly affected by varieties except psedostem circumference. Days to flowering and days to maturity are also phenological determinants of yield including psedostem height at flowering which is almost the time for plant to use all the growth traits to produce their food especially during photosynthesis.

All the yield and yield components studied in this experiment such as bunch yield, number of hands per bunch, number of fingers per hand and finger weight were significantly affected by varieties; whereas, variety had not brought a significant effect on finger diameter. The highest bunch yields of (45.333 t ha^{-1}) and (42.000 t ha^{-1}) were recorded for the banana varieties Dwarf Cavendish and Giant Cavendish, respectively. Therefore, it can be concluded that use of the improved banana varieties such as Dwarf Cavendish or Giant Cavendish is advisable and could be appropriate for banana production in the test area even though further testing is required to put the recommendation on a strong basis.

References

[1]   FAOSTAT (2010). Food and Agricultural Organization of the United Nations. pp. 28-30.

[2]   INIBAP (1992). International network for the improvement of banana and plantain. Annual Report. 1992. Montpellier. France. p. 48.

[3]   Swennen R, Wilson GF (1983). Response of plantain to mulch and fertilizer. Int. Inst. Trop. Agric. Annual Rep. IITA Ibadan, Nigeria p.187.

[4]   Robinson, J.C., 1996. Bananas and Plantains. University Press, Cambridge.

[5]   Viljoen, A. 2010. Protecting the African banana (*Musa spp*): Prospects and challenges. Proceedings of the International Conference on Banana and Plantain in Africa, *Acta Horticulturae*, 879: 305-313.

[6]   Wall, M.M. 2006. Ascorbic acid, vitamin A and mineral composition of banana (*Musa spp*) and papaya (*Carica papaya*) cultivars grown in Hawaii. *Journal of Food Composition and Analysis*, 19: 434-445.

[7]   FAO, 2009. FAOSTAT. In the Food and Agriculture Organization of the United Nations website. http://faostat.fao.org/site/567/default.aspx.

[8]   MoA (Ministry of Agriculture) 2011. Animal and Plant Health Regulatory Directorate Crop Variety Register Issue No. 14 June, 2011. Addis Ababa

[9]   CSA. 2011. Agricultural sample survey 2010/2011 (2003 E.C.). Report on area and production of major crops. Central Statistical Agency of Ethiopia, Addis Ababa, Ethiopia.

[10] Wairegi, L.W.I., van Asten, P.J.A., 2010. Norms for multivariate diagnosis of nutrient imbalance in the East African highland bananas (*Musa* spp. AAA-EA). J. Plant Nutr. (in press).

[11] Van Asten, P.J.A., Gold, C.S., Wendt, J., De Waele, D., Okech, S.H.O., Ssali, H., Tushemereirwe, W.K., 2005. The contribution of soil quality to yield and its relation with other banana yield loss factors in Uganda. In: Blomme, G., Gold, C.S., Karamura, E. (Eds.), Proceedings of a Workshop Held on Farmer Participatory Testing of IPM Options for Sustainable Banana Production in Eastern Africa, Seeta,

Uganda, December 8–9, 2003, International Plant Genetic Resources Institute, Montpellier, pp. 100-115

[12] Gold, C.S., Karamura, E.B., Kiggundu, A., Bagamba, F., Abera, A.M.K., 1999. Monograph on geographic shifts in highland cooking banana (*Musa*, group AAA-EA) production in Uganda. Afr. Crop Sci. J. 7, 223-298.

[13] SAS (2007) Statistical Analysis Systems SAS/STAT user's guide Version 9.1 Cary NC: SAS Institute Inc. USA

[14] Shaibu, A.A., E.A. Majil and M.N. Ogburia. 2012. Yield evaluation of plantain and banana landraces and hybrids in humid agro ecological zone of Nigeria E3 Journal of Agricultural Research and Development Vol. 2(3). pp. 074-079

[15] Dzomeku BD, Armo-Annor D, Adjei-Gwen K, Nkakwa A, Akyeampong E, Banning IS (2006). Evaluation of four Musa hybrids in Ghana. Trop. Sc., 43:176–179.

[16] Dzomeku BM, Bam RK, Adu-kwarteng E, Darkey SK, Ankomah AA (2007). Agronomic and physico-chemical evaluation of FHIA 21 in Ghana. Int. J. Agric. Res., 2:92–96.

7

Small-scale maize seed production in West and Central Africa: Profitability, constraints and options

Awotide Diran Olawale[1, *], Mafouasson Hortense Noelle Tontsa[2]

[1]Department of Agricultural Economics and Farm Management, Olabisi Onabanjo University, Yewa Campus, Ayetoro, Ogun State, Nigeria
[2]Institut De Recherche Agricole Pour Le Developpement (IRAD), Yaoundé, Cameroon

Email address:
w_awotide@yahoo.com (Awotide D. O.)

Abstract: Seed plays a critical role in increasing agricultural productivity. Seed has been described as an essential, strategic, and relatively inexpensive input that often determines the upper limit of crop yields and the productivity of all other agricultural inputs. Given the critical role that seed plays in agricultural production, a key question is how to facilitate the development of a seed system that is capable of generating, producing and distributing new seed varieties that meet the needs of all farmers, in a cost-effective way given the critical role that improved varieties play in increasing agricultural production. The study was conducted in Nigeria and Cameroon in West and Central Africa respectively. A multi-stage sampling technique was used in this study to select 167 maize seed producers. Descriptive and quantitative techniques were employed in the analysis of the study data. Descriptive analytical tools such as frequency tables were used to describe the socio economic characteristics of respondents and options in maize seed production. Normalised profit function analysis was used to determine the profitability of seed production and importance indices were used to rank seed production constraints. Finally regression analysis was used to determine the factors affecting the profitability of maize seed production. The survey conducted in the two countries revealed that there is no formal maize seed production system in Cameroon. Maize seed in Cameroon came from either the government agencies or from the farmers. However, in Nigeria, there is formal maize seed production system. Evidence from the study has shown that maize seed industry in WCA (Nigeria and Cameroon in particular) has not developed remarkably. According to study, a seed system that would integrate large scale and small scale seed companies with the individual seed producers by way of integrating seed producers as out growers or contract growers to the seed companies seems to be the best option. Finally, government agencies could assist the informal sector by providing foundation seed, extension advice on seed production, processing, treatment and storage and legal framework that permits seed marketing. This would facilitate the growth of small-scale entrepreneurs in the informal sector. This is very relevant in Cameroon where there were no small-scale formal seed producers. The evidence provided in this study could lead to the sustainability of maize seed production in WCA where seed companies exist but struggling to survive and could facilitate the establishment of private seed enterprises in Cameroon where none existed. Based on the findings, numerous policy recommendations are proposed.

Keywords: Maize Seed Producers, Profitability, Constraints Analysis, Options for Integration, West and Central Africa

1. Introduction

There is general agreement among national governments and foreign aid donors that the food security situation in most developing countries is worsening. In the African region, almost half the population is being considered to be food insecure. To achieve food security, a country must be able to grow sufficient food. Since most developing countries rely on their agricultural production for their food security, it follows that food insecurity is mainly due to deficient agricultural production and low productivity. A main reason for this situation is that seed and planting material of adapted varieties required by farmers are not always available when needed, a situation often referred to as seed insecurity. There have been consistent efforts by the various national governments to put in place policies and programmes that ensure adequate seed supply systems (Omaliko, 1998).

Seed can play a critical role in increasing agricultural productivity. Seed has been described as an essential, strategic, and relatively inexpensive input that often determines the upper limit of crop yields and the productivity of all other agricultural inputs (Maredia and Howard, 1994; Langyintuo, 2005). Given the critical role that seed play in agricultural production, a key question is how to facilitate the development of a seed system that is capable of generating, producing and distributing new seed varieties that meet the needs of all farmers, in a cost-effective way given the critical role that improved varieties play in increasing agricultural production (Maredia and Howard, 1998).

However, the key to the availability of the seed will be the profitability and riskiness of seed production relative to alternative uses of farmers' limited land and labor resources. Accessible market outlets and high output–input price ratios are the key farmer incentives for sustainable adoption of modern inputs and increased production of seed and grain for the market. If farmers are to produce and supply improved maize seed and grain on a sustainable basis, the market should offer higher grain prices, and seed prices that are high enough to more than offset the costs of seed production, which are usually higher than the cost of grain production.

In many developing countries, small farmers are not considered efficient contract seed growers, and some important crops grown by them are of limited commercial interest to seed companies (Venkatesan, 1994). This limits the diffusion and use of improved varieties and quality seed by small farmers, thus contributing to low productivity. Seed industry in West and Central Africa (WCA) is plagued with myriads of problems. These include collapse of many large scale farms; non-profitability and capacity under-utilization by existing private seed companies; unawareness of the problems of pricing and production costs; seed quality problem; fake seed and flagrant use of grains as seeds in vital projects; and problems of rural sales outlets. Since country-level case studies could provide useful information concerning the economics of smallholder seed organizations, this study attempts to examine profitability of and constraints to seed production with respect to maize. These are considered the two key problems facing seed industry in WCA. Thus, the objectives of the study are to:

1. conduct econometric analysis of maize seed profitability
2. identify the constraints of maize seed production in the study area,
3. identify the criteria used by maize seed farmers to select seed for production
4. develop strategies and options for the integration of informal and formal maize seed producers to promote farmer incentives for maize seed production.

2. Maize Production in WCA

Maize production ranks second among the major cereal grains and it is a major cereal crop in West and Central Africa (WCA), currently accounting for a little over 20% of domestic food production in Africa (Manyong et al. 2000). Its importance has increased as it has replaced other food staples, particularly sorghum and millet (Smith et al. 1994), and it has also become a major source of cash for smallholder farmers (Smith et al. 1997). However, farm holdings are mainly small scale and studies have shown that drought, low soil fertility, insect pests, *striga* and maize streak virus are the major constraints to maize production in Africa. Trends in maize production indicate a steady growth and could be attributed to the expansion of cultivated area. It was observed that about 26 million tons of maize is produced annually in Africa on 20 million hectares of land (Byerlee and Eicher, 1997). During the period 1980 to 1997, maize production in Africa grew at an average annual rate of about two percent per annum (Abamu, 2001).

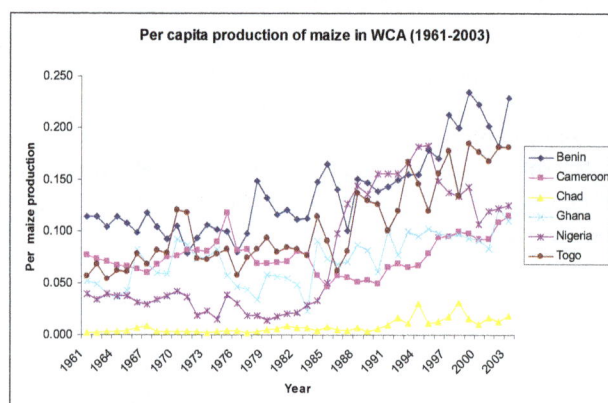

Figure 1. *Per capita production of maize in WCA (1961-2003)*

Maize yields in WCA have been highly variable as a result of climatic factors as well as price variability (Abalu, 2001) and by many biotic stresses (Fakorede et al. 2003). Some of the factors responsible for the low yield in farmers' field are weeds, pests, diseases, soil fertility and crop management practices (Fakorede et al. 2001). Figure 1 and Table 1 show the trend on the per capita maize production in WCA for the period 1961-2003. The figure and the table show the fluctuation in maize production over time. This may be due to price variability (Abalu, 2001) and many abiotic stresses (Fakorede et al. 2003).

Table 1. *Average per capita production of maize in WCA*

Period	Benin	Cameroon	Chad	Ghana	Nigeria	Togo
1961-1965	0.111	0.072	0.004	0.044	0.038	0.061
1966-1970	0.104	0.069	0.005	0.073	0.035	0.086
1971-1975	0.096	0.091	0.004	0.076	0.027	0.085
1976-1980	0.115	0.075	0.004	0.048	0.020	0.078
1981-1985	0.132	0.067	0.007	0.058	0.031	0.090
1986-1990	0.136	0.054	0.005	0.074	0.132	0.107
1991-1995	0.156	0.070	0.016	0.096	0.169	0.131

Period	Benin	Cameroon	Chad	Ghana	Nigeria	Togo
1996-2000	0.208	0.097	0.018	0.095	0.134	0.166
2001-2003	0.205	0.106	0.017	0.106	0.123	0.177

Source: Computed from faostat data

3. Maize Seed Industry

Despite the importance of improved seed for bettering the welfare of small-scale farmers, access to this invaluable technology can be constrained by many factors, including an underdeveloped seed industry (Gemeda, et al. 2001). According to Pray and Ramaswami (1991) a seed industry essentially consists of all the enterprises that produce or distribute seed and has four components: 1) plant breeding and research, 2) seed production and multiplication, 3) processing and storage, and 4) marketing and distribution. However, the industry's performance depends on the efficiency of each component, and each component possesses different economic and technical characteristics that determine the roles that public and private organizations play within the industry (Gemeda, et al. (2001). The performance of the seed industry necessitated the development of an ideal seed system. An ideal system ought to ensure adequate supply of quality seeds for modern varieties at affordable prices in the right time. Hossain, et al. (2003) noted that an ideal seed system needs supportive institutional and policy conditions for active participation of all key entities and for strengthening the public-private interface to play their basic roles in an efficient way.

Seed system development can be viewed as a dynamic process of matching the supply to the changing demand for seeds. The seed system passes through several phases as it evolves from a traditional to advance system. Maize industry in Africa is undergoing rapid changes (Nambiro et al. 2002) and its development seems to follow the same path (Morris et al. 1998). In the early stages of the seed industry, only the public sector can make the necessary heavy investment for research, development, and marketing of seed. However, when the sector expands and develops, seed production and distribution becomes increasingly interesting for the private sector. In the final stages, the private sector can take over increasing parts of the research too. The nature and pace of these changes have varied among countries, reflecting differences in stages of development and the structure of production from one country to the next, as well as differences in the economic, political and institutional climates (Nambiro et al. 2002).

Seeds of the varieties and land races of our food crops provide the biological basis for food security. There has been a number of complementary efforts by public institutions and private seed companies to provide seed to farmers. Despite these efforts and arrangements, the availability and adoption of improved seed varieties to rural farmers is still very low in Nigeria (Fakorede, 1995) and South Africa (Langyintuo, 2005). The low adoption rate was attributed to relatively high cost of hybrid (compared to open pollinated varieties (OPV)) and its suitability for optimal maize growing conditions –

good water regimes and high soil fertility (Langyintuo, 2005). As a result, farmers in many African communities are yet to have access to improved seeds. Most seeds planted by farmers come from local sources including farmers' own crop, neighbors, and relatives, or from local markets (Cromwell et al. 1992; Jaffee and Srivastava 1994; Louwaars and Marrewijk 1999). This has resulted in yield potential losses especially under the poor seed management system adopted by the farmers. Most small scale, resource poor farmers prefer the farmer seed systems to the market source of seed supply as it ensures local seed security, both a function of availability and access although quality may be poor. Seed availability generally refers to the amount of seed harvested, timeliness of the harvest and the sustainability of the supply. Seed access on the other hand relates to the quality (desired variety with the right genetic and sanitary criteria for producing a crop), equity in distribution (to all farmers) and the ability to have from external sources (through financial means) (Langyintuo, 1995).

In Africa, seed parastatals were not effective in meeting the needs of smallholders, private seed firms have not yet filled the gap and smallholder access to improved varieties has worsened in a number of countries following economic reforms (Maredia et al. 1999). This observation and situation suggests the existence of problems in African's seed system. These are in spite of the agricultural seed policy, the legal framework, and the institutional arrangements put in place to ensure success of the seed system. Not many countries have adequately addressed the question of providing farmers with access to good quality planting material, particularly of the modern varieties. In most countries, government policies relating to the regulation of seed production and price-setting inhibit the emergence of private initiatives in seed production and distribution (Venkatesan, 1994).

A well-functioning seed system is defined as one that uses the appropriate combination of formal, informal, market and non-market channels to efficiently meet farmers' demand for quality seeds. The seed system is composed of organizations, individuals and institutions involved in different seed system functions, i.e., the development, multiplication, processing, storage, distribution and marketing of seeds. The seed system includes both informal and formal sectors. The formal sector, is composed of public and private organizations with specialized roles in supplying new varieties. However, the formal seed sector alone is not sufficient to meet farmers' seed needs, and hence cannot produce a sustainable increase in production and productivity to realize Africa objectives of food self-sufficiency and food security. For example, the estimate of the percentage of farmers who purchase seed produced by formal institutions such as parastatal seed organizations and private seed companies in Sub-Saharan Africa ranges from 5% to 10% (Venkatesan, 1994). This suggests that a large percentage of farmers use their own saved seed or seed obtained from other farmers in their communities.

The informal sector is composed of individual farm households, each carrying out most seed system functions on its own, with little or no specialization. Literature is replete on the development of viable farmer seed systems (Brush et al. 1982; Almekinders et al. 1994; Jaffee and Srivastava, 1994; Ghijsen, 1996; Musa and van der Mheen-Sluijer, 1997; Musa and Rusike, 1997; Rohrbach, 1997; Tripp, 1997; Monyo, 1998).

4. Research Methodology

4.1. Study Areas, Sampling Procedure and Sample Size

The study was conducted in Nigeria and Cameroon in West and Central Africa respectively. In Nigeria, maize seed production is concentrated in the Northern part of the country. As a result, Kaduna state from the North was chosen. In like manner, seed production is concentrated in the North and central division of Cameroon. However, because of the population of maize seed producers, two divisions were chosen. A multi-stage sampling technique was used in this study. In the first stage, two divisions in Cameroon and one state in Nigeria were selected. In the second stage of the sampling technique, five communities/villages were selected. From each village, 50 seed producers were initially intended to be chosen using snowball sampling technique. However, in practice, 82 maize seed producers in Cameroon and 85 maize seed producers were used for the study and thus, constituting the third stage of the sampling. In all, 167 questionnaires were distributed for data collection. Questionnaire administration was undertaken in both countries by local enumerators selected from the agricultural development sector. The enumerators were trained in the use of the questionnaires which was piloted in the local language in non-target villages prior to the survey.

4.2. Method of Data Analysis

Descriptive and quantitative techniques were employed in the analysis of the study data. Descriptive analytical tools such as frequency tables were used to describe the socioeconomic characteristics of respondents and options in maize seed production. Normalised profit function analysis was used to determine the profitability of seed production and importance indices were used to rank seed production constraints. Finally regression analysis was used to determine the factors affecting the profitability of maize seed production.

4.3. Model Formulation/Analytical Technique: Normalised Profit Function

Normalised restricted profit function was used to determine the profitability of maize seed farmers. The analytical framework is presented below.

Using the output price as the numeraire, the normalised profit function $(\pi*(q,Z))$ can be written in a generalized form as:

$$\pi*(q,Z) = F\ [X_1*(q,Z)......,X_n*(q,Z)] - \sum_{j=1}^{m} q_j X^*{}_j(q,Z) \qquad (1)$$

Where q_j represents the normalised factor prices, F is a well-behaved production function, X is the vector of variable inputs and Z is the vector of fixed inputs used in the production process. Starting with any well-specified normalised restricted profit function, direct application of Hottelings – Shepherd's Lemmas to the function yields the corresponding factor demand and output supply equations.

$$\partial \pi*(q,Z)/\partial q_j = -X_j*\ j = 1,\ ......,\ m \qquad (2)$$

Multiplying both sides by $q_j/\pi*$ gives a series of m factor share equations.

$$[\partial \pi*(q,Z)/\partial q_j] = -X_j*q_j/\pi* = \alpha_j^*\ j = 1,.....,m \qquad (3)$$

Equations (1) and (3) form the theoretical basis for the specifications of the model.

For this study, the specification of the systems of equations of the normalized profit function equation is given as

$$\text{In } \pi* = \text{In A}* + \alpha*\ L + \delta*D + \sum_{i=1}^{2}\theta_1*Inw_i + \sum_{i=1}^{2}\beta*InZ_i + \psi_i \qquad (4)$$

Where $\pi*$ is the normalized profit defined as revenue less variable costs normalized by the price of maize seed (P).

A* = the intercept

X_1 = the number of hours of labor used including family and hired labor.

X_2 = the quantity of seeds used in maize seed production (kg)

W_1 = the wage rate normalised by the price of maize seed.

W_2 = the price of maize seed.

Z_1 = the capital inputs and is the sum of the costs of various implements used in maize seed production.

Z_2 = the land input in hectares.

D = the dummy variable taking the value of unity for out-growers and zero for non out-growers.

L = Technical linkage is dummy variable taking the value of unity for growers that have links with institutions and zero otherwise.

α, θ_i, and β_i are parameters to be estimated

ψ_i is error term

4.4. Regression Analysis

Ordinary Least Squares (OLS) estimation was used to explain the variation in profit between farmers using profit $(\pi*)$ as the dependent variable and a range of socio-economic and demographic factors of growers as the explanatory variables as follows:

X_1 = Age of farmer in years

X_2 = Education of farmer expressed in number of school years

X_3 = Ecological zone (1 = single rainfall regime and 0 = double rainfall regime)

X_4 = Household headship (1 = male and 0 = female)

X_5 = Membership of association (1 = yes and 0 = no)

X_6 = Earnings from off farm activities in monetary value

X_7 = Access to extension services (1 = yes and 0 = no)

X_8 = Number of field days attended by the respondents

X_9 = Seed production experience in years

X_{10} = Household size

The following equation describing the relationships between gross margin and the explanatory variables was therefore postulated, where δi are regression coefficients and e an error term:

$$\pi^* = \delta_0 + \delta_1 X_1 + \delta_2 X_2 + \delta_3 X_3 + \delta_4 X_4 + \delta_5 X_5 + \delta_6 X_6 + \ldots\ldots + \delta_{10} X_{10} + e \qquad (5)$$

4.5. Constraints Analysis – Importance Indices

Importance indices were constructed to identify the relative importance of constraints in maize seed production. Maize seed producers were asked to rank the identified constraints to maize seed production on an ordinal scale (1 being assigned to the most important). The use of important index in constraint analysis is replete in the literature (Jose and Valluru, 1997; Alimi, 2001; Alimi et al. 2004).

5. Results and Discussion

5.1. Formal Maize Seed Production System in Nigeria and Cameroon

The survey conducted in the two countries revealed that there was no formal maize seed production system in Cameroon. Maize seed in Cameroon came from either the government agencies or from the farmers. However, in Nigeria, there was formal maize seed production system. The National Seed Programme in Nigeria has recognized three tiers of participants in the industry: 1) Large scale seed companies, small/medium seed enterprises, and 3) community seed system (informal sector). There were few seed producing companies in Nigeria. The industry is gradually transforming from a marketing enterprise to a producing/processing and marketing one. A major constraint in the development of seed industry in the country was lack of adequate manpower. During the survey period, not more than ten percent of the required trained personnel were available in the country.

5.2. Large Scale Seed Companies

At the moment, the companies that fall in this category include Premier Seeds Nig. Ltd., (Zaria), Alheri Seeds Nig Ltd (Zaria), Savannah Seeds and Livestock Ltd., (Jos), Nagari Seeds Ltd., (Zaria), Maslaha seeds (Gusau), and Seed Project, (Kaduna).

5.3. Small/medium seed Enterprises

In this category, Agricultural Development Programmes (ADP) is expected to identify suitable out-growers with adequate potentials to develop into small-scale seed enterprises. Such growers are assisted with fertilizer, agro-chemicals, credit and other production inputs at full cost recovery. Furthermore interested growers who would like to float full scale seed enterprises are assisted with training and provision of seed processing equipments on hire purchase, custom processing or outright purchase.

5.4. Informal Maize Seed Production System

The community seed programme is aimed at making good quality seed available to farmers at very cheap prices and within the shortest possible distance. Agricultural Development Programmes (ADPs) were expected under this porgramme to select 100 Village Extension Agents (VEAs) who would identify a contact farmer each. Each contact farmer would be assisted by the ADP and National Agricultural Seed Council (NASC) to establish 1 ha per contact farmer of the most preferred crop and variety in his community. Seeds harvested from these plots were given such publicity that would enable easy distribution within the community of origin. Farmers' seed production essentially refers to growing a crop of which part is saved as seed for own use. Saving the best grains from consumption, their storage and planting developed over centuries into structured local seed systems.

5.5. Socioeconomic Characteristics of Maize Seed Farmers

Table 2. Summary description of maize seed farming household characteristics

Characteristics	Dominant indicator	Mean value (Pool)	Mean value (Nigeria)	Mean value (Cameroon)
Age	43% were aged between 40-50	42.4	43.7	41.0
Formal education	62% had more than 6 years	8.7	7.3	10.1
Household size	52% had between 1-6 people	8.5	10.9	6.0
Experience in maize	40% had between 11-20 years	15.9	16.2	15.5
Experience in seed	78% had between 1-4 years	3.5	4.1	2.9
Farm size	53% had between 1-2 ha	1.8	2.5	1.1
Maize seed output (kg)	42% had more than 3000 kg	4034	6985	976
Maize seed profit ($)	50% made up to $2400 per year	3716	6128	1217
Access to credit	83% did not have access to credit	-	-	-

Source: Field survey, 2008

Summary statistics of some important socio-economic variables are presented in Table 2. The results showed that the mean age of farmers in the study area was 42 years. This suggests that the farmers are relatively young (WHO, 1991). This gives hope for promising future of maize seed production in the Nigeria and Cameroon. Table 2 also shows that the respondents on the average had nine years of formal education. The literacy level was high among the sampled farmers. This would have positive effect on their choice of inputs and the utilization of existing inputs as well as their willingness to adopt improved technologies. Mean profit per production season made by maize seed farmers was $3716. From the descriptive statistics of the socio-economic profile of the farmers, it could be observed that 78 percent of the maize seed farmers had 1-4 years of experience. Furthermore, 50 percent of farmers made up to $2400 per year on maize seed production in spite of the fact that more than 80% did not have access to credit.

The results of the estimated normalised profit function are presented in Table 3. The coefficients of labor input, seed input, land area and capital were significant at between one percent and five percent levels of significance in the pooled model. As expected, the coefficient of labor price (wage rate) and seed price are negatively signed indicating that the estimated profit function is convex in input prices. As costs they lower the profit made by the rice farmers. The coefficient of capital input was positive. However, in Nigeria, seed input, capital input and farm size were significant at 1%, 5% and 1% respectively while in Cameroon, all the variables considered were significant with the exception of seed price. One striking point to note is that in all the models, seed price was not significant. Finally, in Cameroon, there were no out-growers and the growers had no link with any institution. In Nigeria, where they existed, the two variables were not significant in the normalized profit function model.

5.6. Maize Seed Production Profitability

Table 3. Normalized profit function form maize seed farmers in Nigeria and Cameroon

Variable	Pooled data		Nigeria		Cameroon	
	Coeff.	t-value	Coeff.	t-value	Coeff.	t-value
Constant	4.887	10.20***	7.972	27.61***	4.923	6.57***
Labor input	-0.929	-6.99***	0.295	0.98	-0.846	-6.68***
Seed input	-7.316	-5.66***	-8.549	-4.77***	-4.510	-4.04***
Wage rate	-0.091	0.55	-0.137	-1.24	-0.440	2.57**
Seed price	-0.270	-0.15	-0.308	-1.06	-0.227	1.28
Capital	0.309	6.80***	0.064	2.56**	0.224	2.57**
Farm size	0.681	11.58***	0.847	23.54***	0.347	5.28***
Out growers	0.096	0.52	-0.104	-1.39	-	-
Linkage	-0.133	-0.68	0.094	1.12	-	-

Source: Field survey, 2008 *** significant at 1%, ** significant at 5%

5.7. Factors Influencing Profitability of Maize Seed Producers

Table 4. Factors affecting profit made by maize seed farmers in Nigeria and Cameroon

Variable	Pooled data		Nigeria		Cameroon	
	Coeff.	t-value	Coeff.	t-value	Coeff.	t-value
Constant[+]	4117.35	2.65***	5026.32	1.79*	1439.14	2.81***
Ecological zone	-0.537	-5.80***	-	-	-	-
Age	-0.129	-1.50	-0.21	-0.15	-0.161	-1.19
Household headship	0.094	1.40	-	-	0.045	0.406
Formal education	-0.069	-0.98	-0.031	-0.26	1.474	0.145
Household size	0.182	2.01*	0.041	0.31	1.199	0.234
Membership of association	0.142	2.12*	0.082	0.842	-0.198	-1.70*
Earnings from off farm	-0.175	-2.37*	-0.261	-2.59**	-0.131	-1.03
Access to extension	0.005	0.61	0.079	0.71	-0.282	-2.29**
Numbers of field days	0.174	2.40*	0.536	4.98***	0.292	2.49**
Seed production experience	0.070	0.967	0.075	0.717	-0.125	-1.06

Source: Field survey, 2008 *** significant at 1%, ** significant at 5%, * significant at 10%, + unstandardized

The Ordinary Least Square estimates (OLS) of the parameters of the determinants of household farm income from maize seed production in Nigeria and Cameroon is presented in Table 4. Regression results from the pooled model showed that ecological zone specified as a dummy variable, where zero denotes bimodal and one monomodal was negative and was significant at 1 percent level

suggesting profit of maize seed farmers is negatively related to bimodal. This explains why maize production is pronounced in the north than in the south. The negative coefficient of age implies that older farmers made more profit than the younger farmers. The coefficient for age variable was not a significant determinant of farmers' income. Headship was specified as a dummy variable, where one denotes male-headed household and zero otherwise. The coefficient of headship was positive and was not significant even at 10 percent level. The variable of education showed negative relation with farmers' income and not a significant determinant. The coefficient of household size was positive and was a significant determinant of farmers' income suggesting that maize farmers with large household size tend to make more profit in maize seed production. Membership of farmers association was significantly and positively related to farmers' profit in the study area.

Income generated from off-farm was statistically significant at 10% but negatively related to farmers' profit. Reason for the negative relation is not clear. It is expected that money generated from off-farm activities should boost the farmers' potentials to buy necessary inputs for maize seed production. However, if much of the farmers' time is devoted to off-farm activities this may jeopardize his seed production effort and thus has negative relation with profit made. Extension visits was not statistically significant but positively related to profit made from maize seed production. The reason for the positive correlation between extension visits and profitability is very clear. Frequency of visit by extension agents may increase farmers' knowledge of improved technologies and therefore increases profit made from maize seed production. On the contrary, field days attendance was statistically significant and positively related to maize seed profit. Finally, seed production experience was positively related to farm income but was not significant in determining farmers' income.

5.8. Constraints to Maize Seed Production in Nigeria and Cameroon

The perceived constraints to maize production in the two countries are presented in Table 5. The constraints were ranked from 1 (most important) to 6 (least important). Score of zero denotes no constraint. Table 5 reveals that about 68 percent of the respondents considered lack of access to credit facilities to be a constraint of least importance while 2.4 percent considered lack credit as the most important constraint. For seed, 28 percent of the sampled seed growers considered unavailability of foundation seed as the most important constraint while 13 percent considered it as the constraint of least importance. The variation in the percentages across scores for the other constraints is clearly shown in the table.

Table 5. Perceived constraints to maize seed production (percentage)

Scores	Credit	Labor	Market	Fertilizer	Seed	Land
0	2.4	8.5	9.8	2.4	7.3	9.8
1	2.4	13.4	3.7	1.2	28.0	17.1
2	0.0	26.8	8.5	6.1	23.2	25.6
3	6.1	22.0	17.1	7.3	14.6	22.0
4	7.3	9.8	13.4	19.5	3.7	11.0
5	13.4	11.0	15.9	26.8	9.8	7.3
6	68.3	8.5	31.7	36.6	13.4	7.3
Total	100.0	100.0	100.0	100.0	100.0	100.0

Source: Field survey, 2008

Table 6 shows the constraints to maize seed production based on the sample of seed producers in Nigeria and Cameroon. According to the importance index, unavailability of foundation seed ranked first followed by unavailability of land. In this study, unavailability of land refers to inability of willing seed producers to access farm land for maize seed production. On the other hand, lack of access to credit facilities ranked last in order of importance. In the final analysis, the three most important constraints to maize seed production based on a sample of maize seed farmers in WCA were unavailability of foundation seed, unavailability of land, and unavailability of labor.

Table 6. Importance index of the constraints to maize seed production

Constraints	Total value of score	Index	Rank
Lack of access to credit facilities	432	5.3	6th
Unavailability of labor	228	2.8	3rd
Market uncertainty for seed produced	324	4.0	4th
Unavailability of fertilizer	383	4.7	5th
Unavailability of foundation seed	186	2.3	1st
Unavailability of land	212	2.6	2nd

Source: Field survey, 2008

5.9. Farmers' Criteria in Choosing Varieties

Farmers use many but similar criteria in selecting the maize varieties they grow. Table 7 shows the main criteria farmers apply in choosing maize varieties they grow and the importance of each criterion. Importance of the choice of maize varieties was given a quantitative score using a scale of zero to five, with 'very important' given the score=5, "low importance" a score =1 and 0 when the criterion was not mentioned. The scores were averaged for the countries, which allowed ranking the criteria in order of importance (Table 7).

Table 7. Farmers' choice of maize variety by country

Characteristics	Nigeria	Cameroon	Mean score
Maturity period	2.0	4.2	3.1
Taste	1.2	3.2	2.2
Fertilizer requirement	2.0	2.7	2.4
Resistance to diseases	1.7	4.2	3.0
Grain quality	2.2	4.3	3.3
Cooking quality	2.1	3.2	2.7
Market value	2.8	3.9	3.4
Adaptability to poor soils	2.5	2.4	2.5

Source: Field survey, 2008

The most important criteria across the countries were, in that order: market value, grain quality, maturity period and resistance to disease. However, the scores of the criteria varied across the countries. Cooking quality, adaptability to poor soils, fertilizer requirement, and taste were the second group of criteria. These criteria received between 2.2 and 3.4 on the score of importance. Resistance to insects and other pests as a criterion for selection of maize varieties is very useful in practice if the attribute is combined with the most important criteria farmers apply in variety selection, thus adding value to the varieties (Odendo, et al. 2002).

5.10. Options and Prospects of Integration

Nigeria seemingly possesses great potentials for seed business, due to its land area, population and the status of the country in the West African Sub-regional trade. A flexible maize seed system is therefore crucial to effectively respond to these challenges. For a sustainable maize seed production system, the formal seed production sector in most countries should encourage the informal sector, which can meet the seed needs of a wide spectrum of farmers. The formal sector can continue producing hybrids and other high value seed along with the informal sector. Since, in most countries, informal farmer-to-farmer spread of seeds is the single most frequently used source of seed by farmers, it is necessary for governments to recognize the informal sector as an important low-cost source of quality seed, and to use it as a vehicle for providing resource-poor farmers with improved seed of modern varieties at affordable prices.

For a sustainable maize production, there is need to provide farmers with access to good seed within close distance, in time and at affordable prices. One workable option is to evolve a seed industry with both the formal and informal seed systems providing quality seed to the consumers. This will increase the awareness of seed consumers of seed quality and price, and lead to an overall increase in seed use and consequently to agricultural production. Based on the study, the following options were identified:

1. Research institutions providing foundation seed and other inputs to selected farmers through extension services. After harvest, farmers sell certified seed and reimburse the inputs costs to extension services.
2. Potential contract growers are identify by national extension services. The farmers are provided with foundation seed to produce certified seed. In this case, technical advice is given to seed producers by researchers and extension agents.
3. Foundation seeds are given to farmers by researchers. Farmers purchase and apply fertilizer and other inputs. Scientists provide technical assistance to farmers.
4. Non-governmental organizations (NGO's) organize and supply farmers with foundation seed and other inputs for production of certified seed. After seed sales, half of the initial funds are deducted and provided to extension services to encourage seed production by other farmers.
5. Foundation seeds are provided by private seed companies to identified out-growers to produce maize seed. Private seed companies provide incentives such as fertilizer and other inputs to facilitate the production process. After harvest, maize seeds are processed by seed companies and sold as certified seed. The options guarantee maize seed produced by the informal sector.

According to the study, a seed system that would integrate large scale and small scale seed companies with the individual seed productions as presented in option five will provide the best option. The integration can be achieved if the seed producers serve as out-growers or contract growers for the seed companies. The seed companies may also provide technical and financial incentives to the out growers to accelerate and sustain the seed system. In addition, accelerated seed certification and seed quality control will be ensured and there will be sure market for the seed produced by the informal sector. Presently in Nigeria, this option is being adopted by some seed companies in order to meet the needs of their numerous clients, though the integration is not as complimentary and interactive as it should be. This option may seem practicable, a number of analysis have shown that there is a wide diversity in how the two sectors interact (Lanteri and Quagliotti, 1997; Maredia and Howard, 1998).

Strategies to promote informal and formal seed sector integration in Nigeria and Cameroon are:

1. Use of private seed companies as seed producing agencies for large-scale government special seed requirements. This will promote informal seed producers since they will serve as contract growers to the seed companies.
2. Government should organize special programmes to promote awareness of the benefits and use of improved seeds.
3. Government agencies can assist the informal sector by providing foundation seed, extension advice on seed production, processing, treatment and storage and legal framework that permits seed marketing. This will facilitate the growth of small-scale entrepreneurs in the informal sector. This is very relevant in Cameroon where there are no small-scale formal seed producers.
4. Government agencies could provide public goods that are essential to the functioning of both formal and informal sectors including research targeted to maize seed production that are not of interest to the private sector.

5.11. Conclusion and Recommendations

The study gave insight into the activities of the small-scale seed producers and the profitability and constraints. Evidence from the study has shown that maize seed industry in WCA (Nigeria and Cameroon in particular) has not developed remarkably. However, maize seed production is profitable. There still exist problems associated with non-availability of foundation seeds, unavailability of land for potential maize seed growers (especially in Cameroon) and unavailability of required manpower in the production process. The important criteria for chosen maize seed variety for planting are market value, grain quality, maturity period and resistance to pests and diseases. According to the study, a seed system that would integrate large scale and small scale seed companies with the individual seed production by way of integrating seed producers as out growers or contract growers to the seed companies seems to be the best option. Finally, government agencies can assist the informal sector by providing foundation seed, extension advice on seed production, processing, treatment and storage and legal framework that permits seed marketing. This will facilitate the growth of small-scale entrepreneurs in the informal sector. This is very relevant in Cameroon where there are no small-scale formal seed producers. The evidence provided in this study could lead to the sustainability of maize seed production in WCA where seed companies exist but struggling to survive and could facilitate the establishment of private seed enterprises in Cameroon where none exist. Based on the findings, the following policy recommendations are proposed:

1. Government agencies can encourage the growth of informal maize seed producers by providing them with access to NARS (spell out for initial mention) bred foundation (and/or breeder seed) and given extension advice on seed production, processing, treatment and storage.

2. Appropriate mechanisms should be put in place to strengthen public and private extension programs to increase farmer knowledge about the benefits of using new seed and transmitting information about farmer preferences to researchers will also help increase the demand for new seed.

3. Concerted effort should be made to remove compulsory seed certification and restrictive trade licensing requirements. This permits the production of quality seed by smallholders and sale among neighboring farmers. In addition, seed companies would be able to involve smallholders in contract seed production more easily.

4. Public research and extension agencies also need to consider how to use subsidies to strengthen ties to subsistence farmers who may be unable to purchase seed through the market but could benefit significantly from access to improved varieties.

Recommendations for further study

Interplay between the legal framework and the seed system in the study area since a very restrictive legal framework involving mandatory varietal notification and seed certification cannot help the growth of the informal sector.

Acknowledgment

This research work was supported by a grant from the Investment Climate and Business Environment Research Fund, jointly funded by TrustAfrica and IDRC.

References

[1] Abalu, G.I. (2001). "Policy Issues in Maize Research and Development in sub-Saharan Africa in the Next Millennium". In Badu-Apraku, B., Fakorede, M.A.B., M. Ouedraogo, and R.J. Carsky, (eds.), 2001. *Impact, Challenges and Prospects of Maize in West and Central Africa*: Proceedings of a Regional Maize Workshop, IITA-Cotonou, Benin Republic, 4-7 May, 1999. WECAMAN/IITA.

[2] Alimi, T. (2001). "Economic Rationale of Integration in Poultry Production System". *Lesotho Social Science Review*, 7(2):138-156.

[3] Alimi, T., Idowu, E.O. and Tijani, A.A. (2004). "Optimal Farm Size for Profitable and Sustainable Certified Maize Seed Production Enterprise in Oyo State, Nigeria". *Botswana Journal of Economics*, 1(2):135-146.

[4] Almekinders, C .J. M., Louwaars, N. P. and de Bruijn, G. H. (1994). "Local Seed Systems and their Importance for an Improved Seed Supply in Developing Countries". *Euphytica,* 78: 207-216.

[5] Byerlee, D and C. Eicher, "1997). "Africa's Emerging Maize Revolution". Boulder, Colorado, Lynee Rienner Publishers.

[6] Cromwell, E., E. Friss-Hansen, and M. Turner (1992). "The Seed Sector in Developing Countries: A Framework for Performance Analysis". Working Paper 65, Overseas Development Institute, London.

[7] Fakorede, M.A.B, B. Badu-Apraku, A.Y. Kamara, A Menkir and S.O. Ajala. (2003). "Maize Revolution in West and Central Africa: An Overview". In. Badu-Apraku, B., Fakorede, M.A.B., M. ouedraogo, R.J. Carsky, and A. Menkir (eds.), (2003). *Maize Revolution in West and Central Africa*: Proceedings of a Regional Maize Workshop, IITA-Cotonou, Benin Republic, 14-18 May, 2001. WECAMAN/IITA.

[8] Fakorede, M.A.B, B. Badu-Apraku, O. Coulibaly and J. M Fajemisin (2001). "Maize Research and Development Priories in sub-Saharan Africa in the Next Millennium". In Badu-Apraku, B., Fakorede, M.A.B., M. Ouedraogo, and R.J. Carsky, (eds.), 2001. *Impact, Challenges and Prospects of Maize in West and Central Africa*: Proceedings of a Regional Maize Workshop, IITA-Cotonou, Benin Republic, 4-7 May, 1999. WECAMAN/IITA.

[9] Fakorede, M.A.B. (1995). "IITA – Private Seed Companies/NGOs Liaison Activities". A Report of the Current Status and Future Opportunity for a Greater Efficiency in Collaborative Efforts to Improve Maize Production. Pp71.

[10] Gemeda, A., G. Aboma, H. Verkuijl, and W. Mwangi. (2001). "Farmers' Maize Seed Systems in Western Oromia, Ethiopia". Mexico, D.F.: International Maize and Wheat Improvement Center (CIMMYT) and Ethiopian Agricultural Research Organization (EARO). Pp 42.

[11] Ghijsen, H. (1996). "The Development of Varietal Testing and Breeder's Rights in the Netherlands". Pages 223-226. *In:* Integrating Seed Systems for Annual Food Crops (van Amstel, H., Bottema, J., Sidik, M and van Santen, C., eds.) Bogor, Indonesia: CGPRT Centre.

[12] Jaffee, S. and J. Srivastava (1994). "The Roles of the Private and Public Sectors in Enhancing the Performance of Seed Systems". The World Bank Development. The World Bank for the Construction and Development. *The World Bank,* 9(1).97–117.

[13] Jose, H.D. and Valluru, R.S.K. (1997). "Insight from the Crop Insurance Reform Act of 1994". *Agribusiness*, 13(6). 587-598.

[14] Langyinto, A. (2005). "An Analysis of the Maize Seed Sector in Southern Africa". A paper presented at a Rockefeller Foundation Workshop on Biotechnology, Breeding and Seed Systems for African Crops. Nairobi, Kenya 24-27 January, 2005. Pp25.

[15] Lanteri, S. and Quagliotti, L. (1997). "Problems Related to Seed Production in the African Region". *Euphytica*, 96:173-183.

[16] Louwaars, M.P. and G.A.M. Marrewijk (1999). Seed Supply Systems in Developing Countries, CTA.

[17] Manyong, V.M., J.G. Kling, K.O. Makinde, S.O. Ajala, and A. Menkir. (2000). "Impact of IITA-improved Germplasm on Maize Production in West and Central Africa". *Impact*, IITA, Ibadan, Nigeria.

[18] Maredia, M., and J. Howard, (1994). "Facilitating Seed Sector Transformation in Africa: key fingings from the literature". Global Bureau, Office of Agriculture and Food Security. USAID. *Policy Systhesis*, 33.1-6.

[19] Maredia, M., J. Howard, and D. Boughton, with A. Naseem, M. Wanzala, and K. Kajisa (1999). "Increasing Seed System Efficiency in Africa: Concepts, Strategies and Issues". MSU (Michigan State University) International Development Working Paper No. 77.

[20] Morris M. L., J. Rusike and M. Smale. (1998). *"Maize Seed Industries: A Conceptual Framework"*. In Morris M. L. (ed.) *Maize Seed Industries in Developing Countries*. Boulder, Colorado: Lynne Rienner Publishers. pp. 35- 54.

[21] Monyo, E. S. (1998). "Analytical Review of Regional Impacts of SMIP: 15 Years of Pearl Millet Improvement in the SADC Region". Paper presented at Sorghum and Pearl Millet and Development in Southern Africa: Stakeholders Review and Planning Conference. 27-31 Jul. 1998. Harare. Zimbabwe. SADC/ICRISAT Sorghum and Millet Improvement Programme. Bulawayo Zimbabwe.

[22] Musa, T. M., and van der Mheen-Sluijer, J. (1977). "Review of the State of Art in the Field of Local Level Seed Supply Systems". Harare. Published by SADC/GTZ Project Promotion of Small Scale Seed Production by Self Help Groups.

[23] Musa, T. M, and Rusike, J. (1997). *"Constraints on Variety Release, Seed Production and Distribution of Sorghum, Pearl Millet, Groundnut and Pigeonpea"*. *ICRISAT Southern and Eastern Africa Region Working Paper* No 97/02 (SADC/ICRISAT-SMIP). Matopos.

[24] Nambiro, E, H. De Groote and W. O. K'osura (2002). "Market Structure and Conduct of the Hybrid Maize Seed Industry, a Case Study of the Trans Nzoia District in Western Kenya". In Friesen D.K. and A. F. E. Palmer (eds.). Integrated Approaches to Higher Maize Productivity in the New Millenium. Proceedings of the 7th Eastern and Southern Africa Regional Maize Conference, Nairobi, Kenya, 11 - 15 February 2002. Mexico, D. F.: CIMMYT, pp. 474-479.

[25] Odendo M., H. De Groote, O. Odongo and P. Oucho. (2002). "Participatory Rural Appraisal of Farmers' Maize Selection Criteria and Perceived Production Constraints in the Moist Mid-altitude Zone of Kenya". *IRMA Socio-Economic Working Paper* No. 02-01. Nairobi, Kenya: CIMMYT and KARI.

[26] Omaliko, C.P.E. (1998). "Nigeria Seed Industry and its Potential Role in Food Security within the West and Central African Sub-region". Proceedings of the International workshop on Seed security for food security. Florence, Italy, 30 November – 1 December 1997. FAO 1998.

[27] Rohrbach, D. D. (1997). "Farmer to Farmer Seed Movements in Zimbabwe: Issues Arising". Pages 171- 179. *In:* Alternative Strategies for Smallholder Seed Supply. Proceedings of the International Conference on Options for Strengthening National and Regional Systems in Africa and West Asia..10-14 Mar. 1997, Harare, Zimbabwe. Rohrbach, D. D.. Bishaw, Z, and van Gastel A. J. G.; (eds.). International Crops Research Institute for the Semi-Arid Tropics.

[28] Smith, J., A.D. Barau, A. Goldman, and J.H. Mareck. (1994). "The Role of Technology in Agricultural Intensification: the Evolution of Maize Production in the Northern Guinea Savanna of Nigeria". *Economic Development and Cultural Change*, 42. 537–554.

[29] Smith, J., G. Weber, V.M. Manyong, and M.A.B. Fakorede (1997). "Fostering Sustainable Increases in Maize Productivity in Nigeria". Chapter 8 *in* Africa's Emerging Maize Revolution, edited by D. Byerlee, and C.K. Eicher, Lynne Rienner Publishers, London, UK.

[30] Tripp, R. B. (1977). "Between States and Markets-Innovations for Small Scale Seed Provision". Pages 195- 210. *In:* Alternative Strategies for Smallholder Seed Supply. Proceedings of the International Conference on Options for Strengthening National and Regional Systems in Africa and West Asia..10-14 Mar. 1997, Harare, Zimbabwe (Rohrbach, D. D.. Bishaw, Z, and van Gastel A. J. G.; eds.). International Crops Research Institute for the Semi-Arid Tropics.

[31] Venkatesan., V. (1994). Seed Systems in Sub-Saharan Africa: Issues and Options. Discussion Paper No. 266. Technical Department, Africa Region. Washington, D.C.: World Bank.

8

A potential fast growing tree for Agroforestry and Carbon Sequestration in India: *Anthocephalus cadamba* (Roxb.) Miq.

Arvind Bijalwan[1, *]**, Manmohan J. R. Dobriyal**[2]**, Bhartiya J. K.**[3]

[1]Indian Institute of Forest management (IIFM), Nehru Nagar, Bhopal-462 003, M.P., India

[2]Department of Silviculture and Agroforestry, ASPEE College of Horticulture and Forestry, Navsari Agricultural University, Navsari - 396 450, Gujarat, India

[3]Saranda Forest Division, Chaiwasa, Jharkhand, India

Email address:

arvindbijalwan276@gmail.com (A. Bijalwan)

Abstract: *Anthocephalus cadamba* commonly known as Kadamb tree In India is a large tropical tree with straight cylindrical bole belongs to family Rubiaceae. *A. cadamba* tree is fast growing in nature and can grow in different parts of India. Considering the high demand of wood in India; *A. cadamba* is one of the promising and potential trees, being grown on the farm land in the form of Agroforestry. The wood of *A. cadamba* is multipurpose in nature having white to creamy white and straight grain with fine to medium texture wood which is used in variety of services such as ply-wood, pencil making, match splints, pulp wood for paper, packing cases, toys, wooden shoes, flooring, carving and crates etc. The fast decomposition rate of *A. cadamba* is also make it more compatible for the emerging agroforestry systems in various parts of India and considered to be very useful tree in agroforestry and Carbon Sequestration.

Keywords: Anthocephalus Cadamba, Agroforestry, Productivity, Fast Growing, Plantation, Carbon Sequestration, Intercropping

1. Introduction

To meet rising population demand for food and wood in India there is intense pressure on cultivable land and existing forests. The escalating demand of food can be attained either by increasing the farm area or the productivity, however, to increase the farm area has limited options therefore, enhancing the productivity of agricultural field with integration of trees, as agroforestry is the only economic and viable option. In the order to meet the requirement, particularly for wood and tree derived produce the fast growing species are playing major role to increase the productivity. On the other hand the increasing concentration of Carbon in the atmosphere is creating difficulty to the biological entities which needs to be minimized, where fast growing tree are playing role in Carbon Sequestration.

At present the thrust has been given to the exotic species in India like Eucalyptus, Populur, Casurina, Robinia, Mangium etc in Haryana, Punjab, Western Uttar Pradesh and parts of Uttarakhand along with some other states plantation under agroforestry are mainly comprised Eucalyptus and Poplar. Moreover the tree species like Acacia, Casurina, and Teak etc are being encouraged in agroforestry but still adoption is average only. These trees has limitation for the most of the agro-climatic regions of India particularly Madhya Pradesh, Assam, West Bengal, North Eastern states and Maharashtra. However, the tree namely *Anthocephalus cadamba* can grow well in these states, therefore, *A. cadamba* tree can be taken into the agroforestry and farm forestry model of most of the states of India to minimize the gap between demand and supply for wood and wood based industries.

In India the Kadamb tree (*A. cadamba*) is highly regarded as a sacred tree to the Lord Krishna and included in the Indian religions and cultures. The Lord Krishna and Radha encompassed their love play in the generous and endearing scented shade of Kadamb tree. Kadamb scientifically know as *Anthocephalus cadamba* (Roxb.) Miq. Syn. *Neolamarckia cadamba* var *A. chinensis* is a large tropical tree with broad crown and straight cylindrical bole having average height about 15 meters belongs to family rubiaceae. In the favourable climatic condition this tree attains a height of 20 meters or more with a clean bole of about 9 meter and a diameter of 40 to 60 cm (Figure 1 & 2). The bark is generally smooth in the initial stage which turns to darker in the older stage with exfoliating nature. The tree is aesthetic and deciduous in nature but sometimes evergreen and semi-evergreen (semi-deciduous) in nature. The flowers of the tree are small and orange colour in dense terminal globose heads, 2.5 to 5 cm in diameter however the fruit is a pseudocarp, is a globose, orange fleshy mass of closely packed capsules with each bears a numerous minute angular seeds [1]. The tree is near to leafless during in the hot season. The flowers are bisexual, small in size, orange in colour, highly scented in globose heads appear mainly May to July however the fruits ripen August to October and fall in January to February. The tree starts flowering and fruiting an early age even at an age of 6 to 7 years (Figure 3 & 4).

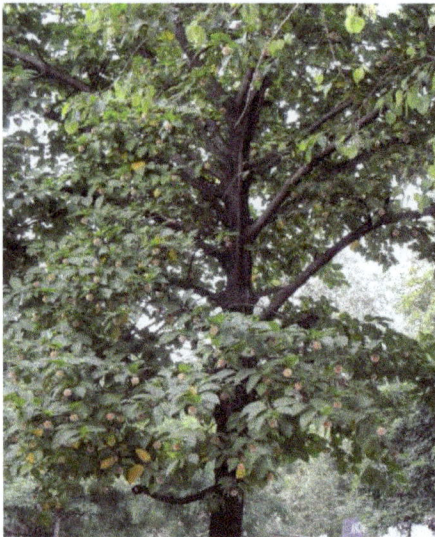

Fig. 1. *Anthocephalus cadamba* Tree

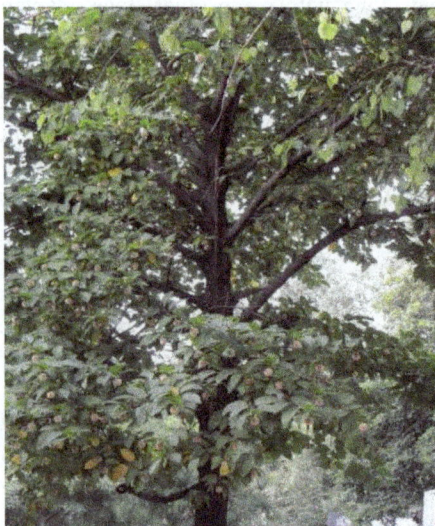

Fig. 2. *Anthocephalus cadamba* Tree

Fig. 3. *Leaves and Flowers of A. cadamba*

Fig. 4. *Flower of A. cadamba*

2. Distribution and Uses

This tree is native to South and Southeast Asia, including India, Indo-Malayan region, Jawa, Sumatra, China, Indonesia, Malaysia, Bangladesh, Sri Lanka, Comboida Papua New Guinea, Philippines and Singapore. In India it occurs in the Sub-Himalayan tract from Nepal eastward to West Bengal and Assam. In other part of India it found in Bihar, Chattisgarh, Madhya Pradesh, Andhra Pradesh and evergreen forest of Karnataka to Kerala [2], the species is also observed in Haldwani Division of Uttarakhand rarely growing on moist and swampy land [3]. It is an ingredient of southern tropical semi evergreen forests, secondary north Indian tropical moist deciduous forest and tropical fresh water swamp forest [4]. Due to the fast growing nature and complementary to annuals in variety of soils it is most suitable species to integrate and mass planting in agroforestry. In India this tree species is being used in plantation, particularly in agroforestry and farm-forestry practices. This species is considered to be increasingly important for wood based industries. In Arunachal Pradesh it is well adopted in agro-forestry.

The wood of *A. cadamba* is white to creamy white, odourless, Lightweight (545 kg/cum at 12% moisture content) the sap wood and heartwood has no clearly differentiated. The wood has straight grain with fine to medium texture and low lusture. The wood of *A. cadamba* is moderately strong and can be treated with preservatives easily, the treated timber becomes durable, and however, the sowing, peeling and other workable operation in this wood are very easy. The *A. cadamba* wood used in the variety of services such as ceiling boards, light construction work, ply-wood, pencil making, match splints, flooring, pulp wood for paper, packing cases, toys, wooden shoes, tea-chests, carving and crates etc. *A. cadamba* wood is also used for inexpensive furniture if properly seasoned and treated. The tree is also suitable for aesthetic and ornamental purpose and planted along roadsides. The fruits, leaves and barks of *A. cadamba* got medicinal properties whereas seeds have anti-poison medical properties. The root bark is also used for natural yellow dye. The *A. cadamba* tree has heavy leaf shedding properties which increases the organic carbon in the soil. The fruits and inflorescences of *A. cadamba* reportedly edible and the fresh leaves are fed to cattle. The scented orange flowers attract Honey- bees as pollinators, moreover the *A. cadamba* flowers are an important raw material used in the Indian perfumes with sandalwood as base and a good bee forage in apiculture.

A. cadamba is light demander in nature but the young plant need protection from sun. It grows generally with a maximum temperature 37.5^0 to 47.5^0 C and minimum 0^0 to 15^0C. The seedlings are frost sensitive and liable to damp off during excess moisture in the soil. The species can grow in the mean annual rainfall of 1500 to 5000mm [1] however; sometimes the *A. cadamba* can also grow in xeric condition with rainfall of as low as 200mm. The altitudinal range of growing this species varies from 300 to 1000m [5] however; sometimes it can go up to 1400m. Kadamb prefers deep, moist well drained loamy soil of alluvial origin. It also comes up in the sandy soils of Brahmaputra valley. The species is considered suitable for soil conservation, agroforestry, jhum land reclamation etc. Due to its heavy leaf shedding nature, Organic Carbon content of the soil can be increased.

Young plants of *A. cadamba* are very susceptible to browsing by the animals. The species has good coppicing ability to be multiplied by vegetative methods. The natural regeneration in *A. cadamba* has observed in favourable condition (light, moisture and well drained soil). Its seeds are very minute has good germination but seeds & seed sprouts swashed into heap along with silt and rain, hence natural regeneration is difficult. Due to small seed size and sensitiveness of seedling the artificial regeneration is considered to be good in *A. cadamba*, though the seedling with ball of earth round the root system is considered good.

The mature fruits are orange brown in colour generally available in August to September which are collected from ground and heaped under shade. The most suitable time of collection of seed is middle to later part of August. The mass of the fruits can be allowed for rotting for three to four day and then pulp of the fruits is washed off and the seeds are dried. The fruits are then rubbed to form paste-like slurry and passed through a 0.50mm sieve with proper shaking. The average dry weight of a fruit is 11.5 gm and wet weight 50gm whereas the a *A. cadamba* fruits yields 456 mg of pure seeds (Luna, 1996). The *A. cadamba* (Kadamb) seeds are minute, trigonal or irregular shaped and measured about 9, 00,000 to 27, 00, 000 seeds/kg (Figure 5 to 8).

Fig. 5. *Fruit of A. cadamba*

Fig. 6. *Rubbed fruits of A. cadamba*

Fig. 7. *Slurry of A. cadamba*

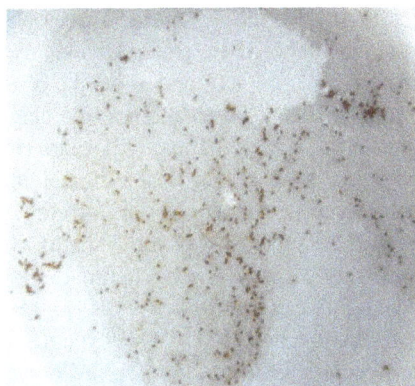

Fig. 8. *Seeds of A. cadamba*

3. Nursery Technique

Tha *A. cadamba* seeds do not require any pre-sowing treatment, but need special care during watering, due to minute nature of seeds. The seeds are generally broadcasted on the raised beds @ 130 gm/m sq and watered properly. About 70% of germination is observed within 15 days and pricked out into polythene bags where they are retained till the monsoon season arrived. Some time the treatment of 5000 PPM of Indole Butyric Acid (IBA) during rooting and survival in air layering considered valuable, the coppicing is also a method used for regeneration in *A. cadamba*. Seeds can be sown in galvanized or wooden trays filled with sand and soil and treated with fungicide. The seed sowing on tissue paper for later transplanting of Kadamb seed was also recommended by [6].

4. Planting Technique

It has observed that *A. cadamba* is best raised by planting out entire seedlings or polythene bag raised seedlings during June-July at the start of monsoon when they are about 4-5months old [1]. The natural seedlings collected from the forest (wilding) in August-September when about 30-40 cm height with balls of earth also gives 90 to 100% success. In Assam, the species has been planted in forest areas which were rendered too open to stock after complete removal of middle canopy layers in the parts [1]. Spacing trials indicate that closer spacing of 1.83m X 1.83m gives the heights yield and the widest spacing of 3.66m X 3.66m the lowest; moreover the spacing appears to have direct effect on the survival of plants [1]. The survival percentage observed to be highest for wider spacing and vice-versa [7]. One more study recommend that 8.5m X 8.5m spacing gives the heights out turn of biomass than either of the wider and closure spacing [8]. In Arunanchal Pradesh, the spacing adopted is 4.5m X 4.5m or 5.0m X 5.0m and in West Bengal it is generally 2.0m X 2.0m [1]. The closer spacing requires little tending and is liable to push through weeds provided protection against grazing animals is assured.

5. Rate of Growth

The species is very fast growing especially in the initial years. As per Troup [9], the *A. cadamba* trees planted in an avenue at Rajabhathkhowa in Buxa Duars, West Bengal, India reached a girth of 61cm and a height of 9m in four years. In Assam, it is noticed to attain a tree height of 6.5m and tree diameter 12.7cm at the age of 4 years [10]. Rate of growth is very fast in the earlier years, the height increment averages 3m per annum for the first six or eight years after which the growth becomes slower up to the age of 20 years and thereafter very slow [1]. The optimal rotation of *A. cadamba* plantations can be worked as 10 to 15 years. The Mean Annual Increment (MAI) of 19 years old Kadamb plantation was recorded to be 6.25m^3/ha in Cooch Bihar [11] however in favourable conditions the trees can attain a diameter of as high as 50cm in about 12-15 years which is good for productivity and Carbon Sequestration.

6. Pests and Diseases

The young plants and coppice shoots are readily browsed by cattle, goats and deer. Porcupines chew the bark as also the bioson and Sambher bark the trees [1]. Among the insects, *Aristobia approximator* feeds on the bark, *Dihamnus cervinus* bores in the stem, and *Dirades adjutaria* defoliates over extensive areas [2].

7. Role in Agroforestry and Carbon Sequestration

A. cadamba is promising plantation trees for humid and sub-humid tropics. The block plantations of species are mainly grown for industrial uses besides selected areas as agroforestry tree. Pruning in *A. cadamba* plantations is unnecessary as the species shows natural pruning with dead branches falling off. The rotation period (harvesting time) depends upon the production purpose. For pulpwood and matches, harvesting can start 4–5 years after planting. For example, a match factory in North Sumatra is growing *A. cadamba* on a 4-year rotation under optimal management, which includes fertilisation. For wood production, felling of trees can start approximately from the age of 10 years. In the Philippines, economic rotations applied in plantations are 5 years for pulpwood and 7 years for the combination of pulpwood and sawn timber [12]. It was estimated [13] that the total carbon sequestered in farm forestry with species such as *Eucalyptus sp.*, *Populus deltoides*, *Tectona grandis*, *Anthocephalus chinensis* trees to be around 16,400 t/yr. It was reported [14] that application extracts of leaves and twigs of A. cadamba did not inhibit the growth of turmeric plants and showed no allelopathic effect. This research conducted in the 3.9-year old A. cadamba stands having average shade level of 73.7%. Production of turmeric at the age of 6 months after planting ranged between 7.4 and 11.9 tons per ha and 9.9-16.4 tons per ha at the age of 8 months.

Due to heavy leaf shading during autumn- spring and easy decomposition of leaves make it suitable alternate for wheat-paddy system in most of the parts of India. There is already high demand for species for pencil industry, plywood and match splints and *A. cadamba* can be extensively adopted in farm forestry in humid tropics for industrial and income generation. Most of the agroforestry in humid tropics concentrated by *Casurina equaetifolia*, *Acacia mangium* and *Tectona grandis*. Recently *Melia dubia* and *Khaya senagalensis* are promoted in agroforestry where *Anthocephalus kadamba* has more potential in terms of growth, yield and economics. *A. cadamba* sporadically also used as shade tree in tea and coffee plantations. They are resistant for insect pest even in humid tropic conditions.

A. cadamba has minimum shade effect and no alleopathic effect on the agricultural crop advocates its suitability for the for agroforestry purpose. The wider adaptability of the tree from semi arid to humid tropics and foothills of Himalayas also make it preferable option for adoptions in most of agroecological zones of India. Being indigenous fast growing early succession tree it can be promoted in new areas or fresh flood plains of most of the river bank cultivation and Tarai areas. Even in marshy areas or puddle crops like paddy, acorus, mint etc. it is best tree combination where most of the tree species affected by pink disease or root rot. It can be suitable in Chattisgarh, West Bengal, Tamilandu, Bihar, Orissa in India where paddy or medicinal crop of acorus is cultivated either sole or in combination on large areas and repetitively. There is need to further selection of better CPTs of *A. cadamba* for wider adaptability and fast growth to compete with other exotic trees. Though it is better than *Gmelina arborea* in terms of compatibility with crops and resistance to disease and pest mainly in humid area but growth of Gmelina is fast. A good plus tree selection in natural forests and in private plantations is being done by Institute of Forest Genetics and Tree Breeding (IFGTB), Coimbatore, still the availability of superior plants of ' *A. cadamba* ' to the tree growers of India is limited. In the present context where there is scope of Tree Outside Forest (TOF) such fast growing tree species are important for the climate change era for the sequestration of carbon.

There is need to extend the *A. cadamba* tree adoption in farms through extension and build the linkages for industrial and other uses of tree for better markets and economic returns from the crop. *A. cadamba* has potential to yield good returns in agroforestry especially in sub humid and humid tropics (Figure 9 & 10). Though scattered adoption by farmers for the *A. cadamba* is there but systematic agroforestry system research is required by monitoring and evaluation of multiplication trials of elite germplasm of species developed by different institutes. Considering the importance of *A. cadamba*, it is being cultivated in various parts of India including various regions of Bihar state like Katihar, Purnia, Madhepura, Araria, Supoal, Saharsa, Muzaffarpur, Samastipur, Vaishali, Khagaria, Chapra, West champaran, East champaran, Begusaria and Darbhanga under agroforestry systems.

Fig. 9. *A. cadamba Plantation in Agroforestry*

Fig. 10. *A. cadamba Plantation in Agroforestry*

8. Economics

The tree is being cultivated in the gardens and field of the many parts of India. In the West Bengal region this tree is grown in plantation for long times [1], however at present this tree is being used under agroforestry in many parts of the country. As per the estimates done by Institute of Forest Genetics and Tree Breeding (IFGTB), Coimbatore (India), 2012 in Tree Growers fare in an 8 years old *A. cadamba* tree with a height of 10m and a girth of 100cm will give 11 Cu Ft of wood. The average price of one Cubic Fit is about Rs. 250/- comprises to Rs. 2750 per tree. Considering the 400 trees/hectare gives a total of Rs. 11 lakh/hectare at 8 year of age which provides a net benefit of Rs. 9000/month besides additional income from the intercropping as well as ample sufficient amount of Carbon.

Acknowledgement

The authors are grateful to the Director, IIFM, Bhopal, India; Vice Chancellor and Principle, College of Horticulture and Forestry, Navsari Agricultural University, India and the anonymous reviewers for support and guidance. The special thanks to the Luna, R.K. 1996 for book "Plantation trees of India" and Krisnawati, H., Kallio, M. and Kanninen, M. 2011 *Anthocephalus cadamba* Miq.: ecology, silviculture and productivity, CIFOR, Bogor, Indonesia for consultation of information about the article. The farmers cultivating *A. cadamba* India are also acknowledged for allowing us to take photographs from their filed.

References

[1]   Luna, R.K. 1996. Plantation trees, IBD Publisher, Dehradun, India.

[2]   Annon, 1985. Troup's Silviculture of Indian Trees. Vol VI. Controller of Publications, Delhi.

[3]   Osmaston, F.C. 1927. Nursery and plantation Notes of Bihar and Orissa. Governmanr Printing.

[4]   Champion, H. G. and Seth, S. K. 1968. A Revised Survey of Forest Types of India, Govt. of India Press, New Delhi, p. 404.

[5]   Martawijaya, A., Kartasujana, I., Mandang, Y.I., Prawira, S.A. and Kadir, K. 1989 Atlas kayu Indonesia Jilid II. Pusat Penelitian dan Pengembangan Hasil Hutan, Bogor, Indonesia.

[6]   Venator,C.R et al. 1972. Extraction and germination of Cadamb seed. Research Note. No. ITF14. Institute of Tropical Forestry, Puerto Rico.

[7]   Singh, S.P. and Lal, P. 1982. Effect of different spacing treatments on yield from *Anthocephalus chinensis* plantations. Indian Forester, 108 (12): 734-740.

[8]   Rai, S.N. and Sarma, C.R. 1991. Effect of planting spacement on diameter growth of *Anthocephalus chinensis*. Indian Forester, 117 (12): 1029-1031.

[9]   Troup, R.S. 1921. Silviculture of Indian trees. Clarendon Press, Oxford.

[10] Ghosh, R.C. 1977. Handbook of Afforestation Techniques. Controller of Publication, Delhi.

[11] Guhathakurtha, P. and Banerjee, A.K. 1970. The rate of growth of some species in North Bengal. West Bengal Forest Department.

[12] Soerianegara, I. and Lemmens, R.H.M.J. 1993 Plant resources of South-east Asia 5 (1): Timber trees: Major commercial timbers.   Pudoc   Scientific   Publishers,   Wageningen, Netherlands.

[13] Singh TP (2003) Potential of Farm Forestry in Carbon Sequestration. Indian Forester 129: 839-843.

[14] Lubis, Muhammad Ripqi 2014. Turmeric (Curcuma Domestica Val.) Plants Under Agroforestry Stands Jabon (*Anthocephalus Cadamba* Miq.), M.Sc. Thesis- IBP Bogor Agricultural   University, http://repository.ipb.ac.id/handle/123456789/68365.

9

The Changes in the Natural Woody Vegetation in Some Yemeni Villages: Basics for Restoration Policies and Afforestation Programs

Anwar A. Alsanabani

Department of Horticulture and Forestry, Faculty of Agriculture, Sana'a University, Sana'a, Yemen

Email address:

a.sanaban@gmail.com

Abstract: The aim of the study is to detect the changes in the natural woody vegetation (NWV) of rural areas of Yemen and analyze the patterns of these changes. Three villages around Sana'a city were selected. To detect the changes, satellite images of different dates (2004 and 2012) for each village were obtained from the Yemeni Center for Remote Sensing. The result showed an increase of 53%, 49% and 90% for Anagah, Dhbir Khairh and Bait Hambus respectively. The differences among years were significant using a paired- samples t test. The study declined the general consensus by experts who consider that land vegetation cover is declining. Identification of plants' species that exist in the area showed a low biodiversity of only 6 species where two *Acacia* species as well as the shrub *Lycium shawii* represent 95%. Furthermore, comparing NWV among the villages and within plots presented valuable information for plantation strategies such as selecting trees with regenerative criterion and seeding each barren land with some regenerative trees. The study also noted the possible negative influence of industrial expansion and signified the importance of developing land-use plans to protect the natural vegetation.

Keywords: Woody Plant Cover, Natural Vegetation, Renewable Resources, Woodland, Yemen

1. Introduction

Earth existing natural biological capital is declining by deforestation which is proceeding in an alarming rate over the globe [1]. Over the last three centuries, earth has lost more than 1200 million ha of forests and wood lands also grassland and pastures have diminished by about 560 million ha [2]. The demand for resources to support the continuous growth of human population increases each year and consequently intensifies the deforestation process. According to FAO [3] 13 million ha of forest and woodland is lost each year.

The consequences of deforestation are most severe in developing countries, especially those that are located in arid and semiarid zones. Yemen is an example, where poverty of rural communities has forced individuals to overexploit the natural vegetation cover in order to secure income to maintain a mere living [4]. In the same time, government's lack of financial resources has hindered their capacity to plan and implement rural development projects. ROY/FAO/UNDP [4] noted that Yemen has considerable difficulties in allocating funds for the acquisition of reliable and timely geo-spatial data which is considered the first step for appropriate planning and implementation. These two factors among other natural and human induced factors have lead some researchers to conclude that by the year 2000, the total available woody biomass of Yemen would be exhausted [5,4]. There has not been any recent literature on regard of the NWV of Yemen since then, but a general perception among scientists that a persistent reduction continues. .

Misana et al. [6] noted to the fact that land cover change is a complex process involving situation specific interactions among a large number of factors at different spatial and temporal scale. In some cases, these factors have lead to positive changes. In recent decades, some researchers have reported an increase in the NWV in certain parts of the world [7,8,9] and ;therefore, opens hope for recovery trends. Finding those areas and understanding the amount and the rate of this positive change can provide valuable information for restoration efforts in different parts of the world.

A keen visual observation by the author of some rural areas surrounding Sana'a city in recent years, also suggests a positive change which fetch a contrary view to the general supposition. Therefore, this study will test the hypothesis which states that there has been an increase in the NWV of some rural areas of Yemen in the recent years. It will further analyze and explain those changes. The study will consequently provide a basis for governmental policy and programs which targets preservation and plantation in order to mitigate environment problems and help in alleviation of rural poverty.

2. Study Areas

Within a 25 Km radius from Sana'a city, there are approximately 27 villages out of which three villages were randomly selected (Figure 1). The following table gives more information about their location:

Table 1. Detailed information about the villages used in the study

Village name	Latitude	Longitude	Altitu'de	Distance from Sana'a
Anagah	15°18'57.7"	44°25'09.8"	2598 m	20 Km
Dhbir Khairh	15°07'60.0"	44°16'05.1"	2408 m	25 Km
Bait Hambus	15°15'55.6"	44°08'45.5"	2783 m	12 Km

Records for the weather conditions for each village are unavailable but the closest area that is documented is Sana'a city which has a mean annual temperature of 18 C and an average precipitation of 200 mm per year and usually has low humidity averaged of approximately 43% [10].

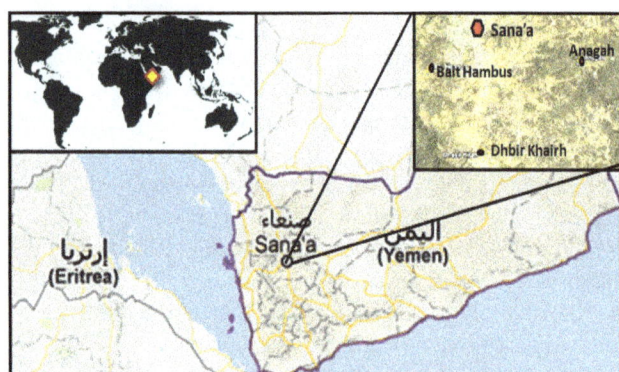

Figure 1. The geographic location of the study area.

3. Methodalogy

3.1. Satellite Images and GIS Analysis

The three villages were observed in different dates for change detection which is defined as the process of identifying differences in the state of an object or phenomenon by observing it at different times [11]. The study looked at the NWV of those three villages in the year of 2004 and the year of 2012 with 8 years interval using QuickBird satellite images with 60 cm resolution and Blue, Green, Red, NearIR bands. For each village two images with the different dates were obtained from the Yemeni Center for Remote Senescing and Geographical Information System. Folega et al. [12] stated that The moderate resolution of satellite images is suitable for mapping the status of land cover features and can enable planners to quantify the pattern of land cover changes that have occurred over time. Due to the small size of the areas, the low density of the vegetation and the importance of the accuracy, the following simple method was adopted.

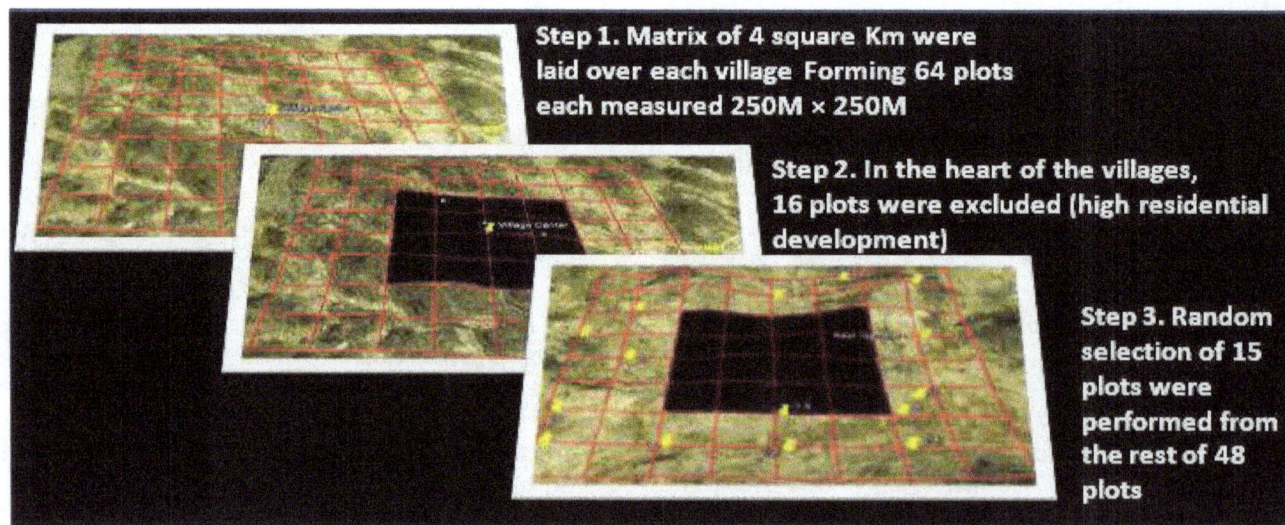

Step 1. Matrix of 4 square Km were laid over each village Forming 64 plots each measured 250M × 250M

Step 2. In the heart of the villages, 16 plots were excluded (high residential development)

Step 3. Random selection of 15 plots were performed from the rest of 48 plots

Figure 2. Methodology steps for randomized data collection using satellite images.

A matrix of 4 square Km was laid over each village and formed 64 plots where each plot measured 250 M × 250 M.

In the heart of the villages were residential development is high, 16 plots were excluded. A random selection of 15 plots were performed from the rest of the 48 plots (Figure 2). Overlapping of two layers for each village one with the year 2004 and the second with the year of 2012 using GIS (ArcMap version 9.3), we were able to trace the fate for each single vegetation and observe any appearance of new vegetation by changing the transparency of the upper layer and the result was recorded (Table 1). To test the accuracy of the recorded result from satellite images of year 2012, a site visit for the plots was performed using GPS device. To estimate the accuracy of the 2004 satellite images, trees and shrubs' ages were observed. Another task which was also performed during the site visit is identifying and counting the types of vegetation that existed in those plots.

3.2. Statistical Procedures

The alternative hypothesis states that there is a difference

in the NWV of some rural areas of Yemen between year 2004 and year 2012. To prove that a difference exists and to evaluate its significance, a paired- samples t test were used in which the means of year 2004 and year 2012 were compared. The tests were two tailed and at alfa level of .05. The SPSS program (IBM SPSS statistics, version 21) was used for the analyses.

4. Results and Discussion

The result of the satellite images showed an increase in woody plants cover between year of 2004 and 2012 (Table 1). The increase was 53%, 49% and 90% for Anagah, Dhbir Khairh and Bait Hambus respectively.

Table 2. Change % of NWV between year of 2004 and 2012 among villages and within plots.

Plot #	Anagah			Dhbir Kharah			Bait Hambus		
	2004	2012	% change	2004	2012	% change	2004	2012	% change
1	6	9	50	33	38	15	75	105	40
2	3	3	0	3	9	200	63	95	51
3	2	3	50	13	23	77	2	15	650
4	4	14	250	20	29	45	10	26	160
5	7	11	57	16	31	94	19	37	95
6	0	0	0	15	31	107	17	42	147
7	0	0	0	20	37	85	11	29	164
8	11	25	127	4	10	150	27	60	122
9	56	64	14	76	69	-9	1	3	200
10	0	0	0	9	20	122	29	43	48
11	5	8	60	3	20	567	0	0	0
12	1	5	400	15	21	40	1	4	300
13	1	0	-100	11	19	73	0	6	600
14	1	7	600	25	19	-24	0	5	500
15	1	1	0	41	77	88	11	36	227
TOTAL	98	150	53	304	453	49	266	506	90

When t test was performed, the result showed that these increases were significant at alfa level of (.05) for each village independently and for all villages together (Table 2). This result is in the contrary to the general consensus by some Yemeni experts and scientists who consider that land vegetation cover is declining. Their supposition is supported by previous studies. Millington [5] indicated the negative impact of fuel-wood collection on the woody biomass and predicted that by the year 2000, the total available woody biomass in Yemen will have been exhausted. Another study estimated that the total fuel-wood consumption in Yemen was 5 million m3 (3 million tons) in 1982 and projected that fuel-wood consumption would reach 8.5 million m3 by the year 2000, at which time the fuel-wood supplies will have disappeared [4]. The researchers reached this conclusion because Yemenis used to relay on wood as their main source of energy. In the early years of this century, gas prices were affordable and in the same time there was a scarcity of wood; therefore, a shift toward gas energy was made. The shift relieved the stress on the NWV and a notable recovery took place.

Table 3. Average count of NWV, % change and t-test results @ α =.05.

	2004	2012	% Change	Paired Samples t-Test	
				Df	P-value
Anagah	6.5	10	53	14	0.008
Dhbir Kharah	20.3	30.2	49	14	0.002
Bait Hambus	17.7	33.7	90	14	0.003
All Villages	14.8	24.6	64	44	0.000

The provision of alternative energy source proved to be practical solution for deforestation and the Yemeni government should take it to account when it sets prices for gas. A solution for deforestation which had been proposed generally by some researchers [13] and specifically to Yemeni rural areas by others [4].

Another study of the vegetation cover between years of 1990 and 2000 which covers all Sana'a governance, stated a decline by 34% [14]. The study used NDVI technique and included agriculture crop cover while ours only focused on natural trees and shrubs cover. The possible explanation is that both studies are true for their certain timeframe studied. Meaning that there had been a decline before year of 2000 and an increase afterward. A fact that can be explained by the shift in the energy demand from woodfuel to gasfuel. A decreased trend in early years followed by an increase in recent years has been reported by some researchers. For example, Doner [9] who showed a decrease of forest trees between 1978 and 1987 and an increase between 1990 and 2000 in Gumushane, Turkey. The increase took place in Turkey earlier than Yemen probably because it is more developed and the energy shift could have been earlier.

A comparison of wood cover within plots and among villages presents a valuable implication for plantation and preservation programs. Anagah has the least NWV cover in 2004 and consequently has an increase of only 52 count by 2012 while Bait Hambus has an increase of almost five fold more than Anagah. This is because plots that showed zero cover remain in most cases zero after 8 years while those that have some plants which have reproductive habits gave more plants cover over time. This fact explains why some huge

distances of Yemeni landscape are empty from any trees and illustrate the importance of seeding all barren land with some regenerative trees.

Only two plots in Dhbir Khairh showed a decrease in number of plants between year of 2004 and 2012. This decrease was explained from observing satellites images by industrial and agriculture development that have reached to the area and consequently have cleared the NWV that had existed. This observation is important and denotes the possible impact of the future industrial and agriculture expansion. It also urges the concerned officials to establish land use plans for the whole area to protect NWV, especially watershed areas which are the most sensitive.

The study area has only 6 different species of NWV where both *Acacias* and *Lycium shawii* represent 95% (Table 3), a fact that can be generalized based on author's ground observation for the high mountain region of Yemen. This indicates the lack of biodiversity in the NWV of the region.

Table 4. *Existing plant species and their count per village and their percentage from whole NWV.*

	type	Dhbir Khairh		Anagah		Bait Hambus	
		2012	%	2012	%	2012	%
Acacia gerrardii	Tree	131	29	0	0	177	35
Acacia origena	Tree	170	38	0	0	215	42
Tamarx aphylla	Tree	0	0	53	35	0	0
Ziziphus spina-christi	Tree	13	3	3	2	0	0
Ficus carica	Tree	0	0	11	7	1	0
Lycium shawii	shurb	139	31	83	55	113	22
TOTAL		453	100	150	100	506	100

Selection and Introduction of new species for plantation program can help in improving biodiversity. The FAO reported three species that used to exist in Yemeni landscape and had been eliminated due to high exploitation [15]. Those can be a good choice for reintroduction.

Ziziphus spina-christi is another important indigenous species which exists in scars number and yet has a great value as an agroforestry tree [16]. Since this important plant already existed; therefore, proved practically to be suitable for the region. Furthermore, its ability to be multiplied by simple method of direct seeding [17], makes it another good suggestion for plantation program.

The study also illustrated the importance of tree reproductive character as a criterion for plant selection. Anagah which has *Tamarx aphylla* and *Ficus carica* which are both less reproductive showed less vegetation cover while Dhibr Khairh and Bait Hambus which have both *Acacias* that are highly reproductive showed better vegetation cover.

5. Conclusions

The result of the study proved a recovery in the NWV in the sample areas studied, and provide new notion which replaces the long held view of decline. One of the main causes of this change is the shift that took place in some parts of Yemen from the use of wood for fuel to the use of gas for

fuel at a time where fossil fuel energy was made affordable and wood was scarce. The rate of the recovery process depended on two factors. The first is the amount of NWV cover that still exists. Those areas that still have a large NWV cover were able to recover easily, while those areas that have no NWV cover remain as they were. The second factor is the species type and its ability to regenerate. An important consideration which had been ignored in previous plantation effort and had resulted in a poor outcome as found in this study. Plots that have single or multiple exotic plants which were unable to regenerate in Yemeni environment, remain the same number after more than 8 years. Plantation programs are a must and lay an ethical obligation upon governmental and non-governmental organization. The study also denotes the importance of land-use plans for areas surrounding big cities in Yemen to protect intensive NWV as well as the importance of policy initiatives for sustainable use of this important natural resource.

The study were restricted to small area due to resources limitation. Yet its finding is important and provoke the attention of officials to allocate funds for the acquisition of reliable and timely satellite images for the whole Yemen. A large scale analysis that could confirm the consistency of those findings over the whole region should be also funded. Although Cover changes can be detected by remote sensing but in order to get a deep understanding of their causes, a socio-economic analysis should be carried out. A further study is also suggested to cover this topic.

Acknowledgement

Yemeni Center for Remote Senescing are highly acknowledged for providing the satellite images and special thank to Dr. Khalid Khanbari and Eng. Ibrahim Al-Samawi for their technical help.

References

[1]   Arekhi S and Jafarzadeh AA (2012). Deforestation modeling using logistic regression and GIS (Case study: Northern Ilam forests, Ilam Province, Iran). Afri. J. of Agri. Res. 7(11)1727-1741

[2]   Fabiyi O. (2011) Change actors' analysis and vegetation loss from remote sensing data in parts of the Niger Delta region. J. of Ecol. and the Nat. Environ. 3(12):381-391

[3]   FAO (Food and Agriculture Organation of the United Nations) (2005). State of the World's forests. FAO, Rome.

[4]   ROY/FAO/UNCCD/UNDP (2000). The National Plan of Action to Combat Desertification in the Republic of Yemen, Sana'a, Yemen.

[5]   Millington AC (1988). Woody Biomass Resource Assessment. Sana'a, Yemen.

[6]   Misana SB, Sokoni C, Mbonile MJ (2012). Land-use/cover changes and their drivers on the slopes of Mount Kilimanjaro, Tanzania. J. of Geog. and Reg. Plan. 5(6):151-164.

[7]   Poyatos,R. Latron J. and P. Llorens (2003) Land Use and Land Cover Change After Agricultural Abandonment The Case of a Mediterranean Mountain Area (Catalan Pre-Pyrenees) Mountain Research and Development 23(4): 362-368

[8]   Herrmann SM. Anhamba A and CJ. Tucker (2005) Recent trends in vegetation dynamics in the African Sahel and their relationship to Global Environmental Change climate. Bio One. 15(4):394-404

[9]   Doner F (2011). Using Landsat data to determine land use/land cover changes in Gümüshane, Turkey. Sci. Res. and Ess. 6(6):1249-1255.

[10] Al-Korasani MA (2005). Guide for agriculture weather in Yemen. The Yemeni agri. res. and ext. auth. Dhmar, Yemen

[11] Olaleye JB, Abiodun OE, Asonibare RO (2012). Land-use and land-cover analysis of Ilorin Emirate between 1986 and 2006 using landsat imageries. Afri. J. of Environ. Sci. and Tech. 6(4): 189-198

[12] Folega F, Zhao X, Batawila K, Zhang C, Huang H, Dimobe K, Pereki H, Bawa A, Wala K, Akpagana K (2012). Quick numerical assessment of plant communities and land use change of Oti prefecture protected areas (North Togo). Afri. J. of Agri. Res.. 7(6): 1011-1022.

[13] Githiomi JK, Mugendi DN, Kung'u JB (2012). Analysis of household energy sources and woodfuel utilisation technologies in Kiambu, Thika and Maragwa districts of Central Kenya. J. of Hort. and Forest. 4(2): 43-48

[14] Herzog M (1998). The Natural Forests of Yemen. Rheinfelden, Switzerland

[15] Elsiddig E, Luukkanen O, Batahir A, Elfadl M (2004). The Important of Ziziphus spina-christi in the Drylands with reference to Yemen. University of Khartoum. Khartoum, Sudan

[16] Alsanabani A, Al-Thobhani M, Al-Gadasi A (2013). Direct-seeding success of Ziziphus spina-christi in rainy seasons of Yemen in preparation for large scale afforestation efforts. Yemeni J. of Agri. Res. and stud. (27)157-168

10

Biological invasion of imported cabbageworm, *Pieris rapae* (L.), on oilseed *Brassica* in Punjab, India

Sarwan Kumar

Department of Plant Breeding and Genetics, Punjab Agricultural University, Ludhiana-141 004, India

Email address:

sarwanent@pau.edu

Abstract: The infestation of imported cabbageworm/ small white butterfly, *Pieris rapae* (L.) (Lepidoptera: Pieridae) is reported on oilseed *Brassica rapa* ssp. *oleifera,* ecotype brown *sarson* cv. BSH 1 from Punjab, India. Though, it has been reported earlier on vegetable brassicas in India, there is no report of its infestation on oilseed brassica in this part of the country so far. A low infestation of 1.7-3.3 larvae per 10 plants was reported during first to third Standard Meteorological Week of 2014. Since this pest has potential to cause significant damage to oilseed brassica crops, therefore, timely reporting of this pest is important to avoid any future outbreak.

Keywords: Cabbage Caterpillar, Oilseed Brassica, Small White Butterfly

1. Introduction

Rapeseed-mustard is an important group of winter season oilseed crops after soybean (*Glycine max* (L.) and Palm (*Elaeis guineensis* Jacq.) oil. India is one of the leading producers including Canada, USA, European Union, Australia and China. Among the seven edible oilseed crops cultivated in India, rapeseed-mustard (*Brassica* spp.) contributed 26 per cent of the total oilseeds production in the year 2012-13 [23]. It is the second most important edible oilseed after groundnut sharing 27.8 per cent in the India's oilseed economy. The share of rapeseed-mustard to the total edible *rabi* oilseeds production was 78.9 per cent, during 2005-06 to 2010-11 [16]. In India, under the name rapeseed and mustard three cruciferous members of *Brassica* species are cultivated; *B. juncea* (Indian mustard or commonly called *rai*) being the chief oil yielding crop, while three ecotypes of *B. rapa* ssp. *oleifera*, viz. brown *sarson*, yellow *sarson*, *toria* and *B. napus* are grown to a limited extent [2].

The average productivity of India (1176 kg/ha) is about two-third of world's average yield of 1695 kg/ha [26]. There are many reasons for this low yield, among which the damage caused by insect-pests is the one. A number of insect-pests are associated with this group of crops right from sowing till harvest. The pest complex of oilseed *Brassica* in this part of the country is witnessing a change possibly due to climate change, intensive cultivation of high yielding varieties, frequent and indiscriminate application of synthetic insecticides, development of insecticide resistance in insects. The changing scenario of insect pest problems in agriculture as a consequence of green revolution has been well documented [8, 9]. The pest complexes have changed to the extent that some of the insects which had been known earlier to be sporadic, minor or non-injurious to these crops, have become serious pests in certain agroclimatic conditions today, whereas, other insects which were never previously recorded on these crops are now becoming a matter of concern [25]. For example, the large white butterfly, *Pieris brassicae* (L.) (Lepidoptera: Pieridae) which was once a sporadic pest of oilseed *Brassica* has become a regular pest of oilseed *Brassica* [17] in this region and is one among the important pests after turnip aphid, *Lipaphis erysimi* (Kaltenbach) (Homoptera: Aphididae). The objective of the present study is to report a new pest *Pieris rapae* for the first time on oilseed Brassica crops in Punjab province of India.

2. Materials and Methods

Field experiment was conducted during November 2013 to March 2014 at Punjab Agricultural University, Ludhiana (30° 56'N, 75° 52'E, 247 m above mean sea level), India to study the population dynamics of insect-pests infesting oilseed *Brassica* crops. The climate of the area is characterized as sub-tropical and semi-arid with hot and dry spring summer from April to June, hot and humid summer from July to September and cold autumn winter from November to January. The average annual rainfall is about 700 mm, most of which is received during monsoon period from July to September with little showers during winter (crop) season. The soil of the experimental field was loamy sand in texture having pH 7.5. The experiment was laid out in a randomized complete block design with three replications of 10 cultivars. The different cultivars included; *Brassica juncea*: RLC 1 (single '0' i.e. low erucic acid), PBR 91, PBR 210 (both conventional i.e. high in erucic acid and glucosinolates)); *B. napus*: GSC 6, GSC 5 (both '00' i.e. canola), GSL 1, GSL 2 (both conventional); *B. carinata*: PC 5 (conventional), *B. rapa* ssp. *oleifera* ecotype brown *sarson* cv. BSH 1 (conventional); and *Eruca sativa*: T 27 (conventional) (all available from Incharge, Oilseeds Section, Department of Plant Breeding and Genetics, Punjab Agricultural University, Ludhiana – 141 004, Punjab, India except T 27 which is available from Director, Directorate of Rapeseed-Mustard Research, Bharatpur – 321 303, Rajasthan, India). All the cultivars were sown on November 9, 2013 in plots of size 4.2 x 3 m with recommended package of practices [19] except spray of insecticides. Sowing was delayed by 19 days than normal since late sown crop is heavily attacked by insect pests in this part of the country [15]. All the 10 cultivars were sown in well prepared seed beds with the help of seed drill. The row to row and plant to plant spacing was maintained at 30 cm and 15 cm, respectively, for all the *Brassica* species except *B. napus* for which it was 45 cm and 15, respectively. A basal dose of 50 kg N (as urea) and 30 kg phosphorous ha^{-1} (as single super phosphate) was applied at sowing time and 50 kg N ha^{-1} was applied three weeks thereafter at the time of first irrigation. After about four weeks of sowing, when the plants reached true leaf stage, data on the population of different insect-pets were recorded in each cultivar from 10 plants selected at random at weekly intervals. Since, the present paper describes the first record of imported cabbageworm on oilseed *Brassica* in this region, the data on other insect pests are not presented here. Larvae of *P. rapae* were collected and maintained individually in laboratory in a styrofoam cup covered with a fine muslin cloth fastened with elastic bands in a Biological Oxygen Demand (B.O.D.) incubator at 22±1°C till pupation and subsequent adult eclosion.

3. Results and Discussion

A very low incidence of imported cabbageworm, *P. rapae* was recorded on *B. rapa* ssp. *oleifera* ecotype brown *sarson* cv. BSH 1 at experimental farm in Ludhiana, Punjab, India, during 1-3 Standard Meteorological Week (SMW) of 2014, which ranged from 1.7 to 3.3 larvae per 10 plants (Table 1). The larvae were found feeding on leaves of BSH 1. They were hard to locate on leaves since their green colour provided excellent camouflage. Damage to host plants consisted of irregular large holes on the leaf lamina (Fig. 3). During all the observations larvae were found feeding exposed on the leaf surface. Caterpillars were velvetty green with yellow lines on

the body one on the dorsal side and others often broken on the lateral sides (Fig. 1 & 2). To confirm the species identification from adult stage, larvae were reared on the same host till adult eclosion from pupae. Adult butterflies that emerged from pupae had a wing span of about 43 mm. The wings were white with black tips of forewings. There were two black spots on the top of forewing of female while males had only one such spot. A black spot was also present on the black outer margin of hind wing which otherwise is white.

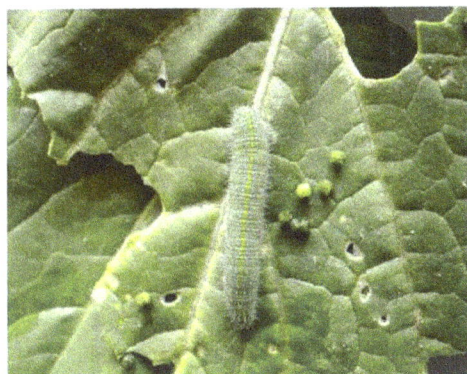

Fig. 1. *Larva feeding on* Brassica rapa *ssp.* oleifera *ecotype brown* sarson *cv. BSH 1 leaf*

Fig. 2. *Dorsal and lateral yellow lines on the larval body*

Fig. 3. *Damage symptoms after larval feeding*

The imported cabbage worm, *Pieris rapae* (L.) (Order: Lepidoptera; Sub order: Papilionoidea; Family: Pieridae; Subfamily: Pierinae; Genus; *Pieris*; Species; *rapae* (Linnaeus 1758); Synonymy: *Artogeia rapae* (L.) (Lepidoptera: Pieridae) is thought to be originated in Europe [21], where it is called the small white butterfly to distinguish it from *Pieris brassicae*, the large white. It has spread around the world in the wake of European colonization. It was first found outside Europe in Canada in 1860 when it was accidently introduced to Quebec which later spread throughout North America [13] and subsequently to Hawaii and Australia as well [3].

Unlike the larvae of *P. brassicae*, the large white butterfly, which are found in gregarious phase during initial stages of their development (first three instars), *P. rapae* larvae were found solitary feeding on the leaves. Though females lay eggs singly, their fecundity is very high and a single female can lay on an average 300-400 eggs which can go as high as 1000/ female [14]. Although, the damage caused by *P. rapae* was slight, but it can be severe in years with high infestation [12]. *P. rapae* is an economically important pest of brassica vegetables the world over. Some of its host plants include: broccoli, Brussels sprouts, cabbage, cauliflower, Chinese broccoli, Chinese cabbage, choy sum, collards, daikon, horseradish, kale, kohlrabi, mustard, mustard cabbage (kaichoy), pak choy, radish and turnip. It may also feed on other plants such as nasturtium, lettuce and sweet alyssum. In addition to these, Shepherd's purse is a weed host of this pest. To date this insect has not been reported to infest oilseed Brassica crops in Punjab province of India though it is reported as a pest of vegetables (cabbage and cauliflower) in other parts of the country [27, 1]. In India, it has been primarily documented as a pest of vegetable Brassica [4, 10]. Even in other parts of the world also, it is documented as a pest of vegetables [5]. It has shown considerable potential to expand its geographical distribution from Europe to other parts of the world and host plant range. It is likely that *P. rapae* can pose a serious threat to oilseed Brassica crops in this geographical region, which can lead to extensive insecticide use potentially causing multitude of undesirable side-effects [28, 7]. Development of an effective management strategy at this early stage of detection is important in order to prevent its damage to these important winter season oilseed crops of India.

Invasive species like *P. rapae* represent a major threat to both natural [6, 22] and agricultural [18, 11, 24, 20] ecosystems. They can reduce crop yields, increase costs (related to their management), and lead to the use of pesticides which ultimately lead to disruption of existing pest management systems [26]. Thus, monitoring and timely reporting of this pest is important to avoid any future outbreak on oilseed Brassica. There is a need of detailed study of its biology, bionomics and population dynamics on different oleiferous Brassica so that management strategy can be devised well in time before it reaches serious proportions.

References

[1]   Bhat D. M., and R. C. Bhagat (2009) Natural parasitism of *Pieris rapae* (L.) and *Pontia daplidice* (L.) (Lepidoptera: Pieridae) on cruciferous crops in Kashmir Valley (India). American Eurasian J. Agric. Environ. Sci. 5: 590-591.

[2]   Bhatia V., P. L. Uniyal and R. C. Bhattacharya (2011) Aphid resistance in *Brassica* crops: challenges, biotechnological progress and emerging possibilities. Biotechnol. Adv. 29: 879-888. doi:10.1016/j.biotechadv.2011.07.005

[3]   Braby M. F. (2000) *Butterflies of Australia.* CSIRO Publishing, Melbourne, vol. 1: 343-344.

[4]   Butani D. K. and M. G. Jotwani (1984) *Insects in vegetables.* Delhi, India: Periodical Expert Book Agency, 356p.

[5]   Capinera J. L. (2001) *Handbook of vegetable pests.* California, USA: Academic Press, 729p.

[6]   Clavero M. and E. Garcia-Berthou (2005) Invasive species are a leading cause of animal extinctions. Trends Ecol. Evol. 20: 110.

[7]   Desneux N, A. Decourtye and J. M. Delpuech (2007) The sublethal effects of pesticides on beneficial arthropods. Ann. Rev. Entomol. 52: 81-106.

[8]   Dhaliwal G. S. and R. Arora (2006) Integrated pest management: concepts and approaches. New Delhi, India: Kalyani Publishers.

[9]   Dhaliwal G. S. And O. Koul (2010) Quest for pest management: From green revolution to gene revolution. New Delhi, India: Kalyani Publishers.

[10] Gupta S. L. (1990) Key for the identity of some major lepidopterous pests of vegetables in India. Bull. Entomol. 31: 69-84.

[11] Haack, R. A., F. Herard, J. H. Sun, and J. J. Turgeon (2010) Managing invasive populations of Asian longhorned beetle and citrus longhorned beetle: a worldwide perspective. Ann. Rev. Entomol. 55: 521–546.

[12] Hern, A., G. Edwards-Jones and R. G. McKinlay (1996) A review of the preoviposition behaviour of small cabbage white butterfly *Pieris rapae* (Lepidoptera: Pieridae). Annals appl. Biol. 128: 349–371.

[13] Howe W. H. (1975) The butterflies of North America. Doubleday, Garden City, 633p.

[14] Jogar K, L. Metspalu, K. Hiiesaar, A. Ploomi, E. Svilponis, A. Kuusik, N. Menshykova, I. Kivimagi and A. Luik (2009) Influence of white cabbage cultivars on oviposition preference of the *Pieris rapae* L. (Lepidoptera: Pieridae). Agronomy Res. 7 (Spl. Issue I): 283-288.

[15] Kular, J. S., A. S. Brar, and S. Kumar (2012) Population development of turnip aphid *Lipaphis erysimi* (Kaltenbach, 1843) (Hemiptera: Aphididae) and the associated predator *Coccinella septempunctata* Linnaeus, 1758 as affected by changes in sowing dates of oilseed *Brassica.* Entomotropica 27: 19-25.

[16] Kumar, A (2012) Production barriers and technological options for sustainable production of rapeseed-mustard in India. J. Oilseeds Brassica 3: 67-77.

[17] Kumar, S. (2011) *Cotesia glomeratus* – a potential biocontrol agent for large white butterfly *Pieris brassicae* in Indian Punjab. *Proceedings of 13th International Rapeseed Congress,* Prague, Czech Republic, June 05-09, 2011. vol. 13, pp. 1141-43.

[18] Olson, L. J. (2006) The economics of terrestrial invasive species: a review of the literature. Agric. Resources Econ. Rev. 35: 178-194.

[19] PAU (2013) Package of Practices for Crops of Punjab: *Rabi* 2013-14. Punjab Agricultural University, Ludhiana, India, pp. 43-53.

[20] Ragsdale, D. W., D. A. Landis, J. Brodeur, G. E. Heimpel and N. Desneux (2011) Ecology and management of the soybean aphid in North America. Ann. Rev. Entomol. 56: 375-399.

[21] Robbins, R. K. and P. M. Henson (1986) Why *Pieris rapae* is a better name than *Artogeia rapae* (Pieridae). J. Lepidopterists' Society 40: 79-92.

[22] Samways M. J. (2007) Insect conservation: a synthetic management approach. Ann. Rev. Entomol. 52: 465–487.

[23] Singh D. (2014) Genetic enhancement of mustard for seed yield and its sustainability. In: V. Kumar, P. D. Meena, D. Singh, S. Banga, V. Sardana and S. S. Banaga, 2014: *Abstracts, 2nd National Brassica Conference on 'Brassicas for Addressing Edible Oil and Nutritional Security'* organized by Society for Rapeseed Mustard Research, Bharatpur, India at Punjab Agricultural University, Ludhiana, India during Feb. 14-16, 2014, p. 18.

[24] Suckling, D. M. and E. G. Brockerhoff (2010) Invasion biology, ecology, and management of the light brown apple moth (Tortricidae). Ann. Rev. Entomol. 55: 285–306.

[25] Taggar, G. K., R. Singh, R. Kumar and P. C. Pathania (2012) First report of flower chafer beetle, *Oxycetonia versicolor*, on pigeonpea and mungbean from Punjab, India. Phytoparasitica DOI 10.1007/s12600-012-0222-8

[26] Thomas, M. B. (1999) Ecological approaches and the development of 'truly integrated' pest management. PNAS USA 96: 5944–5951.

[27] Verma, A. K. (1974) *Systematic studies of the butterflies of Patiala area*. M.Sc. Thesis, Punjabi University, Patiala.

[28] Weisenburger D. D. (1993) Human health: effects of agrichemicals use. Human Pathology 24: 571–576.

11

Effects of biochar derived from maize stover and rice straw on the germination of their seeds

Alie Kamara[1], Abibatu Kamara[2], Mary Mankutu Mansaray[2], Patrick Andrew Sawyerr[1]

[1]Soil Science Department, School of Agriculture, Njala Campus, Njala University, Sierra Leone
[2]Extension Division, Ministry of Agriculture, Forestry and Food Security, Sierra Leone

Email address:
aliekamara@njala.edu.sl (Alie K.)

Abstract: Although there has been an increased focus on the use of biochar for improving soil fertility and mitigating climate change, some biochars have been reported to contain substances that affect germination and seedling growth negatively. It is therefore necessary to evaluate any biochar material for its effect on seed germination before large scale applications. This study was therefore undertaken to assess the effects of (i) biochar derived from maize stover on maize seed germination and (ii) biochar derived from rice straw on rice seed germination. Seeds of maize (*Zea maize* L.) and rice (*Oryza sativa*) were sown separately to soils treated with increasing levels of biochar derived from maize and rice residues respectively. The experiment was conducted using a completely randomized design involving five biochar treatments: 0 g (control), 1.25g, 2.50g, 3.75g and 5.00g each mixed with 300g of a fine sandy loam soil in Sierra Leone. Results of the germination test showed that most of the maize seeds (>80%) germinated by day3 and there was no significant difference in the number of maize seeds germinated on day 7. On the other hand, few rice seeds germinated on day3 (35%) and was significantly greater than the number of the rice seeds (>90%) germinated on day 7. However, even though the number of maize or rice seeds germinated on biochar treated soils was higher than the control, the difference was not significant. Also, no significant differences in root lengths were observed between the control and biochar treatments at day 7 for both plants. However, maize shoot length differed significantly from the control whereas rice shoot length did not. The results showed that sowing seeds of maize and rice on soils treated with biochar derived from their crop residues had no adverse effect on germination. These findings hold great potential for improved and sustainable maize and rice cultivation in Sierra Leone.

Keywords: Biochar, Rice Straw, Maize Stover, Germination

1. Introduction

Crop production on Sierra Leone soils faces challenges due to soil acidity, low water and nutrient retention. For such soils, maintenance or improvement of soil organic matter content is the key to sustainable cropping. One important approach to soil organic matter management is by addition of crop residues. Crop residues when incorporated or left on the soil surface eventually decompose releasing nutrients and carbon dioxide back into the atmosphere [1]. Consequently the benefits of adding organic matter to soil, such as, improved water and nutrient retention are basically short-lived and continuous additions are necessary to maintain productivity.

A more lasting approach to improving soil productivity is by applying carbon into the soil in the form of biochar.

Biochar is a stable form of carbon produced from heating natural organic materials (such as crop residues and other biomass wastes) in little or no oxygen environment in a process known as pyrolysis. Biochar refers to the charred organic matter or charcoal, produced with the intent to deliberately apply to soils to sequester carbon and improve soil properties [2]. Additions of biochar to soil have been reported to increase pH and cation exchange capacity [3, 4, 5] as well as nutrient availability [6, 7].

During pyrolysis, plant biomass undergoes a series of physical and chemical changes. Physical changes in plant biomass during pyrolysis basically involve mass loss and volume reduction without significant changes in the original structure [8]. The shrinkage in volume and mass loss are due to loss of volatile compounds in the original plant material [9] and result in concentration of nutrients into the charred remains or biochar fraction. Thus converting crop residues into biochar and and applying to soils, is a convenient way of returning nutrients to the soil. Furthermore, biochar is resistance to microbial decomposition [10] and therefore soil biochar applications offer the potential to sequester carbon in agricultural lands [11, 12, 13] thereby mitigating climate change.

Generally, biochar as a renewable bio-resource has the potential of positively impacting soil and crop productivity. However some biochars have been reported to contain substances that affect germination and seedling growth negatively [14]. It is therefore necessary to evaluate any biochar material for its effect on seed germination before large scale applications. This study was therefore undertaken to assess the effects of (i) maize stover biochar on maize seed germination and (ii) rice straw biochar on rice seed germination using a soil-based assay.

2. Methodology

2.1. Description of the Study Area

The experiment was conducted in Njala Campus, Njala University in the Moyamba District in Southern Sierra Leone, West Africa. Njala is about 255km from Freetown, the capital city of Sierra Leone and is located at an elevation of 128m above sea level (altitude), on Latitude 8^0N and Longitude 12^0W. The climate is tropical and is characterized by two main seasons; a rainy season (April to October) and a dry season (November to March). The average annual rainfall is about 2500mm and the mean annual temperature ranges from a minimum of 28^0C to a maximum of 33^0C.

2.2. Biochar Production and Processing

Maize stover and rice straw were collected from farms in Mosongo, one of the villages around Njala University Campus. Each crop residue was thoroughly dried in the sun and converted to biochar using the Elsa stove designed for the BeBi Project in Njala University. [The Elsa stove is a low-tech Top-Lit-Up-Draft (TLUD) bioenergy stove designed for use in domestic cooking and production of biochar for application to soils under the Project Agricultural and Environmental Benefits of Biochar Use in ACP Countries – 'BeBi']. The charred material (biochar) was quenched with water (by sprinkling water on the hot char) and dried in the sun for a few days. The dry biochar was crushed and sieved through a 2mm sieve and stored.

2.3. Properties of Biochar and Soil Used in the Study

The soil used in this study had the following characteristics: pH 4.7 (1:1 soil:water ratio), organic carbon 2.1g/kg soil (Walkley-Black), available phosphorus 5.6mg/kg soil (Bray & Kurtz 1), cation exchange capacity 6.2 cmol(+)/kg soil (neutral M NH_4OAc), exchangeable cations 0.28, 0.19, 0.12 and 0.02 cmol+/kg soil Ca, Mg, K and Na respectively, and particle size (70% sand, 16% silt and 14% clay). Maize stover biochar had the following characteristics: pH 8.4, electrical

conductivity (1:5 biochar:water) 3.98 dSm^{-1}. Rice straw, on the other hand, had the following characteristics: pH 8.8, electrical conductivity (1:5) 2.82 dSm^{-1}.

2.4. Germination Test

This study involved five biochar application rates (treatments) in four replications in a completely randomized design (CRD). Each of five biochar treatments (0g, 1.25g, 2.50g, 3.75g and 5.00g) was mixed thoroughly with 300g of a fine sandy loam soil (Mokonde Series) located in the Njala area. The test was conducted for seven days. The soil-biochar mixture was placed separately in five small cylindrical containers of dimensions 10 cm diameter and 5 cm height. The soil in each container was moistened with 140g of water (47%w/w moisture content) and four seeds of maize (Zea maize L.) or rice (Oryza sativa) were planted to 2cm depth in each container. After planting, the pots were covered with a transparent plastic sheet for 48 hours to minimise moisture loss and allowing access to sunlight.

The number of plants that germinated on day 3 and 7 was recorded for each treatment. On day 7 the germination test was terminated and data collected on total plant length, length above ground and length below ground

2.5. Statistical Analysis

Statistical differences among treatments were determined by Analysis of Variance (ANOVA) for Completely Randomized Designs (CRD) and LSD (P<0.05) in the GenStat 12th Edition computer software.

3. Results

3.1. Effect of Maize Stover Biochar on the Germination of Maize Seeds

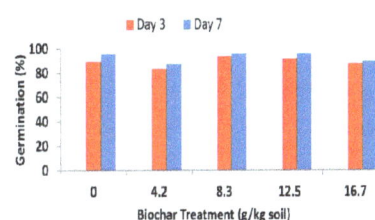

Figure 1. Percentage germination of maize seeds after three and seven days on soil mixed with biochar derived from maize stover

The percentage germination of maize seeds as influenced by increasing levels of application of biochar derived from maize stover on a sandy loam are shown on Fig.2. The pattern of maize seed germination was similar at both 3 and 7 days after planting (DAP). At biochar application rate of 4.2g/kg soil, there was a slight decrease in germination compared with the control but increased slightly again at 8.3g/kg soil biochar application, then decreased slightly again at higher application rates. These variations, however, were not significantly different (p<0.05) among each other at both 3 and 7 DAP. At day 3 most of the maize seeds (>80%) have germinated and did not differ significantly (p<0.05) from the number of seeds germinated at day 7. Thus the presence of biochar neither improved nor adversely affected maize germination.

3.2. Effect of Rice Straw Biochar on the Germination of Rice Seeds

Fig. 2 shows the percentage germination of rice seeds with increasing levels of application of rice straw biochar on a fine sandy loam. At 3 DAP percent rice seed germination was less than the control for most biochar application rates except at 16.7g/kg soil. On the other hand, at 7 DAP there was a slight decrease in percent rice seed germination from the control at biochar application rate of 4.2g/kg soil but increased again slightly above the control at higher biochar application rates. Nonetheless, there was no significant difference ($p < 0.05$) in germination between the control and biochar treatments at 7 DAP. There was also no significant difference among treatments at 3 DAP. On the other hand, the percentage germination of rice seed increased significantly between 3 and 7 DAP.

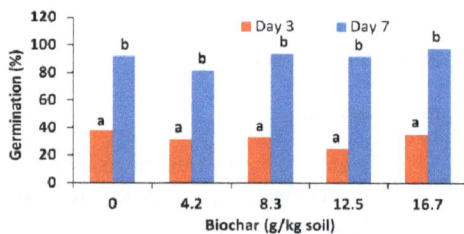

Figure 2. *Percentage germination of rice seeds after three and seven days on soil mixed with biochar derived from rice straw. Bars with the same letter are not significantly different ($P < 0.05$).*

3.3. Effect of Maize Stover Biochar on Shoot and Root Lengths of Germinated Maize Seedlings

Table 1 shows data collected on mean shoot and root lengths of maize seedlings at the end of the germination test on day 7. Except for biochar application rates of 4.2g/kg soil, mean shoot length differed significantly ($p < 0.05$) from the control with increasing rates of biochar application. On the other hand, the control treatment had slightly higher mean root lengths than the biochar treatments. However, the difference between the control and biochar treatments was not significantly different.

Table 1. *Mean shoot and root lengths of maize seedlings after seven days germination on a sandy loam soil treated with increasing levels of maize stover biochar*

Biochar (g/kg soil)	Shoot Length (cm)	Root Length (cm)
0.0	14.01	10.45
4.2	16.17ns	10.37
8.3	17.79*	11.39
12.5	17.56*	10.24
16.7	17.38*	8.89
LSD (0.05)	2.65	2.22ns

ns = not significant (p>0.05), *significant (p<0.05) in comparison to the control

3.4. Effect of rice Stover Biochar on Shoot and Root Lengths of Germinated Maize Seedlings

Mean shoot and root lengths of rice seedlings at the end of the germination test on day 7 are shown on Table 2. Although the mean lengths of rice shoot on soils treated with biochar were greater than the control, the difference was not significant ($p < 0.05$). On the other hand, mean root lengths in soils treated with biochar were found to be less than the control but did not differ significantly ($p < 0.05$).

Table 2. *Mean shoot and root lengths of rice seedlings after seven days germination on a sandy loam soil treated with increasing levels of rice straw biochar*

Biochar (g/kg soil)	Shoot Length (cm)	Root Length (cm)
0.0	11.56	6.92
4.2	12.42	6.21
8.3	12.10	5.74
12.5	11.91	6.01
16.7	12.22	6.81
LSD (0.05)	2.24ns	2.04ns

ns=not significant (p<0.05)

4. Discussion

Seed germination and emergence is critical to crop growth and development. The presence of inhibitory substances [15] can affect germination and emergence and hence plant growth. Some biochars have been shown to contain phytotoxic substances such as dioxins, furans, polyaromatic hydrocarbons, phenolic compounds as well as heavy metals that can harm soil microorganisms, plants and even humans [16].

In this study, percent germination of maize and rice seeds generally increased slightly above the control at higher biochar application rates although the increase was not signififcant. This study revealed that application of maize stover biochar or rice straw biochar did not have any negative impact on the germination and emergence of maize or rice seeds respectively. Other report on related studies [17] have found no significant difference among different biochars (including maize stover biochar) on the germination of maize seeds. Also, biochar application has been reported to enhanced germination of wheat, clover and mung bean seeds [18]. Other studies on forest seed germination showed that biochar enhances seed germination [19].

The study also revealed that after seven days of emergence there was a significant increase in maize shoot length beyond biochar application rates of 4.2g/kg soil. Thus maize stover biochar enhanced seedling emergence at the very early seedling growth (one week). On the other hand, application of rice straw biochar showed no significant difference in shoot length.

Mean root lengths for both maize and rice plants were were found to generally decrease with biochar applications. Thus, whereas there was a significant difference in maize shoot length between the control and biochar treatments, rice shoot length and root length did not differ significantly among treatments.

5. Conclusion

This study has shown that application of biochar derived from maize stover or rice straw to soil has no adverse effect on the germination and seedling emergence of maize or rice seeds respectively. Unlike rice straw biochar, application of maize stover biochar significantly improved maize seedling emergence relative the control.

The study also demonstrated the relevance of conducting seed germination test using a soil-based assay, particularly with soil from the site to which field applications of biochar are to be made in order to ensure good seedling emergence and crop growth. This study may also have relevance to horticulture where biochar may be used for the establishment of nurseries.

Acknowledgements

The authors wish to express their gratitude to the field technicians Joseph S. Domingo and Luseni Jaward, and the laboratory technician Samuel Jaia all of the Department of Soil Science Njala University, Sierra Leone for their assistance in soil sample collection and preparation. We would also like to thank Mr. Tamba in the Agricultural Engineering Department for his assistance in the production of biochar. Special thanks also to Mr. Prince Norman of the Sierra Leone Agricultural Institute (SLARI) for helping in the statistical analysis of rice data.

References

[1] P. M. Fearnside. Global warming and tropical land-use change: greenhouse gas emissions from biomass burning, decomposition and soils in forest conversion, shifting cultivation and secondary vegetation. Climatic Change 46:115–158, 2000.

[2] J. Lehmann, C. Czimczik, D. Laird, and S. Sohi. Stability of biochar in the soil. In: Biochar for Environmental Management: Science and Technology (Eds. J. Lehmann, & S. Joseph), Earthscan Publishers Ltd. 2009.

[3] K. Y. Chan, B. L. Van Zwieten, I. Meszaros, D. Downie, D. and S. Joseph. Using poultry litter biochars as soil amendments. Australian Journal of Soil Research, 46, 437- 444, 2008.

[4] A. Masulili, W. H. Utomo, and Syekhfani. Rice husk biochar for rice based cropping system in acid soil 1. The characteristics of rice husk biochar and its influence on the properties of acid sulfate soils and rice growth in West Kalimantan, Indonesia. Journal of Agriculture Science, 3, 25-33, 2010.

[5] A. Nigussie, E. Kissi, M. Misganaw and G. Ambaw. Effect of Biochar Application on Soil Properties and Nutrient Uptake of Lettuces (Lactuca sativa) Grown in Chromium Polluted Soils. American-Eurasian Journal of Agriculture and Environmental Science, 12 (3), 369-376, 2012

[6] B. Glaser, J. Lehmann and W. Zech. Ameliorating physical and chemical properties of highly weathered soils in the tropics with charcoal: A Review. Biology and Fertility of Soils, 35, 219-230, 2002.

[7] J. Lehman, J. P. Da Silva Jr, C. Steiner, T. Nehls, W. Zech and B. Glaser. Nutrient availability and leaching in an archaeological Anthrosol and a Ferralsol of the Central Amazon basin: fertilizer, manure and charcoal amendments. Plant and Soil, 249, 343-357, 2003.

[8] J. Laine, S. Simoni and R. Calles. 1991. Preparation of activated carbon from coconut shell in a small scale concurrent flow rotary kiln. Chem. Eng. Commun. 99:15–23.

[9] K.Y. Chan,. and Z. Xu. Biochar: Nutrient Properties and Their Enhancement. In: J. Lehmann and S. Joseph (eds.). Biochar for Environmental Management: Science and Technology. Earthscan, London, pp.53-66, 2009.

[10] D. Granatstein, C. Kruger, H.P. Collins, M. Garcia-Perez, and J. Yoder. Use of biochar from the pyrolysis of waste organic material as a soil amendment. Center for Sustaining Agric. Nat. Res. 2009. Washington State University, Wenatchee, WA. WSDA Interagency Agreement. C0800248. (http://www.ecy.wa.gov/pubs/0907062.pdf).

[11] D. A. Laird, R.. Brown, J.E. Amonette, and J. Lehmann. 2009. Review of the pyrolysis platform for coproducing bio-oil and biochar. Biofuels, Bioprod. Bioref., 3:547-562.

[12] C. A. Mullen, A. A. Boateng, N. Goldberg, I. M. Lima, D. A. Laird, and K.B. Hicks. Bio-oil and biochar production from corn cobs and stover by fast pyrolysis. Biomass Bioenergy, 34:67-74, 2010.

[13] G. K. Roberts, B. A. Gloy, S. Joseph, N. R. Scott, and J. Lehmann. Life cycle assessment of biochar system: estimating the enegetic, economic, and climate change potential. Environ. Sci. Technol. 44:827-833, 2010.

[14] K. Jones, A. Stewart. Dioxins and furans in sewerage sludges: a review of their occurrence and sources in sludge and of their environmental fate, behaviour, and significance in sludge-amended agricultural systems. Critical Reviews in Environmental Science and Technology /27: 1-85, 1997.

[15] A.I. Piotrowicz-Cieslak, B. Adomas, D.J. Michal-Czyk. Different glyphospate phytotoxicity to seeds and seedlings of selected plant species. Pol. J. Environ. Stud. 19 (1), 123, 2010

[16] X.D. Cao, L.N. Ma, B. Gao, W. Harris. 2009. Dairy-Manure Derived Biochar Effectively Sorbs Lead and Atrazine. Environmental Science & Technology 43, 3285-3291.

[17] H. F. Free, C. R. McGill, J. S. Rowarth, M. J. Hedley. The effect of biochars on maize (Zea mays) germination', New Zealand Journal of Agricultural Research, 53: 1, 1-4, 2010.

[18] Z. M. Solaiman, D. V. Murphy, L. K. Abbott. 2012. Biochars influence seed germination and early growth of seedlings. Plant and Soil, 353 (1-2), 273-287.

[19] S.J. Robertson, P.M. Rutherford, J.C. López-Gutiérrez, H.B. Massicotte. Biochar enhances seedling growth and alters root symbioses and properties of sub-boreal forest soils. Canadian Journal of Soil Science. 92(2):329-340, 2012. 10.4141/cjss2011-066.

12

Distribution and Indexation of Plant Available Nutrients of District Layyah, Punjab Pakistan

Muhammad Ashraf[1], Fayyaz Ahmad Tahir[1], Muhammad Nasir[2], Muhammad Bilal Khan[3, *], Farah Umer[1]

[1]Soil and Water Testing Laboratory Layyah, Punjab Pakistan
[2]Soil and Water Testing Laboratory Multan, Punjab Pakistan
[3]Soil and Water Testing Laboratory Muzaffar Garh, Punjab Pakistan

Email address:

bilalkhan_arid@yahoo.com (M. B. Khan)

Abstract: During last five years 2008-09 to 2012-13, a study was conducted to assess the fertility and salinity/sodicity status of district Layyah for the provision of guidelines to farmers and researchers for better crop production. Representative soil samples received/collected from farmers' fields were analyzed for texture, electrical conductivity (EC), pH, organic matter (O.M) and available phosphorus (P). A total of 31032 soil samples were collected from all tehsils of Layyah district, (15768 samples from tehsil Layyah, 7650 from Karor Lal Eisen, and 7614 from Chaubara). These soil samples were tested in Soil and Water Testing Laboratory Layyah and fertilizer recommendations were served to farmers according to soil and crop. The results showed that, soil texture of 91.18% soil samples was sandy loam (light), 8.53% loam (medium) and 0.29% clayey (heavy). About 99.42% soil samples had EC values within the normal range (< 4 dS m^{-1}) while 0.68% had (> 4 dS m^{-1}). The pH of 88.43% soil samples was up to 8.5 whereas 11.57% had >8.5. Organic matter content of 94.20% soil sample was poor (<0.86%), 5.25% medium (0.86-1.29%) and only 0.55% adequate (>1.29%). Available phosphorus of 67.75% soil sample was poor, (<7 mg kg^{-1}), 25.00% medium (7.1-14 mg kg^{-1}) while only 7.25% adequate (>14 mg kg^{-1}). Awareness camps, rallies and training programmes can be arranged for farmers regarding the benefits of soil and water testing, balanced use of chemical fertilizers and use of organic agriculture in crop production in improving soil fertility and nutrition status.

Keywords: Soil analysis, EC, pH, OM, P, Layyah, Nutrient Index

1. Introduction

Land is a basic unit, which is used to perform human activities like forestry, horticulture and agriculture. In the biotic ecological meaning land is the source of global bio-diversity by specifying biological territory and genetic material for vegetation, flora and fauna and micro-organisms, above and below Land (FAO, 1997). Relating to water a healthy soil can perform a number of vital functions like store water and nutrients, helps to regulate water flow, and neutralize all kinds of pollutants. Soil production capacity is limited and these limits are laid down by intrinsic characteristics, agro-ecological settings, apply and management (FAO, 1993).

Historically, mainly the enlargement of the various civilizations flourished in the Middle East was dictating by the accessibility of fertile soil and water. Continuous cropping was only feasible where the soils were rich in nutrients or fertility was regenerated through flood-borne sediments. While these days the main constraint to grow crops is inadequate moisture, due to limited water supplies and erratic rainfall, economic crop production is not feasible without an adequate supply of the fundamental nutrients, either from the soil or added as fertilizers. During the past three four decades, when chemical fertilizers began to be used extensively, has established the essential need of fertilizer nitrogen N in all regions specially in drought stressed areas. Likely, the calcareous nature of most soils in the region such that phosphorus P fertility is low, and thus without added P fertilizer sufficient crop yields are not possible. Fortunately, other important crop growth nutrients such as potassium, magnesium, calcium, and sulfur are well supplied in less fertile soils (J. Ryan, 2004).

Most of the soils in Pakistan have poor status of available plant nutrients and cannot support optimum levels of crop productivity (Rafiq, 1996; Ahmed and Rashid, 2003). The primary objective of soil testing is to help making soil test based fertilizer use recommendations. It helps in applying different nutrients in balanced ratio so as to get maximum efficiency of the applied fertilizers and profitable crop production (Motsara, 2002). Soil test measure some fraction of total supply of nutrients in the soil and indicate its available nutrient level. The higher soil test values mean higher level of nutrients and thus the lower will be the need for fertilization and *vise-versa*. There is a network of Soil and Water Testing Laboratories in the country to provide advisory service to farmers on soil and water management (Ahmed and Rashid, 2003).

The Layyah District has an extremely hot climate. Maximum temperature in the summer goes up to 53 Degree Celsius. The temperature in winter is low due to the area's nearness to Koh-Suleman range of mountains. The Chaubara Tehsil is almost barren and consists of forest and sand dunes. It lies between 30–45 to 31–24 degree north latitudes and 70–44 to 71–50 degree east longitudes. The area consists of a semi-rectangular block of sandy land between the Indus and Chenab rivers in Sindh Sagar Doaba. The tehsils of Layyah and Karor Lal Esan are developed agriculturally compared to other tehsils of the distract but still have are large tracks of sand dunes and uncultivated land. The Indus River passes from north to south on the western side of the district and touches Dera Ghazi Khan. Cotton, wheat, sugarcane, gram, watermelon, citrus and green chilly is the main agricultural product of district Layyah. The main objective of this study was to compile information on soil fertility and soil salinity/sodicity status of district Layyah on the basis of soil samples analyzed during the last five years 2008-2013.

2. Material and Methods

This study was conducted in Soil and Water Testing Laboratory,Layyah, Pakistan during 2008-2013. A total of 31032 soil samples were collected from all tehsils of Layyah district, (15768 samples from tehsil Layyah, 7650 from Karor Lal Eisen, and 7614 from Chaubara). These samples were collected from 0-15 and 15-30 cm depths for crops and vegetables while 0-15, 15-30, 30-60, 60-90, 90-120 and 120-150 cm depths for fruit plants and orchards. Samples were air-dried, ground and passed through a 2 mm sieve and analysed for physical and chemical properties. Soil texture was determined by measuring saturation percentage of soils (Malik et al., 1984), electrical conductivity (EC) by preparing 1:10 soil and water suspension (Soil Salinity Lab. Staff, 1954), pH (Schofield and Taylor, 1955), organic matter (Nelson and Sommers, 1982), available P (Olsen and Sommers, 1982) and K (Helmke and Sparks, 1996). The data were subjected to statistical analysis using MS Excel 2007 package. The criteria used for the classification is given in Table 1 as described by Malik *et al 1984*.

Table 1. Criteria of parameters used for classification.

(a) Soil texture

Saturation percentage	Textural class
0-20	Sand
21-30	Sandy loam
31-45	Loam
46-65	Clay loam
65-100	Clay

(b) Soil salinity/sodicity

Status	pH	EC (dS/m)
Normal (salts free)	< 8.5	< 4
Saline	< 8.5	> 4
Saline sodic	> 8.5	> 4
Sodic	> 8.5	< 4

(c) Nutrient status

Status	Organic matter (%)	Olsen P (mg/kg soil)
Poor	< 0.86	0-8
Satisfactory	0.86-1.29	8-15
Adequate	> 1.29	> 15

Source: Malik et al 1984.

3. Results

3.1. Soil Texture

The results (Table 2) showed that 91.18 percent soils in Layyah district were sandy loam and 8.53 percent soils were loam in texture. Heavy textured soils (clay loam) were noticed at few sites (0.29 %). In tehsil Layyah, 88.16 percent soils were sandy loam and 11.49 percent soils were loam. In tehsils Karor Lal Esan and Chaubara, 91.39 and 97.20 percent soils were sandy loam, respectively. In Karor Lal Esan, 8.16 percent soils were loam and 2.79 percent soils were loam in tehsil Chaubara. This shows that soils are quite heterogeneous and variable in texture. Rashid (1993) reported that the soils in Chakwal district were predominantly light textured as sandy loam and sandy clay loam were the dominant textured classes. The dissected old loess and alluvial terraces in the area have predominantly silt and silt loam texture, formed from parent material loess, residual material, old river alluvium and sub recent out wash. About 31% soils are sandy to sandy loam in texture and the remaining are highly eroded (Tager and Bhatti, 2001).

3.2. Dissolved Salts (Electrical Conductivity)

Dissolved salts in soils create hindrance in normal nutrient uptake process by imbalance of ions, antagonistic and osmotic effects. Normally for research purpose, electrical conductivity of soil extract (ECe) is used for total dissolved salts but for assessing soil salinity and sodicity for advisory purpose, a soil-water suspension of EC 1:10 is normally used as described in the manual of Malik et al. 1984. Various workers used the same method for electrical conductivity. However, EC1:10 is converted to ECe by multiplying with the factor Saturation percentage/100 as described by US Salinity Lab. Staff 1954. The data (Table 3) showed that

99.42 percent of soil samples analysed in district Layyah were free from salinity/sodicity. All tehsils showed similar trend i.e. >99 percent soils had total dissolved salts in normal range except a few sites (0.68 %) which were sodic in nature. The reason for low accumulation of salts in soils is that texture of the most of soils is sandy loam to loam and high and sporadic rainfall in monsoon season leaches/washes the salts, if any, from the root zone. These results are in line with the findings of Rehman et al. (2000) and Mahmood et al. (1998).

Table 2. Status of Soil Texture of District Layyah during 2008-2013

Name of Tehsil	Texture		
	Light	Medium	Heavy
Layyah	88.16%	11.49%	0.34%
Karor Lal Eisen	91.39%	8.16%	0.43%
Chaubara	97.20%	2.79%	0.00%
Overall Layyah District	91.18	8.53%	0.28%

Table 3. Status of Soil Salinity/Sidicity of District Layyah during 2008-2013.

Name of Tehsil	EC (dS/m)		pH	
	< 4	> 4	< 8.5	< 8.5
Layyah	99.49%	0.51%	87.60%	10.48%
Karor Lal Eisen	99.09%	0.90%	91.28%	8.71%
Chaubara	99.59%	0.78%	87.29%	12.70%
Overall Layyah District	99.42%	0.68%	88.43%	11.59%

3.3. Soil Reaction (pH)

The results (Table 3) further revealed that 88.43 percent soils at district level had pH <8.5. These soils are also good for agriculture but pH towards higher side (i.e. >8.2) has some limitations for high value crops. Soils having pH >8.5 need special attention and some suitable amendment (acid or gypsum) is to be applied for their reclamation according to the soil gypsum requirement. Such soils in Layyah district are very few (11.59 %). As the pH of soils is alkaline due to the indigenous parent material, calcareousness and low organic matter, this situation is similar in almost all soils. Latif et al. (2008) also reported pH of Chakwal soils in alkaline range of 7.7-7.8. When the average values are taken in to consideration, the area looks free from salinity/sodicity menace.

3.4. Organic Matter

Nitrogen requirements are usually recommended by the Soil Testing Laboratories, based on the estimation of nitrogen released by the SOM contents (Cooke, 1982). Higher organic matter reflects the higher crops yield. The data (Table 4) showed that 94.20 percent soils in Layyah district were poor and only 0.55 percent were adequate with respect to organic matter. Soils in tehsil Layyah were found deficient (92.81 %) in organic matter while 0.87 percent soils in tehsil Layyah were satisfactory in organic matter. The reason for low organic matter in these tehsils is that temperature in summer exceeds 50 °C due to which its decomposition rate is increased. Also farmers generally do not use farmyard manure and remove crops totally (grain plus straw) from

soils leaving it fallow. The trend of green manuring is also not observed. Rashid (1994) reported that OM contents ranged from 0.2- 1.3% in surface soils of Chakwal. The soils of Pakistan are quite low in organic matter. Generally the soils in Punjab contained less than 1% organic carbon (Azam, 1988). The decline in SOM is due to crops grown without or very meager addition of plant and animal manners. When the OM level declined to 40-60% of their original level, the soil productivity was affected, erosion loses of soil surface increased and net mineralization of soil fell below the level required for sustained grain crop production (Doran and Smith, 1987).

Table 4. Status of Organic Matter of District Layyah during 2008-2013

Name of Tehsil	Organic Matter		
	Poor	Medium	Adequate
Layyah	92.81%	6.31%	0.87%
Karor Lal Eisen	91.96%	7.64%	0.39%
Chaubara	99.35%	0.64%	0.00%
Overall Layyah District	94.20%	5.25%	0.55%

3.5. Available Phosphorus

With regards to phosphorus availability to plants, the results (Table 4) showed that 67.75 percent soils of Layyah district were poor in this nutrient. All tehsils had similar trend and were quite deficient in available phosphorus. The reasons for poor available phosphorus is that farmers do not apply phosphatic fertilizers to crops according to recommendations and only nitrogenous fertilizers are applied due to price hike of phosphatic fertilizers. Malik et al. (1984) and Rashid (1994) reported that 75-95% soils in Punjab are deficient in P. They also indicated that 61% soils contained up to 3 mg kg^{-1} and 34% soils had 3-12 mg kg^{-1} P contents.

Table 5. Status of Available Phosphorus of District Layyah during 2008-2013

Name of Tehsil	Available Phosphorus		
	Poor	Medium	Adequate
Layyah	64.56%	29.95%	8.47%
Karor Lal Eisen	62.99%	29.05%	7.94%
Chaubara	79.10%	16.86%	4.03%
Overall Layyah District	67.75%	25.00%	7.25%

4. Discussions and Recommendation

Soil organic matter level and soil fertility status may be increased by green manuring (sesbania, guar, etc.) once in three years. With this practice, the sufficient moisture can be preserved for rabi crops (wheat, canola, etc). Inorganic fertilizers (NPK) should be applied in balanced form according to soil test value and their use efficiency can be increased by band placement for row-sown crops. The low nutrient concentrations might be due to losses through leaching, as these soils were found sandy in nature or due to low soil organic matter because of rapid decomposition at some locations. Farmers from all over the Punjab province can get the fertilizer recommendation for different crops from the website http://www.fertilizeruaf.pk.General

recommendations for the farmers of district Layyah on the basis of results are given in Annexure 1.

Acknowledgements

The author would like to acknowledge the District Government Layyah for providing laboratory facility for analyses purposes to complete this study.

Annexure 1. Nutrients recommendations for different crops in district Layyah.

Sr. No	Name of Crops	Soil Status/Variety Sown/Specific Condition	Nutrients kg/ha			Bags/Acre		
			N	P$_2$O$_5$	K$_2$O	Urea	DAP	SOP
1	Wheat(irrigated)	Poor Soil	128	114	62	1.50	2.00	1.00
		Medium Soil	104	84	62	1.50	1.50	1.00
		Fertile Soil	79	57	62	1.00	1.00	1.00
2	Cotton (BT)	Early Sowing(Feb-March)	341-398	114	91-124	6.00-7.00	2.00	1.50-2.00
		Late Sowing(Apr-May)	200-227	86	91	3.50-4.00	1.50	1.50
3	Sugarcane	Poor Soil	296	170	124	5.25	3.00	2.00
		Medium Soil	227	114	124	4.00	2.00	2.00
		Fertile Soil	170	57	61	3.00	1.00	1.00
4	Rice	Fine varieties	170	101	79	3.00	1.75 as TSP	1.25
		Basmati Varieties	141	101	79	2.50	1.75 as TSP	1.25
5	Maize(Irrigated)	Poor Soil	227	141	91	3.00	2.50	1.50
		Medium Soil	168	114	62	2.25	2.00	1.00
6	Gram	For Tehsil Chaubara	32	84	---	0.50	1.50 as TSP	---
7	Water Melon	For Tehsil Layyah	158	203	62	2.00	2.00 as DAP & 4.00 as SSP	1.00

Source: Rapid Soil Fertility Survey and Soil Testing Institute, Punjab, Lahore (2012).

Annexure 2. Nutrients recommendations for Citrus orchard in district Layyah.

Age of plant	FYM	N	P	K	Zinc
	(kg/plant/year)	(g/plant/ year)			
At plantation	20	0	0	0	0
1 year	0	0	0	0	0
2 years	10	125	0	0	0
3 years	15	250	125	0	0
4 years	20	500	250	0	0
5-9 years	40	1000	500	500	50
> 10 years	60	1500	750	500	50

All P, K and Zn to be applied with FYM in December every year below the canopy of the plant but one meter away from the stem of tree. Fertilizers and FYM to be mixed in the soil with hoeing to be followed by irrigation. Half N to be applied in March and remaining half N to be applied by the end of June or beginning of July.

Source: NFDC, (2003).

Annexure 3. Area, Production and Average Yield of Major Crops of district Layyah (2008-14).

Name of crop	Area/Production/ average yield	2008-09	2009-10	2010-11	2011-12	2012-13	2013-14
Wheat	Area (000 Acre)	491	498	485	471	480	488
	Production (000 Tons)	503.86	515.36	557.55	527.40	519.56	610.512
	Average Yield (Mounds)	27.49	27.73	30.80	30.00	29.00	33.50
Suger Cane	Area (000 Acre)	32	29	22	30	31	35
	Production (000 Tons)	648.55	623.47	648.05	650.57	676.88	770.75
	Average Yield (Mounds)	543	576	570	581	585	590
Gram	Area (000 Acre)	275.52	263.04	270.00	255.18	255.21	255.66
	Production (000 Tons)	87.1	60.30	45.96	-	63.00	47.32
	Average Yield (Mounds)	-	-	4.56	-	6.96	5.02

Source: Statistics Department, Layyah Punjab Pakistan (2014).

References

[1] Ahmed, N. and M. Rashid. 2003. Fertilizer Use in Pakistan. NFDC. Planning and Development division, Islamabad. 141p.

[2] Anon. 2003. Fertilizers and their Use in Pakistan. Training Bulletin. 3rd Ed. NFDC, Islamabad.

[3] Anon. 1954. Diagnosis and Improvement of Saline and Alkali Soils. USDA Handbook No. 60, U. S. Salinity Lab. Staff. Washington, DC, USA. p. 16-17.

[4]   Cooke, G.W.1982. An introduction to soil analysis. *World Crops* 1: 8-9.

[5]   Doran, J.W. and M.S. Smith. 1987. Organic matter management and utilization of soil and fertilizer nutrients. p. 53-71. *In:* Soil Fertility and Organic Matter as Critical Components of Production System. J.J. Mortvedt and D.R. Buxton (eds). Soil Science Society of America, WI, USA.

[6]   FAO. "Land Resources Evaluation and The Role of Land-related Indicators, by W.G. Sombroek", Land Quality Indicators and Their Use in Sustainable Agriculture and Rural Development, land and water bulletin, 5. Rome, [1997].

[7]   FAO. "A Frame work for Land Evaluation", Soils Bulletin, 32. Rome, [1993].

[8]   J. Ryan., "Soil Fertility Enhancement in Mediterranean-type Dryland Agriculture: A Prerequisite for Development. In Challenges and Strategies of Dryland Agriculture", Crop Science Society of America and American Society of Agronomy [2004].

[9]   Latif, R., S. Ali and R. Hayat. 2008. Nitrogen fixation and yield of peanut affected by inorganic fertilizers, variety and inoculums interaction in rainfed areas of Punjab. Soil and Environment 27(1): 77-83.

[10] Mahmood, T., H. Mahmood, M. R. Raja and K. H. Gill. 1998. Soil fertility status of Rawalpindi district. Pak. J. Soil Sci. 14 (1-2): 66-69.

[11] Malik, D. M., M. A. Khan and T. A. Chaudhry. 1984. Analysis Manual for Soils, Plants and Waters. Rapid Soil Fertility Survey and Soil Testing Institute, Lahore, Pakistan.

[12] Motsara, M.R. 2002. Available nitrogen, phosphorus and potassium status of Indian soils as depicted by soil fertility maps. Fertilizer News 47(8):15-21.

[13] Nelson, S.W. and I.E. Sommers. 1982. Total carbon, organic carbon and organic matter. P. 539-80. In: Methods of Soil Analysis. Chemical and Microbial Properties. Agron. No. 9. Part 2, 2nd Ed. A.L. Page (ed.). American Society of Agronomy, Madison, Wisconsin, USA.

[14] Olsen, S.O. and I.E. Sommers. 1982. Phosphorus. p. 403-430. In: Methods of Soil Analysis. A.L.Page (ed.). Chemical and Microbial Properties. Part 2, 2nd Ed. American Society of Agronomy, Madison, Wisconsin, USA.

[15] Rafiq, M. 1996. Soil resources of Pakistan. p. 439-469. In: Soil Science. E. Bashir and R. Bantel (eds.). National Book Foundation. Islamabad, Pakistan.

[16] Rashid, A. 1994. Nutrient Indexing Surveys and Micronutrient Requirement of Crops. NARC, Islamabad.

[17] Rashid, A. 1993. Nutrient disorders of rapeseed-mutard and wheat grown in Potohar area. Micronutrient Project. Annual Report 1991-92. NARC, Islamabad.

[18] Rehman, O., A. A. Sheikh and K. H. Gill. 2000. Available phosphorus and pH status of Attock soils. Pak. J. Agri. Sci. 37 (1-2): 74-76.

[19] Schofield, R.K. and A.W. Taylor. 1955. The measurement of soil pH. Soil Science Society of America Proceeding 19: 164-167.

[20] Soil Salinity Lab. Staff. 1954. Diagnosis and Improvement of Saline and Alkali Soils. *USDA Hand book 60*, Washington, D.C., USA.

[21] Tager, S. and A. Bhatti. 2001. Physical properties of soil. p. 113-144. *In*: Soil Science. E. Bashir and R. Bantel (eds.). National Book Foundation, Islamabad, Pakistan.

13

Evaluation of propagation methods of *Schefflera abyssinica*

Tura Bareke Kifle, Admassu Addi Merti, Kibebew Wakjira Hora

Holeta Bee Research Centre, Oromia Agriculture Research Institute, Holeta, Ethiopia

Email address:

trbareke@gmail.com (T. B. Kifle)

Abstract: *Schefflera abyssinica* is indigenous bee forage tree species promising for honey production. However due to lack of appropriate propagation methods; *S. abyssinica* is not promoted in wide scale plantation. Therefore, the main objectives of this study were to develop and evaluate appropriate seed pretreatment procedures to improve the germination percentage, and assessing the impacts of seed provenances on the growth performance of this plant. Germination trials were conducted in laboratory, and plastic house at Holeta Bee Research Center and Holeta Agricultural Research Center using mature seed collected from mother trees using treatment of smoke solution and soaking seeds in different chemicals. The result indicated, there was significant improvement in germination capacity and vigor of *S. abyssinica* after pre-treated with aqueous smoke solution (p<0.05) particularly at low concentration. Pre-treated seeds of *S. abyssinica* with 1% chlorox, 70% alcohol, imidalm and Ridoml gold chemicals resulted no significant (P<0.05) increases, in the final germination percentage as compared to the controls. Seed provenances affect the germination capacity of *S. abyssinica* and their survival rate. *S. abyssinica* can be propagated by seed by producing seedlings and it can grow alone without need of other tree species as an epiphyte.

Keywords: Bee Forage, Smoke-Water, Germination, Survival Rate, Vigor

1. Introduction

Ethiopia has a high land mass endowed with a great diversity of climate, soil and flora. However, currently due to unwise utilization of the available vegetation resources coupled with lack of knowledge and little consideration to the biology of propagation, indigenous trees and shrubs are being depleted at an alarming rate. The sustainable productivity of ecosystems depends to a large extent on the buffering capacity provided by having rich and healthy indigenous forests (Legesse Negash, 1990 and 1995). Hence, it is essential that they are utmost conserved, propagated and developed to the extent possible. In Ethiopia, there is insufficient knowledge about provenance and genetic variability, and propagation of important indigenous tree species in general and *Schefflera abyssinica* in particular.

Schefflera abyssinica (Hochst. ex A. Rich.) is an indigenous tree belonging to the family of Araliaceae, branched, small/medium to 30 m tall trees and is also sometimes growing as an epiphyte. It grows as an epiphyte mainly on *Acacia abyssinica* and *Olea europea* tree species and finally overwhelms it to become an independent tree in highland areas. It produces creamy-yellowish or creamy-white flowers from March to May.

S. abyssinica grows in Afromontana forest, secondary forests and woodlands within the altitudinal range of 1450–2800 m above sea level.; often occurs in association with *Hagenia abyssinica* (Azene *et al*., 1993; Fichtl and Admassu Addi, 1994).It is also usually found left as scattered tree in farmlands.

S. abyssinica is one of the most important honey trees of the country. It has abundant nectar and pollen. Honeybees produce large quantities of a light and pure white honey which has high demand in the market and could generate high income (Fichtl and Admasu, 1994; Tefera Belay, 2005). However, currently the population of this plant species is highly fragmented and becoming scarce because of the continued forest depletion and its nature. In addition, less attention is given to the propagation of this species which put great pressure on honey production. Therefore, the main objective of this study was to develop appropriate seed pretreatment procedures for attaining maximum germination percentage, as well as assessing the impacts of the growth media on nursery performance of *S. abyssinica*.

2. Materials and Methods

2.1. Project Area

The study area was walmera district, Holeta Bee Research Center, located in Special zones of Oromia around Finfinne. The district is geographically located between latitudes $9^0 03`$ N and longitudes $38^0 30`$ E. The altitude ranges from 2060 - 3380 meter above sea level. The site study is located at an elevation of 2400 meter above sea level. The rainfall pattern is bimodal. The main rainy season is from June to September with a mean annual rainfall of 1150 mm.

2.2. Seed Collection and Processing

Seeds were collected from representative provenances depending on accessibility of the species and the natural distribution of the species (table 1).

Table 1. *Provenances and seed zones*

Species	Provenance	Seed zones
Schefflera abyssinica	Bale –Harenna	24.1
	West Showa-Gedo	20.4
	West Arsi-Munessa	21.1
	Jimma zone –Gera	

Provenances were selected following the tree seed zoning system developed by Azene Bekele *et al*., (1993) for the country. .In this study, the term provenance denotes the original geographic area from which seeds were obtained (Hartmann *et al.,* 1997). To ensure maximum genetic variation within the population, the selected trees were kept at least 100 m apart from each other (FAO, 1975). Mature seeds or fruits were collected from 5 to 10 dominant or co-dominant trees with clear bole, well developed crown and with abundant seeds on each site at the end of May in 2012. Immediately after collection, the mixture of fruits and seeds were packed in perforated sacks or plastic bags and transported to the Holeta Bee Research Center for processing and germination tests.

2.3. Germination Experiment

The germination study was conducted at Holeta Bee Research Center and Holeta Agricultural Research Center. Seeds were pre-treated (soaked) for 6 hours in different concentration of various dilution levels of plant-derived aqueous smoke extracts, and pretreated by chemicals (1% chlorox, 70% alcohol, imidalm and Ridoml gold) treatment.

Aqueous smoke extraction was performed by burning 200 gm of small branches and leaves of various plants (among which, *C. macrostachys, J. procera* and *M. ferruginea* are some) in a 100 mm diameter and 200 mm depth beekeeper's smoker for 30 minutes. The generated smoke was forced through plastic hose fitted to the mouth of the smoker by applying pressure on bellow into a 250 ml Erlenmeyer flask (E-flask)containing 200 ml of double distilled water. The mouth of the E-flask was plugged with a smoke tight rubber material whose center hollowed to allow the entry of plastic hose to the E-flask. The smoke was forced into the flask for

30 minutes. Then the resulted concentrated smoke water was maintained as a stock solution and used to prepare aqueous smoke extracts of different dilution levels. This method of smoke extraction was based on the method used by Kibebew (2007).

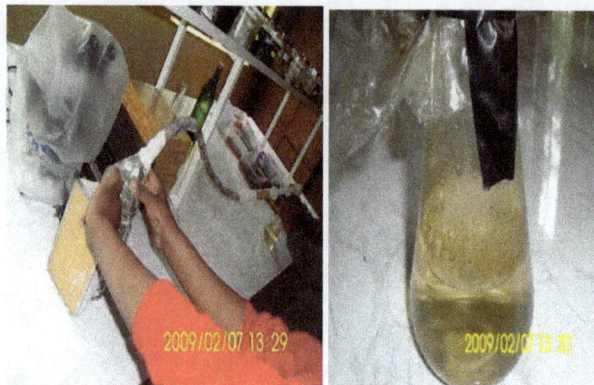

Figure 1. *Extraction of concentrated aqueous smoke solution*

For the studies, seeds obtained from west Arsi (Munessa) were used. Seed pre-treatments were performed employing three dilution levels of aqueous smoke extracts, the concentrations/dilutions levels used were: 1:10, 1:100 and 1:1000 for aqueous smoke extracts.

After soaking seeds in the test solutions for 6 hours, seeds sank to the bottom of each test solution were used for the germination experiments. Also seed pre-treatment were performed for seed collected from Gedo using chemicals for seed were soaked 1 % chlorox and 70% alcohol for 2 minutes washed seeds 3-4 times by water, Imidalm is in powder form and Ridoml gold with water solution for 2 minutes.

A number of pre-treated seeds used for this study was replicated three times, each pre-treated were placed on Whatman`s filter paper in Petri dishes. Then, the Petri dishes were covered with lid and watered as needed based on the moisture conditions of the Petri dishes and this was continued up to the end of the experiment. For the entire experiments, seed germination counts were made every three days after the commencement of seed germination. To facilitate future counts, germinated seeds were removed after recording. The experiments were continued until at least 80% of the replication from each treatment shows no new germination for 2 consecutive counts. A seed is considered germinated at the time when the protrusion of the radicle occurs for the illuminated seeds, and the emergence of the cotyledons for the buried seeds.

2.4. Provenance Variations and Nursery Establishment

Seed germination provenance variation trials were conducted and seeds were sampled for each seed provenance from the seeds sank to the bottom of the water. For provenances study and nursery establishment experiment, a total of three different soil mixtures were used. These were mixtures of local soil, forest soil, and sand in ratios of 2:1:1 respectively. The control contained sand soil only. The selection of these soil mixtures is based on the recommendation made by Legesse Negash (1995) for nursery establishment of various indigenous trees of Ethiopia. These treatments were used by 15cm pot sizes of equal length (20cm) and then transplanted to 30 pot sizes. Thus, for each provenance, 10 replications were used for the tests. Pots were arranged on wooden bench in the plastic house at random and the mouth of each of them was covered with dried grass stalk. The pots were watered once a day until the experiment ended. At the onset of seedling emergence the grass cover was removed to facilitate counting and to prevent bending of the seedlings due to the force applied by the grass stalk. For the experiment seed germination counts were examined every three days and seedlings with two expanded leaves were removed after recording observations. The trials were carried out until at least > 80% of the replication from each treatment showed no new germination for 2 consecutive counts. Finally, the germination responses of seeds were then expressed in terms of germination percentage, mean germination time, germination rate and germination vigor.

Growth status of S. abyssinica in 15cm pot size before transplanted

Growth status of S. abyssinica in 15cm pot size after transplanted

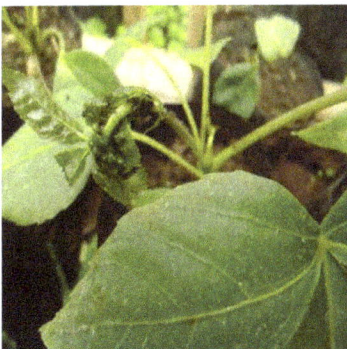

After transplanted to 15 cm pot size S. abyssinica highly eaten by aphids

Growth status *after* transplanting to 30 cm pot size and aphids controlled by manual management

Growth status of S. abyssinica in field condition

Figure 2. *Growth status of S. abyssinica from different pot sizes to field*

2.5. Early Growth and Survival Rate under Field Conditions

The growth performance and survival rate were evaluated under two different sites (Gedo and Holeta).

3. Statistical Calculations and Analysis

The germination responses of seeds were then be expressed in terms of germination percentage, mean germination time, germination rate and germination vigor.

Germination percentage was calculated according to the following formula:

$$\text{Germination Percentage} = (\frac{n}{N}) \times 100\% ,$$

Where:
n = Total number of germinated seeds;
N = Total number of seeds in the sample.

The mean germination time (MGT), mean germination rate (MGR), and germination vigor was determined according to Labouriau and Agudo (1987) as follow:

$$\text{MGT} = \frac{(\sum n_i t_i)}{n} ,$$

Where:

n_i = Percentage of seeds germinated between two consecutive counts;

t_i = Time taken since germination experiment started;

n = Total percentage of seeds germinated.

$$MGR = \frac{1}{MGT},$$

Where:

MGT = Mean germination time

$$\text{Germination vigor} = \sum \left(\frac{G_i}{t_i}\right) / N \times 100\%,$$

Where:

G_i = Number of seeds germinated up to the day under consideration;

t_i = Time taken since the first day of incubation;

N = Total number of seeds

Seedling survival rate (SR) was calculated as follows

$$SR\,(\%) = \frac{\text{no. of seedling alive at the end of the test} \times 100}{\text{Number of seedling transplanted}}$$

3.1. Statistical Analyses were Performed According to the Following Procedures

The effects of smoke solution and distilled water (Control), Chemical treatment and provenance variations on germination of all plants species were analyzed by a one-way ANOVA using SPSS with treatments as factor. Turkey's Honest Significant Difference Test was used for determination of significant differences among mean values for treatments.

4. Result and Discussion

4.1. S. abyssinica Seed Germination Trial Using Aqueous Smoke Solution

The result has indicated that there was significant difference in mean germination percentages and germination vigor among the treatments used (p<0.05), whereas, the mean germination time and rate did not differ significantly (table 2). This indicated that the germination capacity of S. abyssinica increased as the dilution level of the concentration of aqueous smoke solution decreased. Therefore, use of aqueous smoke solution at 0.001ml dilution level had significant effect on increasing the germination capacity of seeds of S. abyssinica. Smoke from a wide variety of biotic sources, including wood, straw, mixtures of dry and fresh plant material and charred wood can result stimulated germination (Brown and Vanstaden, 1997). Drewes et al., (1995) found that high concentrations of smoke-water could be inhibitory to germination but could be leached to promotive levels through irrigation (Delang and Boucher, 1993).Recently, the germination response to smoke is most easily studied using

smoke water, the main germination active compound has been identified as the butenolide, 3-methyl-2H-furo [2,3-c]pyran-2-one, from burned plant-derived smoke (Vanstaden et al., 2004) and cellulose (Flematti et al., 2004) that acts at very low concentrations. Smoke-water may be acting on the seed coat in a way similar to scarification, whereby the passage of water and oxygen into the dormant embryo is made easier (Egerton, 1998).

The germination vigor of 0.001ml dilution of smoke solution was 16.7% which is higher as compared to the rest. The dilution of smoke treatment increased the germination vigor of S. abyssinica at 0.001ml. This result was similar with (Paasonen et al., 2003) that smoke water dilutions improve germination and seedling vigority. As the result indicated that the mean germination time and germination rate of all treatment was not significantly different among the treatments used, therefore aqueous smoke solution has no effect on the germination rate and time required to germinate.

Table 2. *Mean + SE of mean germination percentage (MGP), mean germination time (MGT), mean germination rate (MGR) and germination vigor (GV) of S. abyssinica seed germination trial using aqueous smoke solution*

Treatment	MGP Mean + SE	MGT Mean + SE	MGR Mean + SE	GV Mean + SE
0.1ml	63.3 ± 0.19^d	7.7 ± 0.4^a	0.13 ± 0.02^a	12.8 ± 0.45^d
0.01ml	66.67 ± 0.38^c	8.1 ± 0.02^a	0.12 ± 0.01^a	13.3 ± 0.2^{cd}
0.001ml	83.3 ± 0.4^a	9.3 ± 0.2^a	0.11 ± 0.00^a	16.7 ± 0.4^a
control	73.33 ± 0.19^b	9 ± 0.5^a	0.11 ± 0.00^a	14.6 ± 0.07^b

4.2. S. abyssinica Laboratory Seed Germination Trial Using Different Chemicals

Pre-treated seeds with all the used chemicals resulted in non-significant (P<0.05) increases in the final germination percentage as compared to the controls (table 3). Accordingly, the mean germination percentage of control was better than seeds treated with chemicals. From this it is evident that seeds of S. abyssinica do not require any of the above mentioned chemicals treatments because the mean germination percentage of control was higher than chemical treatment.

Mean Germination Time (MGT) for seed pre-treatments employing all the chemicals was non- significantly (P<0.05) different from control. Mean Germination Time (MGT) of the seed pre-treated with Imidalm and control was 13.2 days, which was the lowest, while for 1% chlorox up to 16.2 days it was the highest and the rest treatment were between those treatments (table 3). Germination vigor and germination rate were non-significantly different (P<0.05) for seed pre-treatments employing all chemicals compared to the controls. Thus, the chemicals used did not show any significant stimulatory effect on final germination percentage, MGT, germination rate and germination vigor (%) of S. abyssinica seeds compared to the control, hence, these treatments did not offer any advantage in increasing the germination capacity, germination rate and vigority of seeds of S.

abyssinica.

Table 3. *Mean + SE of mean germination percentage (MGP), mean germination time (MGT), mean germination rate (MGR) and average germination vigor (GV) of S. abyssinica laboratory seed germination trial using different chemicals*

Treatment	MGP Mean ±SE	MGT Mean ±SE	MGR Mean ±SE	A GV Mean ±SE
1% Chlorox	24±5.16[a]	16.2±1.78[a]	0.04±0.00[a]	1.2±0.25[a]
70% alcohol	19±1.00[a]	14.6±1.06[a]	0.04±0.00[a]	0.95±0.05[a]
Imidalm	25±3.00[a]	13.2±0.21[a]	0.04±0.00[a]	1.25±0.15[a]
Ridoml gold	18±2.58[a]	15.3±0.63[a]	0.04±0.00[a]	0.9±0.12[a]
control	33±4.12[a]	13.2±0.23[a]	0.04±0.00[a]	1.65±0.20[a]

4.3. S. abyssinica Seed Provenance Germination Trial

The results indicated that there was significant difference in mean germination percentage, mean germination rate and mean germination vigority among seed provenances at ($p<0.05$), while the mean germination time was non-significant ($p<0.05$) (table 4). The mean germination percentage was highest for seed collected from Munessa. This indicated that seed provenance has great impact on seed germination capacity of *S. abyssinica*. The germination vigor was also highest for Munessa. This indicated that seed source is great factor that should be considered in germination vigor of *S. abyssinica*. This difference is mainly due to environmental variation. This idea was also supported by Lange (1961) as the variation within plant species was highly affected by environmental factors. Factors such as climatic conditions, abundance and coordinated maturation of pollen grains and thoroughness of pollination can produce variation in seed provenances (Bell *et al.*, 1995; Legesse Negash, 2003). The mean germination rate was highest for Munessa and Gedo whereas low for Harenna and Gera. This showed that seed collected from Munessa and Gedo was highly significant as compared to Harenna and Gera. Therefore, seed source has impact on germination rate. There is no significance difference among seed provenances in terms of mean germination time (MGT). However, MGT was needed short time for seeds collected from Munessa and somewhat long time for Harenna. Therefore seed provenance has non-significant effect on seed germination time of *S. abyssinica*.

Table 4. *Mean + SE of mean germination percentage (MGP), mean germination time (MGT), mean germination rate (MGR) and average germination vigor (GV) of S. abyssinica seed provenances germination trial in green house.*

Treatment	MGP Mean ±SE	MGT Mean ±SE	MGR Mean ±SE	AGV Mean ±SE
Gedo	45.57 ± 2.73[b]	18.47±1.39[a]	0.03 ± 0.00[a]	3.04±0.18[b]
Gera	24.57 ± 2.43[c]	21.64±0.74[a]	0.02 ± 0.00[ab]	1.38±0.16[c]
Munessa	66.72 ± 4.48[a]	18 ± 0.42[a]	0.03 ± 0.00[a]	4.45±0.30[a]
Harenna	5 ± 0.68[d]	23.12±5.81[a]	0.02 ± 0.00[b]	0.2 ± 0.04[d]

4.4. S. abyssinica Seedling Provenances Survival Rate after Transplanting

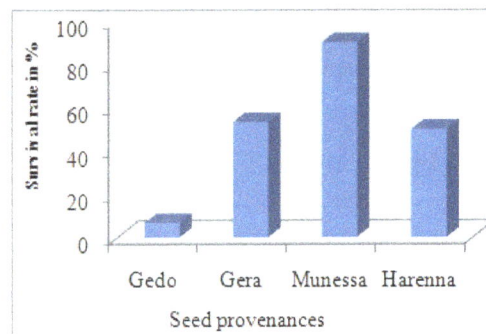

Figure 3. *S. abyssinica seedling provenances survival rate after transplanting*

The survival rate of seedlings of *S. abyssinica* after transplanting was highest for Munessa, whereas, seedlings of Gedo have poor survival rate after germination and transplanting. Seedlings of *S. abyssinica* collected from all provenances were affected by aphids. Particularly seedlings of seed collected from Gedo, Harenna and Gera were highly affected by aphids, whereas, seedlings of Munessa were less affected when compared with other provenances. To control aphids we used manual management only.

5. Conclusion and Recommendation

In conclusion the study revealed that *S. abyssinica* produces large number of seeds and it's expansion, therefore, could be achieved by means of seed propagation. So far it was considered as an epiphyte which grows on another tree species and finally overwhelms it and becomes an independent tree in highland areas. However, the present study clearly indicates that aqueous smoke solution showed potent germination activity of *S. abyssinica* at low concentrations.

Accordingly, there was significant improvement in germination capacity of its seeds after pre-treatment with aqueous smoke solution, especially the low concentration aqueous smoke solution (0.001ml). Seed provenance also has great impact on germination capacity and vigor of *S. abyssinica* and seed collected from Munessa showed good germination capacity, whereas, seed collected from Harenna showed very low. Thus, there is evidence that seed provenances affect the germination capacity of *S. abyssinica* and their survival rate. *S. abyssinica* can be propagated by seed by producing seedlings and it can grow alone without the need of other tree species as an epiphyte. Use of smoke solution is recommended to multiply *S. abyssinica* seeds through seedling

Acknowledgements

The authors are thankful to Holeta Bee Research Center and Oromia Agricultural Research Institute for providing required facilities and logistics. Our sincere thanks are also extended to Holeta Agricultural Research center for allowing their laboratory, Zewdu Ararso, Gemechis Legesse and Dejene Takele for their help on how to control aphids, and Konjit Asfaw and Tesfaye Abera, for their inspiration and support in the implementation and follow-up of the research.

References

[1] Azene Bekele; Birnie, A.; Tengnas, B. (1993): Useful trees and shrubs for Ethiopia: Identification, propagation and management for agricultural and pastoral communities. *Regional Soil Conservation Unit, Swedish International Development Authority, Nairobi.* 474 p.

[2] Bell, D. T., Rokich, D. P., Machesney, C. J. and Plummer, J. A. (1995). Effects of temperature, light and gibberellic acid on the germination of seeds of 43 species native to Western Australia. *Journal of Vegetation Science* 6: 797-806.

[3] Brown, N.., Vanstaden. (1997). Smoke as a germination cue: a review, *plant growth regul.* 22:115-124.

[4] Delange, J., Boucher, C. (1993). Aut-ecological studies on Audouinia capitata (Bruniaceae). 8. Role of fire in regeneration. *South African journal of botany* 59:188-202.

[5] Egerton, W., (1998). A smoke-induced alteration of the sub-test cuticle in seeds of the post-fire recruiter, Emmenanthe penduliflora Benth (Hydrophyllaceae), *J. Exp. Bot.* 49: 1317-1327.

[6] FAO (Food and Agricultural Organization), (1975). Forest Genetic Resources Information. No 4.Forest Occasional Paper (1975/1). Food and Agricultural Organization, Rome.

[7] Fichtl, R. and Admassu Addi, (1994). *Honeybee flora of Ethiopia.* Margarff Verlag Germany.

[8] Flematti, G. R., Ghisalberti, E.L., Dixon, K.W. and. Trengove, R.D. (2004). A compound from smoke that promotes seed germination. *Science* 305:977.

[9] Hartmann, H. T., Kester, D. E., Davies, J. F. and Genève, R. L., (1997). *Plant Propagation Principles and Practices.* Sixth edition, Prentice-Hall of India Private Limited, New Delhi-110 001, 2002.

[10] Kibebew Wakjira (2007). Seed Germination Physiology and Nursery Establishment of Croton macrostachys Hoch t. Ex Del. MSc Thesis. Addis Ababa University, School of Graduate studies, Addis Ababa, Ethiopia.

[11] Labouriau, L. G and Agudo, M., (1987). The physiology of seed germination in Salvia hispanica L. *Anais da Academia Brasileira de ciencias.* 59: 37 – 56.

[12] Lange, A. H., (1961). Effect of sarcotesta on the germination of papaya seed. *Bot. Gazette.* 122(4):305-311.

[13] Legesse Negash (1990). Ethiopia's Indigenous Forest Species and the Pervasive Effects of Deforestation. SINET Newsletter, Vol.14, No. 2.

[14] Legesse Negash (1995). *Indigenous trees of Ethiopia: Biology, Uses and Propagation Techniques.* Printed by SLU Reprocentralen, Umeå, Sweden. ISBN 91-105, pp. 285.

[15] Legesse Negash. (2003). In situ fertility decline and provenance differences in the East African Yellow Wood (*Podocarpus falcatus*) measured through in vitro seed germination. *Forest Ecology and Management* 174: 127-138.

[16] Paasonen, M., Hannukkala, A., Ramo, S., Haapala, H., Hietaniemi, V., (2003). Smoke-a novel application of a traditional means to improve grain quality. In: Nordic Association of Agricultural Scientists 22nd Congress, Turku, Finland.

[17] Tefera Belay (2005). Dynamics in the Management of Honey Production in the Forest Environment of Southwest Ethiopia: Interactions between Forests and Bee Management: MSC. Thesis. Wageningen University, Netherlands.

[18] Vanstaden, J., Jager, A., Light, M., and Burger, B., (2004). Isolation of the major germination cue from plant-derived smoke. *S. Afr. J. Bot.* 70:654–659.

14

Effect of irrigation at different growth stages on yield, water productivity and seed production of onion (*Allium cepa* L. CV BARI piaz-1)

Dilip Kumar Roy, Sujit Kumar Biswas, Abdur Razzaque Akanda, Khokan Kumer Sarker, Abeda Khatun

Irrigation and Water Management (IWM) Division, Bangladesh Agricultural Research Institute (BARI), Gazipur, Bangladesh

Email address:
droy49@gmail.com (D. K. Roy), sujitbari@yahoo.com (S. K. Biswas), razzaquebari@gmail.com (A. R. Akanda), ksarkerwrc@gmail.com (K. K. Sarker), abeda62@yahoo.com (A. Khatun)

Abstract: The study was executed in the experimental field of Irrigation and Water Management Division (IWM), Bangladesh Agricultural Research Institute (BARI), Bangladesh to investigate the effect of irrigation on onion seed yield. There were six irrigation treatments: T_1= Irrigations at vegetative, bolting, flowering and seed formation stages, i.e., no stress, T_2= Stress at vegetative stage, T_3= Stress at bolting stage, T_4= Stress at flowering stage, T_5= Stress at seed formation stage, and T_6= Irrigations at vegetative and flowering stages (Farmers practice). Bulb to seed method was used for this study. Recommended doses of fertilizers for "BARI piaz-1" were applied for all treatments. Measured amount of irrigation water was applied at different growth stages according to the treatment combinations. Data on yield attributing characters, yield, and seasonal water use were recorded. Results showed that umbel diameter and 1000- seed weights were significantly influenced by different irrigation treatments. Irrigation treatments did not show any significant effect on other yield contributing characters studied. The highest yield (1110.89 kg/ha) was observed from the treatment receiving irrigations at four different growth stages while the treatment in which stress was imposed at flowering stage produced the lowest seed yield (897.70 kg/ha). Water productivity was observed highest (0.45 kg/m^3) in treatment T_5 while the lowest (0.38 kg/m^3) was observed in treatment T_4. The highest benefit-cost ratio (3.84) was obtained from treatment T_1 while the lowest one (3.14) was found in treatment T_4. The results suggested that irrigation at the flowering stage is critical for onion seed production.

Keywords: Irrigation, Growth Stages, Water Productivity, Onion, Seed Production, *Allium Cepa*

1. Introduction

Onion (*Allium cepa* L.) belongs to the family Alliaceae is one of the most important spices in Bangladesh [1]. Onion ranks top in respect of production and second in terms of area among the spices crops grown in Bangladesh [2]. The green leaves and flowering stalks are also edible. It is an indispensable part of the Bangladeshi diet and is commonly used both by rich and poor but domestic production does not achieve even 15% of the annual requirement [3]. Low productivity of onion in Bangladesh could be attributed to limited availability of quality seed and lack of appropriate hybrids [1, 4, 5]. The onion's seed size and weight affect the final yield [6] and improved seed contributes substantially to

enhance crop yield as high as 30% [7].

Onion is a biennial crop in the temperate zones. Proper management practices are needed for onion seed production. Onion bulbs are produced in all districts of Bangladesh while seed production of onion are limited to a small number of farmer in Faridpur, Natore and Rajshahi districts [8]. Although the climate of Bangladesh is congenial to the production of high quality onion seed, farmers are not making use of modern technology for its production [5]. In Bangladesh, onion is grown as annual crop in winter season and its seeds are produced from bulb. Due to non-adaptability of exotic cultivars in agro-climatic condition of Bangladesh exotic cultivars generally do not bear seed production [9]. Therefore, proper management practices and production technology should be developed for successful production of onion bulb and seed from local cultivars of Bangladesh. About 8, 94,000 m. tons of onion are produced from an area of 1, 28,745 hectares [2] respect to an annual demand of about 14, 50,000 m. tons [10]. Onion is mainly used as spices in Bangladesh. The price of onion seed remains high in the season of onion cultivation. Seed is the basic and essential input for any crop production. According to Thompson [11], high quality seed is a critical input on which the effectiveness of all other inputs depends. Seeds production is a vital part in onion growing. Steady supply of good quality seeds is a prerequisite for the successful accomplishment of high production of acceptable onions as fresh bulbs or dehydrated forms either for local consumption or for export.

Onions require frequent irrigations. Soil moisture is an important factor that influences seed yield of onion. Because they extract very little water from depths beyond 24 inches; most of the water is from the top 12 inches of soil. Thus upper soil areas must be kept moist to stimulate root growth and provide adequate water for the plant. Hawthorn [12] found that high soil moisture in the seedling year performed high seed yields. Borgo et al. [13] reported that water stress during bulb sprouting and beginnings of the anthesis (period when onion flowers are fully open and functional) reduce the number of umbels and flowers per plant. However, in practice, the soil surface should not be continuously wet because it will predispose the crop to infection of root rot disease or damping off. Literature indicated that the use of irrigation could improve yield and quality of seed to a great extent [14-16]. Ali et al. [1] suggested that earthing up with 3-4 times irrigation is more effective for onion seeds production in Bangladesh.

Irrigation scheduling based on developmental stage or deficit irrigation is the technique of applying water on a timely and accurate basis to the crop, and is the key to conserving water and improving irrigation performance and sustainability of irrigated agriculture. Water availability and cultural practices may influence not only the interrelationships between seed yield and its components but also the seeds quality characteristics. A lot of work has been conducted on onion bulb production but a little information is available on onion seeds production in Bangladesh. Yield and quality of onion seed are greatly affected by soil moisture content during growth and development. Therefore, it is essential to find out the critical growth stage(s) of onion for quality seed production. Therefore, the present study was undertaken to evaluate the effects of irrigation regimes on onion seed yield and determine the critical growth stage(s) of onion for seed production.

2. Materials and Methods

The study was conducted during the *rabi* season of 2010-2011 at the experimental field of IWM Division, BARI, Bangladesh. The experimental site was located at 25°58' N

latitude and 83°58' E longitude. The soil was silty clay loam with field capacity and bulk density, 29% and 1.44 gm/cc, respectively. Onion variety "BARI piaz-1" was used to study the effect of irrigation levels on onion seed production. There are two methods of onion seed production, viz. seed to seed method and bulb to seed method. Seed to seed method is unpopular, since all the varieties are not suitable for annual seed production due to poor bolting habit and lower seed yield. Moreover, the seed production in this method is not suitable for further multiplication. For this reason, bulb to seed method was used for this study. The experiment was laid out in a randomized complete block design with three replications. The treatments were: T_1 = Irrigations at vegetative, bolting, flowering and seed formation stages; T_2 = Irrigations at bolting, flowering and seed formation stages; T_3 = Irrigations at vegetative, flowering and seed formation stages; T_4 = Irrigations at vegetative, bolting and seed formation stages; T_5 = Irrigations at vegetative, bolting and flowering stages; T_6 - Irrigations at vegetative and flowering stages (Farmers practice).

Well-decomposed cow dung (10 ton/ha) was applied 3 days before sowing of onion bulb. Recommended dose of fertilizer (150-100-180 kg/ha N-P-K) in the form of urea, triple super phosphate and muriate of potash was applied. In addition to N-P-K, gypsum, zinc and borax were also applied at the rate of 20-3.7-1.7 kg/ha. Nitrogen was applied in two equal splits, the first along with phosphorus and potash at the time of soil preparation while the remaining half was applied during apparent flowering and umbel formation of crops. The unit plot size was 3m x 2m. Onion bulbs were planted on 06 December, 2010 in 25 cm apart rows maintaining plant to plant distance of 20 cm. Bulbs were set upright and at a depth of 2.5 cm at 60 bulbs per plot. Measured amount of water (according to the treatments) was applied to each plot at several intervals to maintain the soil moisture content in the root zone up to field capacity. The crop was kept weed free by manual hoeing. Ten plants from each plot were selected randomly at harvest for collection of data on growth, yield components and yield.

2.1. Estimation of Irrigation Water

The irrigation water was applied to bring the soil moisture content at the root zone to field capacity taking into account the effective root zone depth. Before each irrigation, soil moisture was determined by digital moisture meter and Gravimetric method. Measured amount of water was applied to all treatments. The effective root zone of onion was considered as 30-40 cm depending on the growth stage [17]. The depth of water was determined using the following equation:

$$d = \sum_{i=1}^{n} \frac{M_{fci} - M_{bi}}{100} . A_i . D_i$$

Where,

d = net amount of water to be applied during an irrigation, cm

M_{fci} = field capacity moisture content in the ith layer of the soil, per cent

M_{bi} = moisture content before irrigation in the ith layer of the soil, per cent

A_i = bulk density of the soil in the ith layer

D_i = depth of the ith soil layer, cm, within the root zone, and

n = number of soil layers in the root zone D.

2.2. Determination of Effective Rainfall

Effective rainfall means useful or usable rainfall [17]. Effective rainfall was estimated by using the United States Department of Agriculture (USDA) Soil Conservation Method [18]. The equations are as follows:

$P_{effective} = P_{total} (125-0.2 \times P_{total})/125$ for $P_{total} < 250$ mm

$P_{effective} = 125 + 0.1 \times P_{total}$ for $P_{total} > 250$ mm

Where,

$P_{effective}$ = Effective rainfall, mm

P_{total} = Total rainfall, mm

2.3. Harvesting and Seed Extraction

In onion seed crop, the timing of harvest is complicated by the asynchronous pattern of seed growth and development within and between umbels. Furthermore, there is a tendency for seeds to shed soon after physiological maturity as a result of capsule dehiscence. Therefore, selection of optimum

harvest time must balance the increase in the number of physiologically mature seeds in umbels over time with the decrease in seed number caused by capsule dehiscence. Although all the seed heads did not mature simultaneously, there was only one cutting. This was made sufficiently early to avoid shattering of seeds but on the other hand sufficiently late to obtain well ripen seeds. The seed heads were cut 10-15 cm of stem attached. The seed heads were then dried in the sun, threshed, and cleaned for seed collection. Data were analyzed statistically following MSTAT-C package program and the mean differences were evaluated by Least Significant Difference (LSD) following [19].

3. Results and Discussion

3.1. Effect of Irrigation Water on Growth Parameters and Yield of Onion Seed

The data on yield and yield attributing characters of onion seed production are incorporated in Tables 1a, 1b, 2 and 3. Data presented in tables 1a and 1b indicate that irrigation water had no significant effect on growth parameters like plant height, number of leaves per plant, number of flower stalk per plant, length and diameter of scape, number of total and effective florets per umbel etc. There was a significant difference in the diameter of umbel due to different amount of irrigation water applied.

Table 1a. Effect of irrigation water on the growth parameters of onion for seed production

Treatment	Plant height, cm	No. of leaves/ plant	No. of flower stalk/ plant	Length of scape, cm	Diameter of scape, mm	Diameter of Umbel, mm	No. of florets/ umbel	No. of effective florets/ umbel	% Effective fruits set per umbel
T_1	47.27	17.17	2.50	75.50	12.42	66.57	252.47	220.60	87.63
T_2	43.83	17.33	2.67	70.20	11.58	62.37	240.10	216.90	90.28
T_3	46.77	19.17	2.90	72.27	11.52	61.36	218.37	203.83	93.24
T_4	49.10	17.30	2.40	77.96	12.55	65.03	209.83	187.33	88.80
T_5	48.47	17.93	2.63	74.93	12.00	65.93	203.90	186.17	89.40
T_6	49.63	15.27	2.37	77.00	12.40	67.37	217.40	195.10	91.43
CV (%)	4.51	14.13	17.76	3.93	4.36	2.48	13.99	14.50	3.25
LSD (0.05)	NS	NS	NS	NS	NS	2.92	NS	NS	NS

*Treatments: T_1 = Irrigations at vegetative, bolting, flowering and seed formation stages; T_2 = Irrigations at bolting, flowering and seed formation stages; T_3 = Irrigations at vegetative, flowering and seed formation stages; T_4 = Irrigations at vegetative, bolting and seed formation stages; T_5 = Irrigations at vegetative, bolting and flowering stages; T_6 = Irrigations at vegetative and flowering stages (Farmers practice). CV means Coefficient of Variation; LSD indicates Least Significant Difference; NS in the last row indicates "Not Significant" while a value in this row indicates significant difference at P = 5%.

The number of effective florets per umbel is a very important component contributing to final seed yield. Table 1a revealed insignificant variations in the number of effective florets per umbel due to application of different amount of irrigation water. Four irrigations (treatment T_1) produced the highest (220.60) number of effective florets per umbel while treatment T_5 (stress imposed on seed formation stage)

produced the lowest (186.17) number of effective florets per umbel. Table 1b revealed that irrigation water had significant effect on the quality of produced seeds. The heaviest seed (2.82 g/1000-seed) was obtained from treatment T_1 and the lowest weight (2.15 g/1000-seed) was obtained in treatment T_4. The results are in good agreement with Ali et al. [1].

Table 1b. *Effect of irrigation water on the yield parameters of onion for seed production*

Treatment	No. of seeds per umbel	Seed yield per umbel, g	Seed yield per plant, g	Seed yield, Kg/ha	1000 – seed weight, g
T_1	788.77	2.22	5.55	1110.89	2.82
T_2	743.83	2.10	5.35	1069.76	2.79
T_3	739.63	1.95	5.00	999.03	2.64
T_4	743.67	1.60	4.49	897.70	2.15
T_5	751.70	1.84	4.81	962.24	2.45
T_6	791.33	1.90	4.67	933.63	2.41
CV (%)	5.47	12.24	15.95	15.98	9.30
LSD (0.05)	NS	NS	NS	NS	0.43

*Treatments: T_1 = Irrigations at vegetative, bolting, flowering and seed formation stages; T_2 = Irrigations at bolting, flowering and seed formation stages; T_3 = Irrigations at vegetative, flowering and seed formation stages; T_4 = Irrigations at vegetative, bolting and seed formation stages; T_5 = Irrigations at vegetative, bolting and flowering stages; T_6 = Irrigations at vegetative and flowering stages (Farmers practice). CV, Coefficient of Variation; LSD, Least Significant Difference; NS in the last row indicates "Not Significant" while a value in this row indicates significant difference at P = 5%.

The yield of onion seed varied among the treatments but the variation was statistically insignificant (Table 1b). The maximum seed yield (1110.89 kg/ha) was found in treatment T_1 and the minimum (897.70 kg/ha) was found in treatment T_4. Detailed examination of the yield components helped elucidate why higher yield was observed in the treatment experienced no stress at any growth stage (treatment T_1). The higher number of effective florets per umbel containing relatively higher number of seeds with heavier weights appeared to have contributed to the higher seed yield in the treatment with four irrigations, i.e., no stress. Table 1b revealed that the treatment experienced stress at flowering stage produced the lowest seed yield (897.70 kg/ha). This was attributed to the fact that adequate watering conditions early in the season led to the development of an abundant leaf cover and a shallow root depth which cannot withstand water stress at later stage [20]. From the results it is clear that flowering stage of onion seed production may be considered critical. However, the role played by each of the yield contributing factors is actually the result of complex interactions with all the others and, as such, is difficult to interpret. Pejić et al. [21] reported the similar trend and concluded that onion yield was significantly influenced by irrigation.

Table 2. *Days to 60%, 80% and 100% scape initiation, bursting and flowering*

Treatment	Scape Initiation			Bursting			Flowering		
	60%	80%	100%	60%	80%	100%	60%	80%	100%
T_1	49.33	51.33	56.67	71.00	73.00	79.00	82.33	85.33	88.33
T_2	49.33	51.33	56.33	71.00	73.00	79.33	82.67	85.00	92.00
T_3	49.00	51.33	56.67	71.67	73.00	79.33	82.33	85.00	89.00
T_4	49.33	51.00	57.00	71.00	73.33	79.33	82.00	86.00	90.00
T_5	50.33	52.00	56.33	72.00	74.33	79.00	82.67	85.33	88.67
T_6	49.67	51.67	56.00	72.00	74.00	79.67	82.33	85.33	89.33
CV (%)	1.43	1.73	2.40	0.87	0.77	0.64	0.69	0.98	2.36
LSD (0.05)	NS	NS	NS	NS	NS	NS	NS	NS	NS

* Treatments: T_1 = Irrigations at vegetative, bolting, flowering and seed formation stages; T_2 = Irrigations at bolting, flowering and seed formation stages; T_3 = Irrigations at vegetative, flowering and seed formation stages; T_4 = Irrigations at vegetative, bolting and seed formation stages; T_5 = Irrigations at vegetative, bolting and flowering stages; T_6 = Irrigations at vegetative and flowering stages (Farmers practice). CV, Coefficient of Variation; LSD, Least Significant Difference; NS in the last row indicates "Not Significant" while a value in this row indicates significant difference at P = 5%.

Table 2 summarizes the effect of irrigation on the number of days taken to scape initiation, bursting and flowering. From table 2, it was clear that days to scape initiation, bursting and flowering were insignificant among the treatments. Days to scape initiation was found highest (57 days) in treatment T_4 while the lowest (56 days) was found in treatment T_6. The longest time (79.67 days) for bursting was found in treatment T_6 whereas the shortest time (79 days) was found in treatment T_1. The highest time (92 days) to flowering was found from treatment T_4 while the lowest time (88.33 days) to flowering was observed in treatment T_1. Results of the current study are in accordance with Ali et al. [1] who mentioned that days to scape initiation are affected by irrigation water application.

3.2. Water Requirement and Water Productivity

Total water used, based on water requirement, and water productivity, based on yield (kg/ha) per unit water applied, were presented in Table 3. The total amount of irrigation water ranged from 21.52 cm for T_1 to 13.34 cm for T_6 and the number of irrigation events ranged from 4 to 2 in various treatments. The average rainfall occurred during the crop seasons was 105 mm and total was effective since it was much less than the soil moisture deficit. The development of crop can therefore be considered to have largely depended on the irrigation water. Total water use varied with the variation of the amount of irrigation water applied to the plots. Total water use was found maximum (28.77 cm) in treatment T_1

and minimum (20.65 cm) in farmer's practice treatment T_6. The highest amount of irrigation water (21.52 cm) was required in the treatment that received a total four irrigations at all stages. The quantities of water applied during each irrigation event were low under this treatment. Quantity of water applied during irrigation increased when interval between consecutive irrigation events was increased. The water use pattern by different treatments was like that as higher the frequency of irrigation the less the amount of water needed in each irrigation. This was due to the existence of more soil moisture in the treatments, in which intervals are short. The highest water productivity (0.45 kg/m^3) was obtained in treatment T_5 while the lowest (0.38 kg/m^3) was found in treatment T_4.

Table 3. Component of water requirement and water productivity in different irrigation treatments

Treatment	No. of irrigation	Amount of total irrigation, cm	Effective rainfall, cm	Soil moisture contribution, cm	Total water used, cm	Yield, kg/ha	Water Productivity, kg/m^3
1	2	3	4	5	6 = 3+4+5	7	8 = 7/6
T_1	4	21.52	10.50	-3.25	28.77	1110.89	0.39
T_2	3	18.94	10.50	-2.91	26.53	1069.76	0.40
T_3	3	19.32	10.50	-3.28	26.54	999.03	0.39
T_4	3	16.10	10.50	-2.85	23.75	897.70	0.38
T_5	3	15.07	10.50	-3.24	22.33	962.24	0.43
T_6	2	13.34	10.50	-3.19	20.65	933.63	0.45

* Treatments: T_1 = Irrigations at vegetative, bolting, flowering and seed formation stages; T_2 = Irrigations at bolting, flowering and seed formation stages; T_3 = Irrigations at vegetative, flowering and seed formation stages; T_4 = Irrigations at vegetative, bolting and seed formation stages; T_5 = Irrigations at vegetative, bolting and flowering stages; T_6 = Irrigations at vegetative and flowering stages (Farmers practice).

4. Economic Comparison

Data pertaining to economic comparison is presented in Table 4. All the variable costs except the irrigation costs were similar in all the treatments. The highest net margin (Tk. 657332) and benefit-cost ratio (3.84) were found in treatment T_1 while the lowest benefit-cost ratio (3.14) was obtained from treatment T_4. This may attributed to the fact that treatment T_1 experienced no water stress during the growing season and thus produced maximum seed yield and highest net margin and benefit-cost ratio. On the other hand, the lowest net margin and benefit-cost ratio of treatment T_4 may be attributed to the stress imposed on the flowering stage which is the critical stage for onion seed production.

Table 4. Economic analysis of onion seed production under different irrigation treatments

Indicators	Treatments					
	T_1	T_2	T_3	T_4	T_5	T_6
Variable costs (Tk./ha)						
Land preparation	7000	7000	7000	7000	7000	7000
Labor	54000	54000	54000	54000	54000	54000
Fertilizers	25230	25230	25230	25230	25230	25230
Manure	20000	20000	20000	20000	20000	20000
Onion bulb	100000	100000	100000	100000	100000	100000
Pesticide	16000	16000	16000	16000	16000	16000
Irrigation	9150	8053	8214	6845	6407	5672
Total cost (Tk./ha)	231380	230283	230444	229075	228637	227902
Gross margin (Tk./ha)						
Onion seed	888712	855808	799224	718160	769792	746904
Net margin (Tk./ha)	657332	625525	568780	489085	541155	519002
BCR	3.84	3.72	3.47	3.14	3.37	3.28

* 1 Tk. is approximately US$ 0.125

5. Conclusion

The yield of onion seed was reasonably affected by the irrigation regimes. Four irrigations at different growth stages contributed to the increased seed yield of onion due to the availability of adequate moisture in the root zone during the critical growth stages. Depending on the quantity and timing of irrigation, the applied irrigation imparted different degrees of influence on the various components of growth and yield parameters. The yield was the least when stress was imposed on the flowering stage, irrespective of the amount of water applied. Therefore, it can be concluded that flowering stage is the most critical to irrigation water for onion seed production in the study area. Three irrigations including one at flowering stage may be used without any significant yield loss for onion seed production in the study area. So, three irrigations each at vegetative, flowering and seed formation stage may be the optimum and feasible irrigation scheduling for onion seed production under irrigation water shortage situation.

Acknowledgements

The authors would like to thank Irrigation and Water Management Division, Bangladesh Agricultural Research Institute, Bangladesh for providing support and expertise for carrying out the study. The authors gratefully acknowledge the financial support of Bangladesh Agricultural Research Institute to meet the full research expenses for this study.

References

[1] Ali MK, Alam MF, Alam MN, Islam MS, Khandaker SMAT (2007). Effect of nitrogen and potassium level on yield and quality of seed production of onion. J. Appl. Sci. Res. 3:1889-99.

[2] BBS (2008). Monthly (February) Statistical Bulletin of Bangladesh. Bangladesh Bureau of Statistics, Ministry of Planning, Government of the People's Republic of Bangladesh, Dhaka. 66 p.

[3] Hossain AKMA, Islam J (1994). Status of *Allium* production in Bangladesh. Acta Hort. 358: 33-36.

[4] Tomar BS, Singh B, Hassan M (2004). Effect of irrigation methods on seed yield and seed quality in onion cv. Pusa Madhavi. Seed Res. 32: 72-81.

[5] Bokshi AI, Mondal MF, Pramanik MHR (1989). Effect of nitrogen and phosphorus nutrition on the yield and quality of onion seeds. Bangladesh Hort. 17(2): 30-35.

[6] Gamiely S, Smittle DA, Mills HA, Banna GI (1990). Onion seed size, weight, and element content affect germination and bulb yield. Hort. Sci. 25: 522-523.

[7] Shaikh AM, Vyakaranahal BS, Shekhargouda M, Dharmatti PR (2002). Influence of bulb size and growth regulators on growth, seed yield and quality of onion cv. Nasik Red. Seed Res. 30: 223-229.

[8] Rahim MA, Amin MMU, Haider MA (1993). Onion seed production technology in Bangladesh. Allium improvement Newsletter, USA 3: 26-33.

[9] Rahim MA, Hussain A, Siddique MA (1982). Seed production ability of three onion cultivars. Bangladesh Hort. 10: 31-38.

[10] Anonymous (2006). Action plan for increasing the productivity of spices 2006-2009. National technical working group. Ministry of Agriculture, Govt. of the People's Republic of Bangladesh.

[11] Thompson JR (1979). An Introduction to Seed Technology. Leonard Hill Books Ltd., London, pp. 19.

[12] Hawthron LR (1951). Studies on soil moisture and spacing of seed crops of carrot and onion. USDA, Circular no. 852.

[13] Borgo R, Stahlsehmidt DM, Tizio RM (1993). Preliminary study on water requirements of onion cv. Valcatarce in relation to seed production. Agri. Scientia. 10: 3-9.

[14] Brown MJ, Wright JL, Khol RA (1977). Onion seed yield and quality as affected by irrigation management. Agron. J. 34(8): 260-268.

[15] Shasha ANS, Campbell WF, Nye WP (1986). Effect of fertilizer and moisture on seed yield of onion. Hort. Sci., 11(4): 425-426.

[16] Bhonde SR, Lecehiman R, Srivastava KJ, Ram L (1989). Effect of spacing and levels of nitrogen on seed yield of onion. Seed and Farms, 15(1): 21-22.

[17] Michael AM (1985). Irrigation: Theory and Practice. Vikas Publishing House Private Limited. New Delhi, India p.539.

[18] Smith M (1992). CROPWAT, A computer programme. Irrigation Planning and Management, FAO Irrigation and Drainage Paper 46. Rome. Italy.

[19] Gomez KA, Gomez AA (1984). Statistical Procedures for Agricultural Research. John Wiley and Sons Inc. New York p. 214.

[20] Bazza M (1999). Improving irrigation management practices with water-deficit irrigation. In: Kirda C, Moutonnet P, Hera C, Nielsen DR (Eds.), Crop Yield Response to Deficit Irrigation. Kluwer Academic Publishers, Dordrecht, the Netherlands, pp. 49-71.

[21] Pejić B, Borivoj Pejić J, Milić S, Ignjatović-Ćupina A, Krstić D, Ćupina B (2011). Effect of irrigation schedules on yield and water use of onion (Allium cepa L.). African J. Biotech. 10(14): 2644-2652.

15

Evaluation of culture media for biomass production of *Trichoderma viride* (KBN 24) and their production economics

Kishor Chand Kumhar, Azariah Babu, Mitali Bordoloi, Ashif Ali

Tea Research Association, North Bengal Regional Research & Development Centre, Nagrakata, District – Jalpaiguri, West Bengal – 735 225, India

Email address:

kishorkumarc786@gmail.com (K. C. Kumhar)

Abstract: The Genus *Trichoderma* is of immense importance in agricultural crop protection because of their bio-control potential role against an array of phytopathogens through several modes of action. It is well established with fairly good acceptability, worldwide. The establishment and utilization on a commercial level of any promising isolate may not be successful, unless the cost effective mass production is evident. Present study is aimed at the evaluation of the laboratory media as well as locally available food grains for cost effective mass production of local strain KBN-24 (*Trichoderma viride*) for large scale adoption. Among different lab media, potato dextrose agar (solid medium) and potato dextrose broth (liquid medium) yielded comparatively more biomass of tested strain of *Trichoderma viride*. However, among the different grains rice ranked the first which produced the maximum biomass (148.04 gram) followed by wheat (126.87 gram) where as maize produced the least biomass. Similar trends were recorded on the conidial production and colony forming units (CFUs) in case of potato dextrose agar, potato dextrose broth and rice whole grain. Results indicated that the locally available food grains like rice and wheat were comparatively cheaper and serve as convenient substrates for the mass multiplication of *Trichoderma viride* and their cost economics were also discussed.

Keywords: *Trichoderma Viride*, Mass Multiplication, Culture Media, Production Economics

1. Introduction

In the current era of agriculture, the plant protection paradigm has been shifted towards integrated pest & disease management (IPDM) approach which gained popularity and acceptance to a considerable extent among the agricultural community including researchers. Since injudicious agrochemical usage is responsible for wide range of side effects [1]. Management of plant diseases though biological control agents is ideal and need of hour. This practice (biological control) is successfully established in different crops in different agro-ecological regions. *Trichoderma* is one of the most common fungal bio control agents which is being widely used for the management of various foliar and soil borne plant pathogens [2]. It has been acclaimed as an effective, eco-friendly, cheaper, and reducing the ill effects of synthetic chemicals due to its unique modes of action such as competition, antibiosis and enzymatic. Different species of

Trichoderma are being used to protect tea plantation from some of the plant pathogens of economic importance [3] in many parts of India. However, its local strain from the tea ecosystem in Dooars region of West Bengal, is still needs to explored. Its successful utilization as a potential biocontrol agent at the garden level needs isolation of effective local strain, standard and cost effective mass multiplication techniques. Various substrates including grains may be used for mass multiplication of *T. viride* with variable mass productivity [4]. The success of any biological control agent not only depends on its virulence but also on the successful mass production in laboratory assuring cost effectiveness and its shelf life. Therefore looking towards need for large scale cost effective production of eco-friendly bio-pesticide, present investigation has been carried out to evaluate locally available less expensive substrates for mass multiplication of *T. viride* for sustainable tea cultivation. Solid state fermentation has an edge over submerged (liquid state) fermentation in terms of high volumetric productivity, low cost equipments, much lesser by-product or waste generation and lesser time [5].

2. Materials and Methods

Solid synthetic, liquid synthetic media were procured from Himedia and and food grains (rice, maize and wheat) from the local market (Table 1) and evaluated for their suitability for mass multiplication of *T. viride* (KBN-24) following the methods adopted by earlier workers with slight modifications [6, 7]. Petri plates of 90 mm diameter were used for solid media whereas, for liquid media as well as food grains, 250 ml capacity conical flasks were used adopting CRD (Complete randomised design). Sterilized media was poured in to each Petri plate (20 ml/ plate) and allowed the same for solidification.

In the case of liquid synthetic media 250 ml capacity conical flasks, broth media (100 ml) along with the selected food grains (100g) were taken. Food grains were soaked in distilled water for 15-20 minutes to make them soft, excess water was drained off and 2 % sugar solution was added. The mouth of the flasks was closed using cotton plugs. All conical flasks were autoclaved 121 ^{0}C (15 psi) for 15-20 minutes.

All plates and flasks were inoculated with 5 mm mycelial discs of 4-7 days old pre-cultured *T. viride* using a sterilized cork borer and incubated for 15 days at room temperature. After 15 days, biomass produced was estimated by weighing, haemocytometer observations and plating technique (Fig.1) and the cost of economics was calculated for the different media. Each treatment was replicated three times.

Table 1. *Media of media for T. viride (KBN-24) mass production.*

SN	Culture Media	Media state	Quantity of media used per treatment
1	Potato dextrose agar	Solid	20 ml / Petriplate
2	Sabouraud Dextrose Agar	Solid	20 ml / Petriplate
3	Potato dextrose broth	Broth	100 ml / flask
4	Czapek Dox Broth	Broth	100 ml / flask
5	Rice	Grain	100 gm/ flask
6	Wheat	Grain	100 gm/ flask
7	Maize	Grain	100 gm/ flask

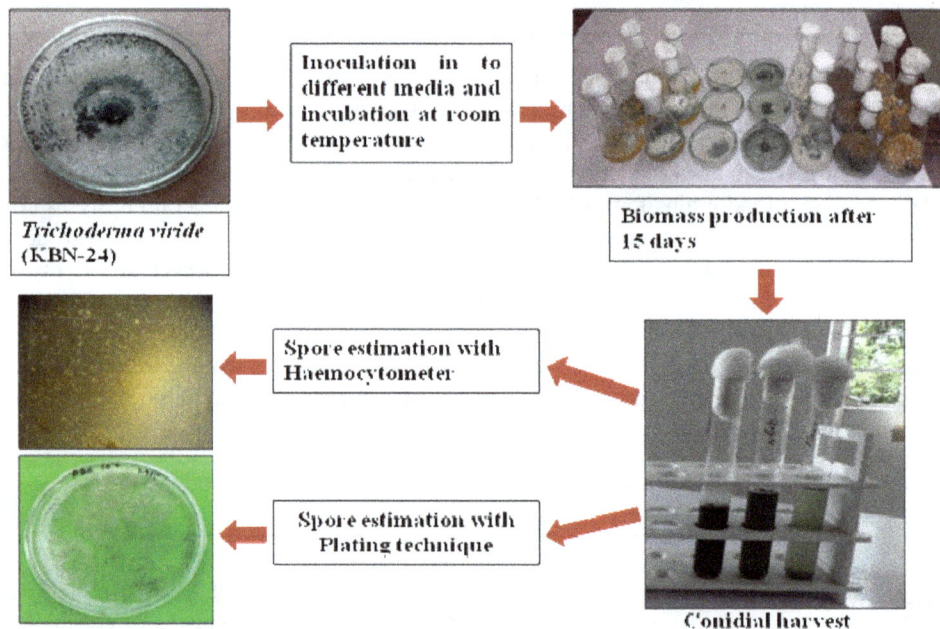

Figure 1. *Experiment design to evaluate the media for biomass production of Trichoderma viride (KBN-24)*

3. Results and Discussion

3.1. Solid Synthetic Media

The results of the two different laboratory media *viz.,* potato dextrose agar (PDA) and Sabouraud dextrose agar (SDA) which were evaluated for mass multiplication are presented in (Table 2). Results indicated that, the potato dextrose agar is better than Sabouraud dextrose agar in terms of biomass produced (Fig. 2). The PDA medium produced 0.54g (fresh weight) per plate whereas; the SDA medium could produce only 0.37g of biomass including both fungal conidia and mycelia.

Table 2. *Biomass production of T. viride (KBN-24) in different media.*

Media	Fresh weight of biomass (gram)*
Potato Dextrose Agar	0.54 ± 0.05
Sabouraud Dextrose Agar	0.37 ± 0.05
Potato Dextrose Broth	96.30 ± 0.36
Czepek Dox Broth	90.38 ± 0.16
Rice	148.04 ± 4.97
Wheat	126.87 ± 5.26
Maize	114.79 ± 5.68
C.D.	10.66
SE(m)	3.48
SE(d)	4.92
C.V.	7.39

*Mean of 3 replications

Table 3. Conidial production of T. viride (KBN-24) in different media.

Media	Water added to harvest conidial biomass	final amount of spore suspension after addition of water (ml)	Conidia / column of Haemocytometer field (at 10X)*	Conidia / ml / Haemocytometer field*
Potato Dextrose Agar	15 ml	15	167	10.00 ± 0.67
Sabouraud Dextrose Agar	15 ml	15	46	3.51 ± 0.38
Potato dextrose broth	0	100	268	2.44 ± 0.14
Czapek Dox Broth	0	100	53	0.54 ± 0.01
Rice	150 ml	150	131	3.19 ± 0.37
Wheat	150 ml	150	240	1.88 ± 0.15
Maize	150	150	419	0.72 ± 0.09
C.D.				1.03
SE(m)				0.34
SE(d)				0.47
C.V.				18.25

*Mean of 3 replications

Figure 2. Mycelial growth and sporulation on different media and food grains.

3.2. Synthetic Broth Synthetic Media

Among synthetic broth media, potato dextrose broth (PDB) performed better than Czapek Dox broth (Table 2 and fig 2) which produced 96.30 gram as compared to second media, with 90.38 gram biomass (fresh weight).

3.3. Food Grains

Table 4. CFU estimation of T. viride (KBN-24) plating technique.

SN	Media	CFUs at 10⁻⁸ dilution*
1	Potato dextrose agar	91.89 ± 23.42
2	Sabouraud Dextrose Agar	59.75 ± 33.30
3	Potato dextrose broth	12.86 ± 2.51
4	Czapek Dox Broth	8.54 ± 1.66
5	Rice	40.62 ± 12.32
6	Wheat	30.69 ± 6.31
7	Maize	29.07 ± 12.64
	C.D.	51.99
	SE(m)	16.98
	SE(d)	24.01
	C.V.	75.28

*Mean of 3 replications

In the case of the three different grains, rice ranked first with 148.04 gram biomass followed by wheat (126.87 gram) and maize (114.79 gram) according to Table 2 and Fig. 2. Conidial quantity assessment was undertaken after 15 days. Potato dextrose agar, potato dextrose broth and rice produced 91.89, 12.86 and 40.62 colony forming units, at 10^{-8} dilution, respectively (Table 4).

4. Determination of Cost Economics

The cost of different media and food grains were considered to work out the production cost. Among the solid synthetic media, potato dextrose agar was found to be cheaper than Sabouraud Dextrose Agar. In liquid media, potato dextrose broth was economic than Czapek Dox Broth (Table 5). Several plant materials such as *Tripxacum laxum,Cymbopogon citrates, Crotalaria anagyroids, Albizzia chinensis, Indigofera stachyodes, Albizzia Montana* and *Derris robusta* were evaluated by earlier workers [8] and reported that *Tripxacum laxum* yielded the maximum cfu of *T. harzianum* (6.81×10^4) followed by *Albizzia chinensis* (4.10×10^4), where as *Derris robusta* produced the least cfu (7.12×10^3). Potato dextrose agar (PDA) and malt extract agar (MEA) were studied for the mycelial growth of *T. viride* and it was noted that it could produce better mycelial growth when PDA was enriched with 2.5 gm glucose and 1.5 gm lactose respectively [9]. It has been reported that large scale mass production of *T. harzianum* can be achieved through liquid fermentation using inexpensive media such as molasses and brewers yeast [10]. Household waste, vegetable waste and other wastes can also be utilized for mass production of *T. candidum* [11]. Vegetable waste, fruit juice waste, sugarcane baggase and rotten wheat grains have been used for mass multiplication of *T. viride* and reported that sugarcane baggase as the best substrate that yielded high amount of

mycelia, spore & higher cfu count [12]. Similarly, in the present investigation, among food grains evaluated, both rice and wheat are found to be cost effective than maize. The production cost per gram biomass in different media and substrates ranged from 0.01 to 15.11 rupees (Table 5). Cost of synthetic media varied from 0.15 to 15.11 rupees whereas mass multiplication with the use of food grains was found to be economic and the cost of production per gram of fresh biomass ranged from 0.01 to 0.08 rupees only indicating the fact that these media could serve the purpose of cost effective manner in the commercial production of this strain.

Table 5. Production economics of different media.

Media	Media Code	Cost ₹/ per Kg (A)	Media quantity required / litre (B)	Cost ₹ / litre media (C=A/1000 x B)	Quantity of media used (ml or g) per treatment (D)	Cost ₹ / treatment (E=C/1000 x D)	Gram Biomass produced (F)	Production cost ₹ / g (G=E/F)
PDA	MH096-500G	4210	39	164.19	20	3.28	0.54	6.07
SDA	M063-500G	4300	65	279.50	20	5.59	0.37	15.11
PDB	M403-500G	4352	24	104.45	100	10.44	90.30	0.12
CDB	M076-500G	3876	35.01	135.70	100	13.57	90.38	0.15
Rice	-	24			100	2.40	148.04	0.02
Wheat	-	16			100	1.60	126.87	0.01
Maize	-	90			100	9.00	114.79	0.08

*Values represent mean of 3 replications

Acknowledgements

We gratefully acknowledge the help rendered by Dr. S. K. Singh, Co-ordinator, Agharkar Research Institute, Pune, India in identifying the strain. The authors are also thankful to Dr. N. Muraleedharan, Director, TRA Tocklai Tea Research Institute, Jorhat, Assam, for his constant support, guidance and encouragement.

References

[1] S. Roy, A. Mukhopadhyay and G. Gurusubramanian. (2011) Resistance to insecticides in field collected population of tea mosquito bug (*Helopeltis theivora* Waterhouse) from the Dooars (North Bengal, India) tea cultivations. J. Entomol. Res. Soc., 13(2): 37-44.

[2] F.C. Dominguesa, J.A. Queiroza, J. M. S. Cabralb and L. P. Fonsecab. (2000) The influence of culture conditions on mycelial structure and cellulose production by *Trichoderma reesei* rut C-30. Enz. Microbial Technol. 26: 394-401.

[3] B. K. Borthakur and P. Dutta. (2011) Disease management in tea, In Tea field management, Tea Research Association, Tocklai Experimental Station, Jorhat -785 008, Assam, India, pp. 182-188.

[4] E. Esposito and M. da Silva. (1998) Systematics and environmental application of the genus *Trichoderma*. Crit. Rev. Microbiol. 24:89-98.

[5] G.S. Kocher, K.L. Kalra, G. and Banta. (2008) Optimization of cellulase production by submerged fermentation of rice straw by *Trichoderma harzianum* Rut-C 8230. Int. J. Microbiol. 5(2): 8230.

[6] N. Subash, J. Viji, C. Sasikumar, and M. Meenakshisundaram. (2013) Isolation, media optimization and formulation of *Trichoderma harzianum* in agricultural soil. Journal of Microbiology and Biotechnology Research.3 (1): 61-64.

[7] Mridula Khandelwal, Sakshi Datta, Jitendra Mehta, Ritu Naruka, Komal Makhijani, Gajendra Sharma, Rajesh Kumar and Subhas Chandra. (2012) Isolation, characterization & biomass production of *Trichoderma viride* using various agro products- A biocontrol agent. Advances in Applied Science Research, 2012, 3 (6):3950-3955.

[8] D. Ajay, J. S. Bisen, M. Singh, R. Saha and B. Bera. (2013) Evaluation of various plant residues in tea plantations of Darjeeling for on farm production of *Trichoderma harzianum*. Indian Journal of research and practices. 1: 1-4.

[9] Prashant Pingolia, Rohini Maheshwari, Narendra Vaishnav, Gajendra P. Sharma and Jitendra Mehta. (2013) Mycelium growth of *Trichoderma viride* (Biocontrol agent) on different agar medium. International Journal of Recent Biotechnology. 1 (1): 43-47.

[10] G. C. Papavizas, M. T. Dunn, J. A. Lewis and J. Beagle-Ristaino. (1984) Liquid fermentation technology for experimental production of biocontrol fungi. Phytopathology. 74:1171-1175.

[11] K. Nagur Babu, and P.N. Pallavi. (2013) Isolation, identification and mass multiplication of *Trichoderma* an important bio-control agent. International Journal of pharmacy and life sciences. 4(1): 2320-2323.

[12] P. J. Chaudhari, Prashant Shrivastava and A. C. Khadse. (2011) Substrate evaluation for mass cultivation of *Trichoderma viride*. Asiatic Journal of Biotechnology Resources. 2(04): 441-446.

16

Relationship between physiological and seed yield related traits in winter rapeseed (*Brassica napus L.*) cultivars under water deficit stress

Gader Ghaffari[1], Mahmoud Toorchi[2], Saeid Aharizad[2], Mohammad-Reza Shakiba[2]

[1]Department of Agricultural Engineering, Payame Noor University, East Azarbaijan Province, Iran
[2]Department of Crop Production and Breeding, Faculty of Agriculture, University of Tabriz, Iran

Email address:
ghaffari314@yahoo.com (G. Ghaffari)

Abstract: Finding the relationship between physiological traits and seed yield components is an important objective in crop breeding programs. Canonical correlation analysis has been adopted to study the strength of association between the physiological traits and seed yield under water deficit stress and to obtain the physiological traits that have the largest effect on seed yield and its components. This study revealed that leaf water potential, relative water content, leaf osmotic potential and chlorophyll index had the largest influence on seed yield and its components under severe water deficit. Under mild water deficit, leaf water potential and relative water content were also important for improving seed yield. Leaf water potential, relative water content, chlorophyll fluorescene and chlorophyll index were had the largest effect on seed yield and its components under well watered condition.

Keywords: Canonical Correlation, Water Deficit Stress, Winter Rapeseed

1. Introduction

Rapeseed is the third most important oilseed cropoily plant in the world after soybean and palm (FAO, 2007). New seed varieties naturally contain 40- 45% of oil which is used as raw material to produce industrial and hydraulic oil, cleaner, soap and biodegradable plastics (Friedt, 2007). After extracting the oil, the remained oil cake, which contains 38-44% high-quality proteins, is used for animal nutrition (Walker and Booth, 2001). Drought and its stress is one of the common environmental stresses which limit farm productions in around 25% of world's land (Mendham and Salisbury, 1995). Access to water is one of the main limitations in realization of full yield and quality of most species and it may erupt during the whole growth stage or in critical conditions (Parry et al., 2005). Plants employ a range of particular responses in order to minimize the effects of water shortage or to increase water absorbing rate (Morison et al., 2008). The effect of water stress is a function of genotype, stress race, weather condition, growth and development stage of rapeseed (Robertson and Holland,

(2004). Water stress in particular stages of rapeseed phonology affects seed qualitative properties such as percentage of oil and protein and the amount of glucosinolates (Strocher et al., 1995). Liang et al, (1992) by evaluating the morphological and physiological responses to water stress showed that *Brassica juncea* is more adaptable to water stress than *B. napus*. The results of Kumar and Singh (1998) indicate that in *Brassica* oilseeds, the cell turgidity is maintained up to 2.4 Mega Pascal leaf water potential by the genotypes with high osmotic adjustment but with low osmotic adjustment, the fall rate in pressure potential was fast accordingly. Also, Valeric et al, (2002) remarked that when the separately planted rapeseed leaves were positioned under high osmotic- laboratory, huge amount of proline flocked in leaves. Zulini et al, (2002) found a significant correlation between Fv/Fm and leaf water potential in stressed plants so when leaf water potential decreases to less than 0.9 Mega Pascals, decrease in Fv/Fm can be observed. Numerous experiments suggest rapeseed yield is influenced by high number of pods per plant or per area unit (Rao and Mendham, 1991). Jensen et al, (1996) reported that the eruption of water stress in vegetative growth and flowering stages didn't have significant effect on each rapeseed weight. However, during water shortage in seed filling stage there is reduction in their weight. It has shown that supplemental irrigation of rapeseed increases the number of pods and seeds per pod by extending flowering stage; and it's because of having many leaves in this stage (Kimber and McGregor, 1995).

In spite of a surge in literature on drought tolerance in crops during the past two decades, clear picture on association between physiological characters and seed yield components is yet to emerge. This is mainly because, most of the studies relied on simple correlation coefficients to analyze the relationships. Simple correlations are inadequate to address this complex issue as, physiological and yield components are neither independent from each other nor among themselves. Therefore one has to consider the correlation between these two sets of variables, simultaneously. Canonical correlation, a well-known multivariate technique, has been established for such situations, where one would like to measure the relationship between two sets of interrelated variables. We have adopted canonical correlation analysis to study the strength of association between the physiological traits and seed yield components under severe water deficit stress, mild water deficit stress and well water conditions. Further, we intended to find the physiological characters that have the greatest influence on seed yield and its components under the three conditions.

2. Materials and Methods

The experiment was conducted under greenhouse conditions in Agricultural Faculty of Tabriz University, IRAN in 2007-2008. Temperature during the day was 23°C-25°C and during the night was 15°C-17°C with 14 hours of light. Also the relative humidity was 50- 60%.

2.1. Plant Materials

The plant material included 12 winter rapeseed cultivars named Zarfam, Okapi, Modena, Licord, Olera, Dexter, Arc-4, Elite, Opera, SLM046, Fornax, and Orient obtained from Agricultural and Natural Resources Research Center of East Azerbaijan province_Iran. Seeds were sown in 8-kilogram flower_ pots with 5 seeds planted in each. Thinning was done at two leaf stage and keep only one plant per pot. Considering that cultivars were winter type, vernalization was done on cultivars. Water deficit stress was imposed from stem elongation to physiological maturity. Gypsum blocks were used in order to control soil moistur. The factorial experiment was done with two factors irrigation at 3 levels: well watered stress (100% FC), mild stress (75% FC), severe stress (50% FC) and 12 winter rapeseed cultivars in randomized complete block design with 3 replications.

2.2. Measured Traits

1. Leaf water potential was measured by Pressure Chamber; model: Soil Moisture Equipment Crop, Sanata Barbara, CA.
2. Relative water content: The method of Morant-Manceau et al, (2004) was used. First the Fresh Weight (FW) of samples was measured. Then, the samples were put in distilled water and after 24 hours the Turgid Weight (TW) was measured and after putting samples in 75°C Oven the Dry Weight (DW) was measured. Finally the percent relative water content was measured by using formula: RWC= FW-DW/TW-DW×100
3. Osmotic potential was measured by Osmometer; model: Osmomat 010, Genotel.
4. Stomata conductance measured by Porometer; model: AP4- Porometer (Delta-T Devices) Cambridge, UK.
5. For chlorophyll fluorescence we used florometer; model: Opti Science, OS-30, USA.
6. Chlorophyll index is determined by Chlorophyll meter; model: SPAD-502, Minolta, Japan.
7. Proline contents were measured by Acid Hydrin method. The plant height, plant dry weight, volume of root, root dry weight, length of siliquae, number of siliquae per plant, seeds per siliquae and 1000-grain weight were measured at the end of growth stage.

2.3. Statistical Analysis

Statistical analysis of individual characters was carried out using standard biometrical procedures. The data of individuals were subjected to ANOVA (to partite the variance) and canonical correlation analysis using PROC ANOVA and PROC CANCORR procedures in the SAS program (SAS institute, 1996). Physiological traits and seed yield related traits were considered as independent (X) and dependent (Y) sets of variables, respectively. An overall test for statistical significance of all the five possible canonical correlations from zero was performed using Wilks Lambda (Gittins 1985). The Wilks Lambda was computed using the formula

$$A = \prod_{i=1}^{s}(1 - C_i^2)$$

Where, Ci is the ith canonical correlation and s = min (p, q), p and q are the number of seed yield related and physiological traits studied. Approximate F-test (as per SAS default) was used for assessing the statistical significance of the Wilks Lambda. The structure correlation (Johnson and Wichern 1998) were calculated using the following formula

$$S_{i(j)k} = \frac{e_{ki(j)} \sqrt{\lambda_{i(j)}}}{\sqrt{\sigma_{kk}}}$$

Where, i (j) k = 1, 2,... (q) depends on the number of characters studied in the seed yield and physiological related sets of variables, respectively; σ_{kk} is the variance of the kth variable and ($\lambda_{i(j)}$ $e_{i(j)}$) are the eigenvalue-eigenvector pairs of the corresponding covariance matrices.   The practical importance of the canonical correlations was obtained by calculating the redundancy measure (RM) according to Sharma (1996) using the formula

$$RM_{Vi.wi} = AV(Y / V_i) \times C_i^2$$

Where, AV (Y/V_i), variance extracted, is the average variance in Y variables that is accounted for by the canonical variate V_i (i^{th} linear combination of the seed yield related characters).

Total redundancy for the Y variables (RM_{y/x}) was computed as

$$RM_{Y/X} = \sum_{i=1}^{q} RM_{Vi.wi}$$

Where, $RM_{Vi.Wi}$ is the redundancy measure as explained above.

3. Results and Discussion

Analysis of variance revealed significant differences among the genotypes for all the traits studied under water deficit conditions. The result of canonical correlation analysis carried out between the seed yield traits and physiological traits showed that first three canonical correlations were significant under SW, MW and WW conditions. The first canonical correlation coefficient was 0.88 under SW, 0.74 under MW and 0.79 WW conditions while second canonical correlation coefficient was 0.51, 0.57 and 0.53 under SW, MW and WW conditions. Squared canonical showed that 0.78 (SW), 0.56 (MW) and 0.64 (WW) of the variability in yield related traits was explained by the first linear combination of the physiological traits. The contribution of these linear composites for the second canonical variate was 0.25, 0.33 and 0.28 under SW, MW and WW conditions, respectively (Table 1).

Table 1. The first four canonical correlations between physiological and seed yield related traits, squared canonical correlation, cumulative proportion, approximate F and significance level under SW, MW and WW conditions

	Canonical correlation			Squared canonical correlation			Cumulative proportion			Approximate F			Value		
	SW	MW	WW	SW	MW	WW	SW	MW	WW	SW	MW	WW	SW	MW	WW
1	0.881	0.748	0.798	0.777	0.560	0.637	0.874	0.595	0.689	2.35	1.76	1.98	0.001	0.023	0.008
2	0.507	0.574	0.529	0.257	0.329	0.280	0.961	0.825	0.842	0.70	1.21	1.14	0.797	0.274	0.337
3	0.315	0.488	0.472	0.099	0.239	0.223	0.988	0.972	0.955	0.42	0.97	1.07	0.932	0.480	0.404
4	0.206	0.235	0.319	0.042	0.055	0.102	1	1	1	0.31	0.41	0.80	0.867	0.799	0.537

The intra-set structural correlation which gives the magnitude and direction of the contribution of variables to the variates within domain is presented in Table 2 and Table 3 for yield related and physiological traits, respectively. Among the yield related traits, SPP (0.82), showed the maximum contribution to the first canonical variate in the SW condition followed by SPS (0.72) and Y (0.73). 1000-GW was the most influential trait in forming the second and third canonical variate (0.56) and (0.66). Under MW condition, SPP and Y still remained the highest contributing traits (0.78) and (0.72) followed by 1000-GW (-0.69). In formation of the second canonical variate, Y (0.55) was the highest contributors. SPP (0.88) SPS (0.74) and Y (0.90) showed the maximum contribution to the first canonical variate in the WW condition. Y still remained the highest

contributing traits (0.34). 1000-GW was the most influential trait in forming the third canonical variate (0.94). Among the physiological traits, CI had the highest contribution (0.87) to the first variate followed by RWC (0.81), LWP (0.77) and LOP (0.67) under SW condition, while, CF and PC were (0.57) and (0.54), respectively. In second variate, PC was the highest contributing character (0.60). Under MW condition, LWP had the highest contribution (0.93) to the first variate followed by RWC (0.63), CI (0.51) and CF (0.42). CF (0.84) and CI (0.54) still remained the highest contributing characters. CF (0.82) had the highest contribution followed by CI (0.81), LWP (0.78) and RWC (0.76) under WW conditions. PC (0.60) was the most influential character in forming the second canonical variate.

Table 2. Structure correlation between physiological traits and their (first four) canonical variates under SW, MW and WW conditions

Trait	Severe-Watered				Mild-Watered				Well-Watered			
	V1	V2	V3	V4	V1	V2	V3	V4	V1	V2	V3	V4
LWP	0.767	0.098	-0.190	0.502	0.931	0.145	-0.134	-0.068	0.781	0.100	-0.359	-0.257
RWC	0.798	-0.318	-0.158	0.026	0.628	0.356	-0.204	-0.071	0.758	0.029	-0.293	0.218
LOP	0.673	-0.120	0.094	0.078	0.245	0.312	0.748	0.272	0.526	-0.327	0.326	-0.326
PC	0.539	0.591	-0.255	0.033	0.283	0.323	0.287	-0.011	0.446	0.593	0.257	-0.257
CF	0.572	0.247	0.364	0.336	0.422	0.844	-0.159	0.056	0.818	-0.215	0.239	-0.239
CI	0.868	0.302	-0.211	-0.222	0.506	0.545	0.461	-0.104	0.816	0.464	0.001	-0.134
SC	-0.173	-0.258	0.572	0.142	0.060	-0.290	0.028	-0.690	0.045	-0.512	-0.041	0.762

The contribution of the linear function of all the physiological characters for expression of CI, RWC, LWP and LOP were 0.77, 0.70, 0.68 and 0.60 percent from the first linear function under SW condition (table 4). Under MW condition, the contribution of the linear function of all

the physiological characters for expression of CI, RWC and CF were decreased to 0.38, 0.47, and 0.32 percent but it was 0.70 for LWP. The contribution of the linear function of all the physiological characters for expression of CI, RWC, LWP and CF 0.65, 0.60, 0.62 and 0.65 percent from the first linear functions under WW condition. Similarly, the contribution of the individual physiological characters to the first three linear functions of the yield-related traits showed 0.72, 0.64 and 0.63 percent of the variability explained by

SPP, SPS and Y, respectively under SW condition (table 5). Under MW environment, the contribution decreased in comparison to SW condition. In WW condition, the results were same as in SW condition.

Table 3. *Structure correlation between seed yield related traits and their (first four) canonical variates under SW, MW and WW conditions*

Trait	Severe-Watered				Mild-Watered				Severe-Watered			
	W1	W2	W3	W4	W1	W2	W3	W4	W1	W2	W3	W4
SPP	0.817	-0.192	0.522	0.146	0.782	0.473	0.338	-0.221	0.876	0.003	-0.277	0.393
SPS	0.718	0.464	-0.088	0.510	0.439	0.428	-0.570	0.546	0.736	0.135	0.201	-0.630
1000-SW	-0.480	0.563	0.663	0.110	-0.694	0.491	0.501	0.157	-0.153	0.293	0.942	-0.046
Y	0.726	0.090	0.390	0.558	0.717	0.549	0.309	0.296	0.896	0.345	-0.089	0.266

Table 4. *Correlation of physiological traits contributing to the first four variates of seed yield related traits under SW, MW and WW conditions*

Trait	Severe-Watered				Mild-Watered				Well-Watered			
	W1	**W2**	**W3**	**W4**	**W1**	**W2**	**W3**	**W4**	**W1**	**W2**	**W3**	**W4**
LWP	0.676	0.060	-0.060	0.103	0.697	0.083	-0.065	-0.016	0.623	0.053	-0.170	-0.082
RWC	0.703	-0.161	-0.050	-0.005	0.470	0.204	-.099	-0.016	0.606	0.015	-0.138	0.069
LOP	0.594	-0.061	0.029	-0.016	0.183	0.179	0.366	0.064	0.420	-0.173	0.026	-0.104
PC	0.476	0.300	-0.080	-0.007	0.212	0.185	0.140	-0.028	0.356	0.314	0.215	-0.082
CF	0.504	0.125	0.115	0.069	0.316	0.485	-0.077	0.013	0.653	-0.114	-0.160	-0.075
CI	0.766	0.153	-0.066	-0.045	0.379	0.313	0.225	-0.024	0.651	0.246	0.000	-0.043
SC	-0.153	-0.130	0.180	0.029	0.045	-0.167	0.014	-0.162	0.036	-0.271	-0.019	0.243

Table 5. *Correlation of seed yield related traits contributing to the first four variates of physiological traits under SW, MW and WW conditions*

Trait	Severe-Watered				Mild-Watered				Well-Watered			
	V1	**V2**	**V3**	**V4**	**V1**	**V2**	**V3**	**V4**	**V1**	**V2**	**V3**	**V4**
SPP	0.720	-0.097	0.164	0.030	0.585	0.272	0.165	-0.052	0.699	0.002	-0.131	0.125
SPS	0.633	0.235	-0.027	0.106	0.328	0.246	-0.278	0.128	0.588	0.072	0.095	-0.201
1000-SW	-0.423	0.285	0.209	0.022	-0.519	0.282	0.245	0.037	-0.122	0.155	0.445	-0.015
Y	0.640	0.045	0.123	0.115	0.537	0.315	0.151	0.069	0.714	0.183	-0.042	-0.085

4. Discussion

Canonical correlation analysis was carried out to identify how physiological characters influence the seed yield traits under SW, MW and WW conditions in winter rapeseed cultivars. Results suggested that CI was the most important character followed by RWC, LWP and LOP under SW condition. In the MW conditions, LWP was the most important character followed by RWC, CI and CF. CF was the most important character followed by CI, LWP and RWC under WW conditions. Pirdashti et al. (2009) observed positive and significant relationship of CI, PC and RWC with seed yield in rice cultivars under drought stress. Pirevatlou et al. (2008) showed positive correlation of CI with seed yield and leaf area index in wheat genotypes under drought stress.

The interrelationships between physiological characters and seed yield-related traits clearly identified the importance of CI, LWP, CF, RWC and LOP under SW, MW and WW conditions. Therefore, the contribution of physiological characters on seed yield under stress condition is more important than well-watered condition. In fact CI, LWP, CF, RWC and LOP can be considered as proper criteria for

selecting cultivars for high yield.

Abbreviation

LWP: Leaf Water Potential, RWC: Relative Water Content, LOP: Leaf Osmotic Potential, PC: Proline Content, CF: Chlorophyll Fluorescence, CI: Chlorophyll Index, LSC: Leaf Stomata Conductance, PH: Plant Height, PDW: Plant Dry Weight, RV: Root Volume, RDW: Root Dry Weight, SL: Siliqua Length, SPP: number of Siliqua Per Plant, SPS: Seeds Per Siliqua, 1000-SW: 1000-Seed Weight, SY: Seed Yield, SS: Severe Stress, MS: Mild Stress, WW: Well Watered.

References

[1]   FAO, 2007. http:// faostat. fao. org/.

[2]   Friedt, W., Snowdon, R., Ordon, F., and Ahlemeyer, J. 2007. Plant Breeding: Assessment of genetic diversity in crop plants and is exploitation in breeding. Progress in Botany, 168: 152-177.

[3] Gittins, R. 1985. Canonical analysis, A review with applications in ecology. Springer-Verlag, Berlin. pp, 56-85.

[4] Jensen, C.R., Mogensen, V.O., Mortensen, G.., Fieldsend, J.K., Milford, G.F.J., Anderson, M.N., and Thage, J. H. 1996. Seed glucosinolate, oil and protein content of field-grown rape (*Brassica napus* L.) affected by soil drying and evaporative demand. Field Crops Research, 47: 93-105.

[5] Johnson, R.A. and Wichern, D.W. 1998. Applied multivariate statistical methods (4th edition). London-Prentice Hall. Englewood Cliffs, pp, 65-85.

[6] Kimber, D.S. and McGregor, D.I. 1995. The species and their origin, cultivation and world production. In: Kimber, D.S. and McCregorceds, D.I. (eds.). Brassica oilseeds. CABI, PP: 1-7.

[7] Kumar, A. and Singh, D.P. 1998. Use of physiological indices as a screening technique for drought tolerance in oilseed Brassica species. Annals of Botany, 81: 413-420.

[8] Liang, Z.S., Diang Z.R., and Wang, S.T. 1992. Study on types of water stress adaptation in both *Brassica napus* L. and *B. juncea* L. Acta Botanica Boreali, Occidentalia Sinica., 12(1): 38-45.

[9] Mendham, N.J. and Salisbury, P.A. 1995. Physiolog of Crop development, growth and yield. In: Kimber, D. and McCregor. D.I. (eds). Brassica oilseeds, CABI, Pp: 11-64.

[10] Morant-Manceau, A., Pradier, E., and Tremblin, G. 2004. Osmotic adjustment, gas exchanges and chlorophyll fluorescence of a hexaploid triticale and its parental species to salt stress. Journal of Plant Physiology, 169: 25-33.

[11] Morison, J.I., Baker, N.R., Mullineaux, P.M., and Davies, W.J. 2008. Improving water use in crop production. Philosophical Transactions of the Royal Society of London. Series B: Biological Science, 363: 639-658.

[12] Parry, M.A.J., Flexas, J., and Medrano, H. 2005. Prospects for crop production under drought: Research priorities and future directions. The Annals of Applied Biology, 147: 217-226.

[13] Pirdashti, H., Sarvestani Z.T., and Bahmanyar, M.A. 2009. Comparison of physiological responses among four contrast rice cultivars under drought stress conditions. Proceedings of World Academy of Science. Engineering and Technology, 37: 2070-3740.

[14] Pirevatlou, A.S., Aliyev, R.T., Hajieva, S.I., Javadova S.I., and Akparov, Z. 2008. Structural changes of the photosynthetic apparatus, morphlogical and cultivation responses in different wheat genotypes under drought stress condition. Genetic Resources Institute. Bako Republic of Azerbaijan, 14: 123-130.

[15] Rao, M.S.S., and Mendham, N.J. 1991. Soil-plant-water relation of oilseed rape (*Brassica napus* and *B. compestris*). Journal of Agricultural Science Cambridge, 197: 197-205.

[16] Robertson, M.J. and Holland, J.F. 2004. Production risk of canola in the semi-arid subtropics of Australia. Australian Journal of Agricultural Research, 55: 525-538.

[17] SAS Institute, Inc. 1996. SAS language guide for personal computers. Edition 6.12, Carry, NC, USA.

[18] Sharma, S. 1996. Applied multivariate techniques. John Wiley and Sons, New York, pp, 245-256.

[19] Strocher, V.L., Boathe I.G., and Good, R.G. 1995. Molecular cloning and expression of a turgor gene in Brassica napus. Plant Mol Biol., 27: 541-551.

[20] Valeric, H.R., Sulpice, R., Lefort, C., Maerskack, V., Emery, N., and Larher, F.R. 2002. The suppression of osmoinduced proline response of Brassica napus L. var. Oleifera leaf discs by polyunsatutated fatty acids and methyl-jasmonate. Plant Science, 164: 119-127.

[21] Walker, K. C. and Booth, E. J. 2007. Agricultural aspects of rape and other *Brassica* products. European Journal of Lipid Science Technology, 103: 441- 445.

[22] Zulini, L., Rubinigg, M., Zorer, R., and Bertamini, M. 2002. Effects of drought stress on chlorophyll fluorescence and photosynthetic pigment in grapevine leaves (*Vitis vinifera* cv. White Riesling). www. Actahort. org / html.

17

Level of Adoption and Factor Affecting the Level of Adoption of Sustainable Soil Management Practices in Ramechhhap District, Nepal

Bikal Koirala[1, *], Jay Prakash Dutta[1], Shiva Chandra Dhakal[1], Krishna Kumar Pant[2]

[1]Department of Agricultural Economics, Institute of Agriculture and Animal Sciences, Rampur, Chitwan, Nepal
[2]Department of Environmental Science, Institute of Agriculture and Animal Sciences, Rampur, Chitwan, Nepal

Email address:
bikal.koirala@gmail.com (B. Koirala)

Abstract: This study investigated the level of adoption and factor affecting the level of adoption of sustainable soil management practices. This research was based on the primary data that was collected in 2012 at Chisapani, Nagdaha and Kathjor VDCs of Ramechhap district of Nepal to analyze the level of adoption and factor affecting the level of adoption of sustainable soil management practices. Pre-tested semi structured interview schedule were used to collect the primary data from 120 farmers, 40 farmers from each VDC by applying simple random sampling technique. The major sustainable soil management practices adopted were found to be improved farm yard manure, improved cattle urine, inclusion of the vegetable and legume in the farming system and use of the bio-pesticide. The level of technology adoption was found to be 79.55% and this shows that the level of adoption was high. The higher level of adoption was due to increase in the production and productivity of the crops and improvement in soil fertility. While considering about the factor affecting the adoption of sustainable soil management practices, the study showed that the five variables namely training, credit, income, livestock standard unit and experience were significantly affecting for higher level of adoption. A unit increase in training, credit, livestock standard unit and experience would increase the probability of level of adoption by 3.48%, 0.83%, 0.45% and 1.1% respectively and a hundred rupees increase in income would increase the probability of level of adoption by 0.0819%.

Keywords: Adoption, Production, Technology, Significantly, Sustainable Soil Management

1. Introduction

Nepal is an agricultural country with 65.6% of economically active population has agriculture as the major occupation (CBS, 2013). The supply of chemical fertilizer from Agricultural Input Company Limited rises to 10328.83 Metric ton in year 2009/10 as compared to 7133 metric ton in year 2008/09 and 142 metric ton of chemical pesticide in year 2011/12. The quantity was increase very highly in year 2010/11 which reached to 110013 metric ton (MoAD, 2012). This reflects the excessive use of fertilizer and its bad impact on soil health. The farmers obtains higher yield but do not replenish the harvested nutrients. This results in degradation of soil (karki and Dacayo, 1990). There is poor management (fertilizer, pesticides, irrigation etc) of soil as per its requirements which seriously have shown the problem of sustainability. Hence in the Nepalese context nutrient

depletion and soil management for optimal crop production is major concern on soil sustainability. The haphazard use of chemical fertilizer and pesticide had resulted in unsustainability of soil and had toxic effect to human beings and animals. Nevertheless large amount of money is being spent on purchasing chemical fertilizers and pesticides from other countries .So adoption of sustainable agriculture is utmost necessary in Nepalese field. Sustainable agriculture system is that agricultural system which is capable of maintaining the productivity and usefulness to society for generations to generations (USDA, 1990). This system has some salient features such as resource conservation, socially just and supportive, commercially competitive and environmentally sounds.

2. Materials and Methods

2.1. Study Area and Sample Size

Three VDCs of Ramechhap district namely Chisapani, Nagdaha and Kathjor were purposively selected to study the impact of SSM practices. Altogether 120 farmers adopting sustainable soil management practices were taken, 40 from each VDC using simple random sampling technique. The field survey was conducted in June, 2012. A co-ordination schema was developed and semi-structured interview schedule was prepared containing both closed and open-ended questions. Observations in the farmer's field were done, focus group discussion was conducted and key informant survey was carried out. The final analysis was done by using computer software Statistical Package for Social Sciences (SPSS), Microsoft Excel and STATA.

2.2. Methodological Approach of Impact Evaluation

Before after approach was used for the impact study of sustainable soil management practice. For the impact assessment pair t test was used to test the impact of adoption of these practices on area and production of the different crops.

2.3. Level of Technology Adoption

First of all adoption score was calculated by doing sum of assigned points to the respondents. The level of technology adoption was calculated by using following formula

$$\text{Level of technology adoption} = \frac{\text{Total score obtained by farmer}}{\text{Maximum possible score}} \times 100$$

2.4. Logit Regression Model

Binary logit regression model was applied for analyzing the factors affecting the level of adoption of sustainable soil management practices which can be expressed as:

Y_i = f(β_i, X_i) = f(Age, Economically active members, Education, Training, Farm size, Experience, Annual household income, credit, Livestock standard unit, gender)

The logit transformation can be expressed as: (Gujrati, 2003).

L_i = ln [p_i / 1-p_i] = Z_i = α + $\Sigma\beta_i X_i$ + ε_i where, X_i = explanatory variables, β_i = parameters to be estimated , ε_i = error term

L_i = logit and [p_i / 1- p_i] = Odd ratios

3. Result and Discussion

3.1. Sustainable Soil Management Practices Adopted by Farmers

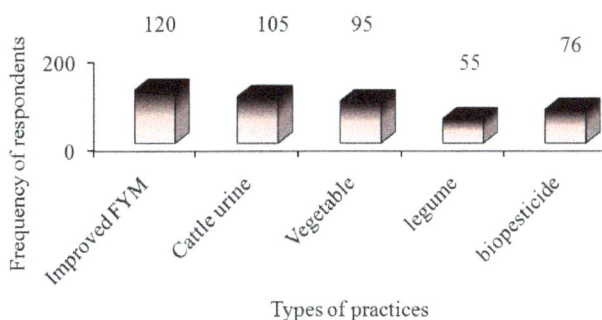

Figure 1. Frequency of different SSM techniques of used by farmers in study area (2012).

For the purpose of determining the techniques adopted by farmers focus group discussion was done. Mainly, five types of techniques were found that were used by farmers. During primary data collection it was found that one farmer had adopted one or more practices. They were improved farm yard manure, cattle urine, inclusion of vegetable in cropping system use of bio pesticide and inclusion of legumes in cropping pattern. Improved farm yard manure and improved cattle urine technique has resulted the less nitrogen loss. Similarly, inclusion of legume had helped in the reduced dose of nitrogen fertilizer to the soil due to the storage of nitrogen in the soil by the nitrogen fixing bacteria. Use of the bio pesticide had used in order to make the food organic. It was shown in Figure 1.

3.2. Impact of Sustainable Soil Management Practices on Area and Production of Different Crops

The impact of sustainable soil management practices on area and production of different crops was also studied using before after approach and pair t test was used for analysis. There was increase in production of rice and maize because they were main staple food in the study area. There was decrease in area and production of finger millet and wheat. This was due to farmers are oriented in vegetable farming and legumes, where previously finger millet and wheat were used to be grown and less attraction mainly towards finger millet. Farmers who used majority of finger millet for preparation of wine had now decreased in the preparation of wine and are used it consumption purpose, so the decrease in finger millet area and production and has not much effect. In gist the increase in production was due to increase in area (either by purchasing, or by sharing and lease hold), adoption of sustainable soil management practices and due to reason of increasing of income particularly by selling vegetables. Due to increase in production the level adoption of these practices is high. Pretty and Hine in 2001 also stated that adoption of sustainable agricultural practices had increased the yield by 50-100% for rain-fed crops. They also stated that by adopting sustainable agriculture by 12,500 households of Ethopia, there is increased in 60% yields. Similarly in a review of 286 projects in 57 countries the agricultural productivity was increased by 79% by means of sustainable agriculture or resource conserving agriculture (Pretty et.al, 2006). The area below was in ropani (Table 1) and the production was in kilogram (Table 2).

Table 1. Impact of sustainable soil management practices on area of different crops in the study area (2012).

	After	Before	Difference	t-value	Sig
Rice area	4.792	4.175	0.617	11.89***	0.000
Maize area	3.45	3.5	-0.05	-1.068	0.856
wheat area	1.254	1.479	-0.225	-5.742	0.500
Finger millet area	3.02	3.362	-0.342	-4.687	0.487
Vegetable area	3.7	1.742	1.958	21.562***	0.000
Legume area	2.304	1.483	0.821	11.746***	0.000

t-value obtained from paired t-test are significant different at 1% level of significance (***)

Table 2. Impact of sustainable soil management practices on production of different crops in the study area (2012).

	After	Before	Difference	t-value	Sig
Rice production	521.5	361.625	159.875	16.727***	0.000
maize production	290	202.458	87.542	17.47***	0.000
wheat production	102.25	105.95	-3.7	-1.153	0.932
Finger millet production	134.14	134.441	-3.301	-1.008	0.842
Vegetable production	24.013	9.754	14.259	21.305***	0.000
Legume production	97.516	58.483	39.033	14.537***	0.000

t-value obtained from paired t-test are significant different at 1% level of significance (***)

3.3. Level of Technology Adoption

Majority of respondents was found higher level of adoption (>75%) which was 75.38% followed by 24.17% of respondents doing 50-75% level of adoption of sustainable soil management practices which is shown in table no. 3. The mean level of adoption was found to be 0.7955 with standard deviation of 0.0543. Bhusal in 2012 also reported that level of adoption of cattle urine use technology (one of the sustainable soil management practices) was high. The higher level of adoption may be due to increase in production of different crops and improvement in soil fertility.

Table 3. Level of adoption of sustainable soil management practices in study area (2012).

Level of adoption	Chisapani	Nagdaha	Kathjor	Total
<50% (Low)	0(0.00)	0(0.00)	0(0.00)	0(0.00)
50%-75% (Medium)	11(27.50)	8(20.00)	10(25.00)	29(24.17)
>75% (High)	29(72.50)	32(80)	30(75.00)	91(75.83)
Total	40(100.00)	40(100.00)	40(100.00)	120(100.00)

Figures in the parentheses indicates percentage

3.4. Factor Affecting the Level of Technology Adoption

Logit model was used for determining the factor that affect the level of adoption of SSM practices. In this age, economically active members, education, training, farm size, experience, income, credit and L.S.U are explanatory variables and level of adoption of practices is an explained variable. The adoption level was categorised into a binary response by the adoption level equal or more than 79%= 1 and 0 otherwise.

Logit regression analysis was done on 120 farmers, who were adopting SSM practices at different level. The wald test (LR chi^2) for the model had good explanatory power at the 1% level. The pseudo R^2 was 0.7727. In order to interpret the model, marginal effects were driven from regression coefficients. The interpretation is shown in table no. 4.

Logit regression analysis shows that out of ten explanatory variables taken, five variables were found to be statistically significant. These variables were training, experience, income, credit and livestock standard unit. It was found that age, economically active members, education, farm size and gender were found to be statistically insignificant. (shown in table 4)

Table 4. Factor affecting the level of SSM practice adoption in the study area (2012).

Variables	Coefficients	P>|z|	Standard error	dy/dx^b	S.E^b
Age	-0.0947	0.277	0.087	-0.0002	0.0004
Economically active members	-0.282	0.76	0.924	-0.0008	0.003
Education	0.421	0.683	1.031	0.001	0.003
Training	4.490**	0.015	1.847	0.0348	0.048
farm size	-0.902	0.117	0.575	-0.002	0.004
Experience	2.150*	0.093	1.279	0.0116	0.022
Income	0.0002**	0.019	0.0001	0.00000819	0.000
Credit	2.749**	0.05	1.444	0.0083	0.014
LSU	1.582*	0.094	0.946	0.0045	0.007
Gender	-0.889	0.383	1.019	-0.002	0.004
Constant	-6.079	0.415	7.453		
Summary statistics					
Number of observation(N)		120			
Log likelihood		-17.513984			
LR chi^2(10)		119.08			
Pseudo R^2		0.7727			
Goodness of fit test		Pearson Chi^2= 38.63 Prob> Chi^2= 0.9986			

Training (dummy) was positively significant (P<0.05) to adopt the practices. The farmer may receive the training from both governmental and non-governmental organizations, the probability to adopt SSM practices increases by 3.48 percent. This might be due to increasing awareness to farmers, realization of advantage of practicing different practices and improving of skills. Bayard *et al.,* 2006 also reported that training will positively influence the adoption level in soil management practices.

The variable experience was positively significant (P<0.1) to affect the level of adoption. A unit increase in the year of farming experience would increase the level of adoption by 1.1 percent. Rezvanfar *et al.* (2009) also reported that increase in experience increase the level of knowledge and increase the adoption of soil conservation practices. This might be due to experienced farmers have high level of knowledge skills in farming activities and well farm management.

Income was the important variable affecting the level of adoption. Annual income was found to be positively significant (P<0.05) and remaining other things constant, a hundred rupee increase in annual income would increase the level of adoption by 0.0819 percent. Higher the level of income they have high risk bearing capacity and there will be increase in investment. So, higher level of income will help to increase the adoption level and in their farm. Onweremadu and Matthews in 2007 also reported, increase in income will increase in adoption level of soil management practices.

The variable credit (dummy) was found to be positively significant (P<0.05). The farmers may have provision of credit from banking sector or cooperatives or other private conditions the probability to increase the level of adoption by 0.836 percent. This might be due to it open the area for investment, better farm management and decision. Jayasawal *et. al.* (2001) also reported that if there is provision of credit, than adoption of any technology would be high.

Livestock holding was one of the important variables in affecting the level of adoption and found to be positively significant (P<0.1). Keeping other things constant, each unit increase in livestock unit will increase the level of adoption by 0.45 percent. This was due to SSM practices are directly related with livestock. Romano *et al.* (2010) also reported that more the livestock there would be more adoption of technology. It was due to the reason that higher level of income and helping the adopters towards food security.

4. Conclusion

Agriculture is the major occupation in Ramechhap district. There is the subsistence oriented agriculture farming with maize, wheat, rice, finger millet and vegetables as the main crops. Before the adoption of practices the use of fertilizers and pesticides had adverse affect on soil resulting on different soil health problems. But after adopting different technique of sustainable soil management practices, there is the reduction of fertilizer and pesticide use and this has resulted in the increase in the production and productivity of the different crops. So the adoptions of these practices are high. This is due to increase in the production and less bad effect in the soil and the material required for these practices are locally available and cost effective. The different factors that affect the level of adoption are training, experience, income, credit and livestock standard unit.

Acknowledgements

I want to express my thanks to Dean, IAAS; Academic Dean, IAAS; Advisory committee, Department of Agricultural Economics, Colleagues, seniors, juniors, respondents of Ramechhap district for valuable and genuine information and my family.

Refernces

[1] Adhikari, J. (2000). Decisions for farm survival: Farm management strategies in the Mid-hills of Nepal. Adriot Publishers, New Delhi, India. 385p.

[2] Bayard, B., C.M. Jolly and D.A. Shannon. 2006. The adoption and management of soil conservation practices in Haiti: The case of rock walls. Agricultural Economics Review: 7(2) 28-39.

[3] Bhusal, D. R.(2012). Impact of cattle urine use technology on farm income of the vegetable growers of Dhading district, Nepal. Thesis submitted to Tribhuvan University, Nepal. 38p.

[4] CBS. (2013). Statistical pocket of Nepal. Government of Nepal, Central bureau of statistics, National Planning Commission Secretariat, Thapathali, Kathmandu.

[5] Karki, K.B. and J.B. Dacayo. (1990). Assessment of land degradation in southern Lalitpur of Nepal. Proceedings of 14th Congress of International Society of Social Sciences, Tokyo, Japan. 38p.

[6] Jayasawal, M. L., B. N. Regmi, T. R. Noor, A Mcleod and J, Best. (2001). Using livestock to improve the livelihoods of landless and refugee affected livestock keepers in Bagladesh and Nepal. DFID. pp. 31-33.

[7] MoAD. (2012). Statistical information on Nepalese Agriculture. Ministry of Agriculture Development, Singh Durbar, Kathmandu, Nepal.

[8] Onweremadu, E.U. and E.C. Matthews. (2007). Adoption level and sources of soil management practices in low- input agriculture. Nature and Science 5(1) 39:45.

[9] Pretty, J. & R. Hine. (2001). Reducing food poverty with sustainable agriculture: a summary of new evidence. UK: University of Essex Centre for Environment and Society. 24p.

[10] Pretty, J.N., A. D. Noble, D. Bossio, J. Dixon, R. E. Hine, F. W. T. Penning de Vries and J. I. L Morison. (2006). Resource-conserving agriculture increases yields in developing countries. Environmental Science and Technology (Policy Analysis) 40(4): 1114-1119.

[11] Rezvanfar, A., A. Samiee and E. Faham. (2009). Analyse of factor affecting adoption of sustainable soil conservation practices among wheat growers. World Applied Science of Journal 6(5): 644-651.

[12] Romano, D., E. Mane, M. D. Errico and L. Ellinovi. (2010). Livelihoods strageies and household resilience to food insecurity: An empirical analusis to Kenya. European Report Development, Rome, Italy. pp. 15-25

[13] USDA. (1990). Debate on sustainability and development. USDA Bulletin, American Congress.

18

Climate Change Impacts, Adaptation and Coping Strategies at Malindza, a Rural Semi-Arid Area in Swaziland

Siboniso M. Mavuso, Absalom M. Manyatsi[*]**, Bruce R. T. Vilane**

University of Swaziland, Department of Agricultural and Biosystems Engineering, Manzini, Swaziland

Email address:

Manyatsi@uniswa.sz (A. M. Manyatsi)

Abstract: The objective of the study was to assess the impacts of climate change faced by rural households in the lowveld of Swaziland, using Malindza as a case study area, and further identify adaptation and coping strategies employed by households. A questionnaire was developed and used to conduct interviews from 160 households randomly selected in four rural communities of the study area. Data were analysed with SPSS software, and reported in forms of tables and figures. More or less all the respondents (99%) were aware of climate change and climate change variability, Sources of information included radios (92.5%), television (5.6%) and agricultural extension officers (2%). The information was however considered inadequate and of short term remedy as it was in the form of daily weather forecast. The perceived effects of climate change included crop failure (99%), loss of livestock (99%) and drying of surface water (99%). Only 9% of the households harvested enough maize to last for a year, and the rest (91%) had to rely on buying maize, exchanging it for labour or receiving food aid. The climate change adaptation strategies practiced included contour ploughing (49%), use of organic fertilisers (29%) and crop rotation (20%). Thirty two percent of the households planted hybrid maize seeds and 15% planted open pollinated maize seeds. Another 26% planted both hybrid maize and open pollinated maize seeds. On the other hand, coping strategies practiced included selling or consuming small livestock and chicken (97%), consuming maize left for seeds (93%) and reducing food intake (23%). It was clear that the effects of climate change in rural areas were severe and needed to be addressed before critical damages like loss of human life manifest. The government should ensure that farmer's knowledge on climate change and variability is increased through education to improve their adaptive capacity.

Keywords: Adaptation, Climate Change, Climate Variability, Coping, Adaptive Capacity

1. Introduction

Swaziland is a landlocked country located on the southern part of the African continent along the geographic coordinates 31°30'E and 26°30'S and has an area of 17,363 km^2,with a population of about 1.2 million inhabitants (Government of Swaziland, 2007a). It shares its boundaries with South Africa and Mozambique (Figure 1). The general climatic characterization of Swaziland is sub-tropical with wet hot summers and cold dry winters. About 75% of the annual rain falls in the period from October to March. The cold dry winters are during the months of April to September (Government of Swaziland, 2010).

The CSIRO and the MIROC models predicts a temperature increase of about 1–1.5 C and 1.5 C respectively and the average daily temperatures of an area may further increase to about 2–2.5^{0}C by 2050 for the country (Manyatsi et al., 2013). The models further estimated a 200 mm reduction in rainfall in most parts of the country. Climate change as a phenomenon has some implications for the developing countries, most of which are negative (Kandji et al., 2006).

Climate variability can be defined as the short-term variations in mean state of the climate and variations in other statistics (such as the occurrence of extremes) on all temporal and spatial scales beyond that of individual weather events. This differs from climate change which refers to a change in the state of the climate that can be identified by changes in the means and/or changes in the variability of its properties, and that persist for an extended period (IPCC, 2007). The effects of climate change includes among other things the rise in sea level, changes in the intensity, timing and spatial distribution of precipitation, changes in temperature and the frequency, intensity and duration of extreme climate events such as drought, floods and tropical storms (IPCC, 2007). The changes in temperature and precipitation have resulted in changes in land and water managements that have subsequently affected agricultural productivity in all developing countries (Kandji et al., 2006).

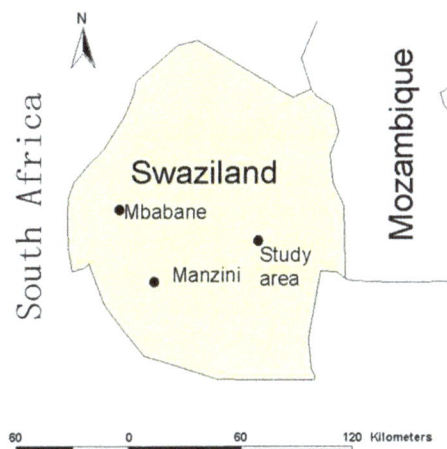

Figure 1. *Map of Swaziland showing location of major towns and study area.*

Evidence from different climate change prediction models suggests that climate change and variability are likely to significantly increase the risk of crop and livestock failures, unless significant adaptation and coping strategies are employed (Chambwera and Stage, 2010). However, rural communities lack the sufficient knowledge, financial capacity and resources to implement the best mitigation practices for sustainable agricultural production. Climate variability and change have brought changes in the timing and length of the growing season and these have affected crop growth in different parts of the country (Manyatsi et al., 2010).

Climate change adaptation and coping strategies distinguishes between farmers' short-term and long-term responses to climate variability, climate change and food insecurity (Mudzonga, 2011). Although farmer's adaptive and coping strategies may not succeed completely, they form the basis of solutions to natural disaster preparedness. Addressing the threat of increased animal and crop failure posed by climate change will require better quantification of climate change effects, greater attention to prioritizing which forms of production systems are more vulnerable, and redoubling farmers efforts in the management of land and water resources, especially in drought stricken areas (Kandji et al., 2006).

There has been a sharp decline in crop production levels and crop diversity in Swaziland, which is affecting the country's economy as it is highly dependent on agriculture (Shongwe et al., 2013). The loss in crop yield in the country was noticeably higher in the lowveld as compared to the other ecological zones. The yield reduction was due to some arable land left uncultivated due to delayed rainfall and the high risk of making loss from agriculture. About 40% of arable land in the lowveld has not been cultivated for over 10

years and rural communities were solely dependent on social interventions for food aids (Manyatsi et al., 2010).

The objective of the study was to determine the climate change impacts, and adaptation strategies in Swaziland, using Malindza, a rural community as a case study.

2. Methodology

2.1. Description of Study Area

The study was conducted at Malindza area, under the Lubombo administrative region of Swaziland (Figure 1). Annual rainfall has varied from 200 mm in 1990 and 2005 to a high of 800 mm in 2000. Since 2011, there has been a continuous decline in the annual rainfall in the area. Since the year 2000, average temperatures increased form 28^0C to a high of 31^0C in 2011 (Government of Swaziland, 2014). The area is also prone to drought and climate variability. Households rely on growing crops such as maize, sweet potatoes, beans, pumpkin, and jugo beans for livelihood. However, due to high temperatures and low summer rainfall, yield, in particular maize are poor in most years. Households in the area also rear livestock such as cattle, goat and poultry.

2.2. Data Collection

The study was conducted at Malindza chiefdom, which has four communities (Lawini, Malindza, Njobo and Sikhuphe). The total number of households in the four communities is 363 with a population of 2,548 inhabitants (Government of Swaziland, 2007a). The list of households was used as sampling frame, and they were allocated identification numbers. A sample size calculator (Creative Research Systems, 2014) was used to determine the representative households for data collection at 95% confidence level and 5% confidence interval. The representative number of households was found to be 160. The households to be sampled were randomly selected. The number of households that were randomly selected from each community is shown in Table 1.

Table 1. *Number of households that were selected for interviews from each community.*

Community	Number of households selected
Lawini	34
Malindza	50
Njobo	57
Sikhuphe	19
Total	160

A questionnaire that was developed by FANRPAN (2013) was modified for use to collect both qualitative and quantitative data. The questionnaire solicited information on household profile, agricultural productivity, understanding of climate change, climate change adaptation and coping strategies. Prior to administering the questionnaire, it was pretested by administering it to 20 households within an adjacent community for face and content validity. The questionnaire was further modified based on the feedback from the pre-test. The questionnaire was administered through interviews in each of the selected households, and the respondent was the head of each household, or any adult person who was available at the time of conducting the interviews.

2.3. Data Analysis

The collected data were coded and entered into a Statistical Package for the Social Sciences (SPSS) computer software for analysis (SPSS, 2008). Descriptive statistics that were used to analyse the data included percentages and frequencies. The primary data collected from the sampled households was used to establish respondents' perception on the impacts of climate variability and change in the area. Moreover, the data was used to determine different adaptation and coping strategies employed by households in the area. The results were presented in the forms of tables and graphs.

3. Results and Discussions

3.1. Profile of Households

The age of household heads ranged from 17 to 80 years, with 54 years being the mean for the 160 households in the study area. The results further shows that males were the most dominant heads of households as 63% were male headed while 37% were female headed (Table 2). This indicates that in the households, the final decision makers were predominant males and this has an implication on decision making regarding the adoption of climate variation and change adaptation and coping strategies. Some studies reported that female headed households were more likely to take up climate change adaptation measures as opposed to males (Shongwe et al., 2013).

Table 2. Gender of household heads (N=160).

Community	Household gender		Total
	Male	Female	
Lawini	21	13	34
Malindza	31	19	50
Njobo	35	22	57
Sikhuphe	14	5	19
Total N	101	59	160
Total %	63	37	100

The majority (53.8%) of household heads were married with another 31.3% being divorced (Table 3). Married households possess better chances of adapting to climate variability and change as they share ideas and workloads than single, widowed and divorced households (FAO, 2007). Thirteen percent of the household heads were single and 3% of these households were child headed. Many African countries were greatly affected by the HIV/AIDS pandemic and it caused serious damage to the agriculture sector, especially in areas highly dependent on labour for production (UNAIDS, 2002). In Swaziland for example, about 25% of the population was infected with the HIV/AIDS virus (WFP,

2005).

Table 3. Marital status of household heads (N=160).

Community	Marital Status of household head				Total
	Married	Single	Widow	Divorced	
Lawini	23	1	9	1	34
Malindza	19	10	18	3	50
Njobo	33	5	19	0	57
Sikhuphe	11	4	4	0	19
Total N	86	20	50	4	160
Total %	53.8	12.5	31.3	2.5	100

The level of education of households has an impact on the adoption of adaptation and coping strategies, as households with educated household heads were likely to adopt climate change adaptation strategies (Deressa et al., 2011). Farmers' education, access to extension and credits, climate information, social capital and agro-ecological settings have great influence in farmers' choice of adaptation methods to climate variability and change (Manyatsi et al., 2010). About 34% of the household heads had not attended any formal education. Sixteen percent of the of household heads attended primary school but could not complete primary education, while 13% completed their primary education, but could not go any further than that. Twenty percent attended secondary school without completion while only 2% completed secondary school. Only 11% of the population completed high school education, 3% obtained college certificates, while only 1% had attended and completed university education (Figure 2). The fact that the majority of household heads (63%) had not gone beyond primary school (Grade 7) was likely to compromise the understanding of climate change issue, and the adoption of climate change adaptation strategies.

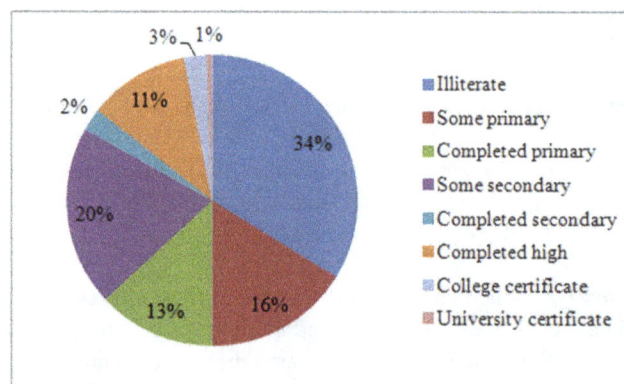

Figure 2. Education level of household heads.

The study area had a very high unemployment rate of 54%. Some adaptation and coping strategies are dependent on the availability of capital to buy farming inputs and implements (UNFCCC, 2007). Seven percent of the respondents relied on subsistence agriculture and could sell excess produce once in a while to buy farming inputs. Child headed households were more vulnerable to the effects of climate change and other shocks (Government of Swaziland, 2007b). Poverty is a key driver of vulnerability to climate change as it directly affects food security and increases community's reliance on natural resources. Child headed and female headed households have less access to credit and are less able to respond to financial constrains in cases of loss of crop due to drought and other hydrological disasters (Young, 2008).

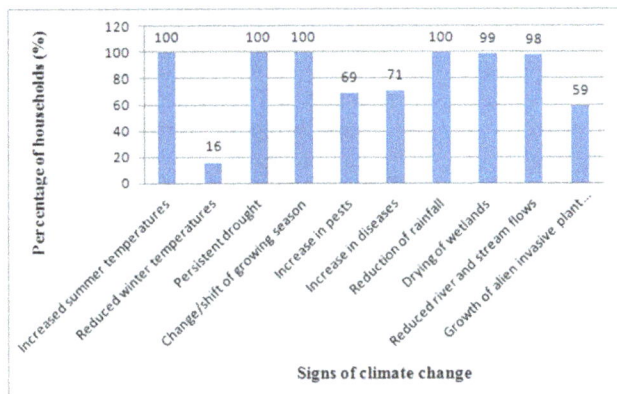

Figure 3. Perceived signs of climate change.

3.2. Knowledge and Understanding of Climate Change

More or less all the households' heads (99%) had heard about climate variability and change but the information was inadequate and of short term remedy as they only receive climatic information in the form of daily weather forecast. Sources of information included radio (92.5% of respondents), television (5.6% of respondents) and extension officers (2% of respondents). Another challenge for the communities was that weather prediction information was provided for major towns and was not relevant to rural areas. Weather forecast should provide medium to long term information to increase the warning times for climatic hazards and minimise climatic risks (Manyatsi, 2010). Proper delivery of information on climate change is important for communities to plan better adaptation and coping strategies. Moreover, the feeling of some respondents was that information relevant to improving agriculture production could be sent through the rapid increasing mobile telephone to reach the otherwise remote and unreachable areas. Encouraging farmers' field days among rural communities could also be effective for the rapid spread of new technologies (Manyatsi et al., 2010).

The perceived signs of climate change are shown in Figure 3. They included increased summer temperatures, persistent drought, shifting of growing season and drying of wetland. The results also concur with findings by other researchers who stated that there have been frequent drought occurrences in Swaziland since 1983. The most severe drought occurred in 1983, 1992, 2001, 2007 and 2008 (Manyatsi et al., 2010). The perceived climate change effects included crop failure, increased soil erosion, loss of livestock and increased risks of fires (Figure 4).

3.3. Crop Production and Productivity

Twenty four percent of the households cultivated the same area that they did some ten years ago. Another 25%

cultivated just more than 50% of their arable land. About 28% did not grow any crop (Table 4).The reasons for reduction in area under crop production or not growing any crops included shortage of farming inputs, family disputes over land, perceived lack and unreliability of rainfall and persistent drought (Figure 5). The largest land under crop production for a household was 4 ha with the highest yield for maize for one household being 1,500 kg (30 bags). The majority of households did not get enough harvest for maize to last them the whole year, as only 9% of the households harvested enough maize to last them for the whole year. About 65% of the households reported that their land productivity was poor due to loss of soil fertility and soil erosion. They attributed that to shortage of farm (kraal) manure, which used to be applied in their fields. The lack of farm manure was attributed to the reduction in the number of livestock in the study area due to drought and shortage of drinking water.

Forty four percent of households used hybrid seeds for maize. The hybrid maize seeds were purchased from input dealers. The agricultural extension officers from the Ministry of Agriculture recommended the varieties to grow. The suppliers of hybrid maize varieties in the country include Pannar and Seed-Co (Government of Swaziland, 2013). Twenty one percent of households planted open pollinated variety and about 35% planted a combination of both open pollinated and hybrid seeds (Table 5). The open pollinated variety was either recycled from previous harvest or obtained from relatives and neighbours. The hybrid seed were more tolerant to drought, as compared to the open pollinated variety. They were however expensive to purchase and needed more inputs in terms of fertiliser. The Ministry of Agriculture has realised the potential of open pollinated maize varieties (OPVs). Three OPV maize varieties; namely ZM521, ZM309 and ZM611, have been developed in an effort to improve maize production in the country (Government of Swaziland, 2011). Households in the area further intercrop maize with pumpkins, and legumes such as beans, groundnuts, and bambara nuts (jugo beans). The wide variety of crops in the field minimises risk of loss due to poor rains, and it adds variety to household's diets. Including leguminous crops also improves soil fertility and improves crop yields with minimal application of inorganic fertilizers.

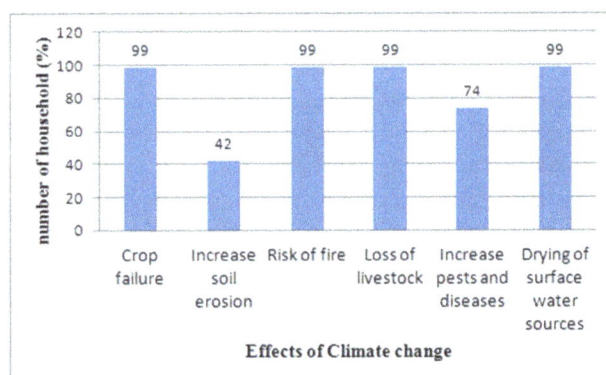

Figure 4. Perceived effects of climate change.

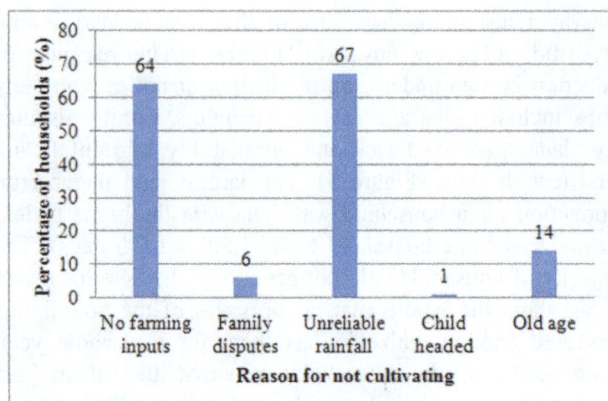

Figure 5. *Reasons for not cultivating some part of the land.*

Table 4. *Proportion of land being cultivated (N=160).*

Community	Cultivated land					Total
	Whole area	More than half	Half	Less than half	None	
Lawini	5	8	6	6	9	34
Malindza	11	15	9	3	12	50
Njobo	13	13	2	8	21	57
Sikhuphe	10	4	1	2	2	19
Total N	39	40	18	19	44	160
Total %	24.3	25.0	11.3	11.9	27.5	100

Table 5. *Variety of maize seeds grown.*

Community	Planted seed variety				Total
	No crop	Hybrid	Open pollinated	Both	
Lawini	8	11	3	12	34
Malindza	12	5	17	16	50
Njobo	21	24	3	9	57
Sikhuphe	3	11	1	4	19
Total	44	51	24	41	160
Total %	27	32	15	26	100

3.4. Climate Change Adaptation and Coping Strategies

Climate change adaptation practices include all the proactive measures or responses to the effects of climates change and variability that rural households use to reduce vulnerability (World Bank, 2003). Adoption of such measures is considered to be very important in poverty reduction especially in highly affected areas like the lowveld of Swaziland. The results indicate that about 50% of the households ploughed their fields along the contours which reduce the risks of soil erosion in cases of extreme rainfall (Table 6). The cultivation along the contours is in line with the Kings Order of 1953 (Manyatsi, 1998). Contour ploughing ensures that most water infiltrates the soil other than being lost as runoff (Manyatsi et al., 2010). Drought tolerant varieties also play a major role in improving crop yield but most farmers lacked sufficient capital to buy the seeds.

The results further showed that 29% of the households used organic fertilizer (mostly animal manure) and 20% practise crop rotation. The reduction in number of farmers using organic fertiliser is attributed by the fact that most farmers had lost their livestock due to drought and poor rainfall that affected pasture and rangelands. None of the household in the study area grew their crops using minimum

tillage practices. Minimum tillage is one of the important adaptation practices which have yielded positive results to the effects of climate change in different parts of the world (Manyatsi and Mhazo, 2014). Benefits of minimum tillage include providing soil cover which helps reduce water evaporation, reduce soil temperatures and increases water infiltration (Ngigi, 2009). Other benefits include reducing soil erosion and weed infestation in the fields, increase soil fertility, improve soil structure and reduce pest and disease infestation. It is important that farmers at Malindza adopt this practice as it seemed to have increased yield and improved crop quality in different parts of the world.

Table 6. *Climate change adaptation practices by households.*

Climate change Adaptation practices by households	Number of households	Percentage of households
Stop grass burning	153	96
Contour ploughing	79	49
Organic fertilizers	46	29
Crop rotation	32	20
Rotational grazing	4	3
Agroforestry	2	1

Table 7. *Coping strategies adopted by households.*

Climate change coping strategies by households	Number of households	Percentage of households (%)
Sell or consume seeds meant for planting next season	149	93
Borrow food or money for food	135	84
Sell or consume livestock	106	66
Change to cheaper family diets	100	63
Have some days without eating at all	36	23
Work overtime or take another job	88	55
Use money intended for investing in small business	27	17
Children discontinue school	7	4
Sell some household possessions	3	2
Sell agriculture tools or implements	2	1
Access to micro finance loans	2	1

Only 1% of the households used agroforestry which also helps in reducing excessive evapo-transpiration thus increasing water use efficiency in plant. Some of the household with livestock used rotational grazing to ensure that their cattle survive the winter (dry) season.

As one of the coping mechanisms some 73% of the households had reduced the size or the number of their meal intake (Table 7). Moreover, 63% of the households in the area opted for food products that cost less in order to cope with shortage of food. About 55% of the respondents worked overtime or took double jobs in order to provide food for the family. Some households further mentioned that their children had been forced to leave school as there was no money to pay fees. The results also show that most households had sold and/or consumed more of either small or large livestock in order to provide food for their family. About 95% of the respondents failed to reserve maize for recycling as seeds for the following season. In another research that was done for an area that had similar climatic conditions to Malindza, it was found that the dominant coping strategies to effects of drought included receiving food rations and farm inputs from Non-Governmental Organisations, Benefiting from government feeding schemes at schools and benefiting from water that was delivered to households by the National Disaster Management Agency (Vilane et al., 2015).

4. Conclusion

The results demonstrated that climate change and climate variability was affecting the Malindza area. Only 9% of the households harvested enough maize to last for the whole season, and the rest had to rely on purchased maize or food aid (which had also declined). The majority of the household heads (54%) were not employed. Over 70% of the households did not plant all the available cropping area due to problems that included lack of farming inputs and unreliable rainfall.

The climate change adaptation strategies included contour ploughing, use of organic fertilizers and crop rotation. On the other hand, coping strategies included reducing the number of meals eaten per day, selling or consuming of small livestock and selling or consuming of maize that was reserved for seeds. One of the most significant impacts of climate change encountered by the rural households at Malindza area was loss of crop yield, especially maize. Farmers were failing to produce sufficient food that could sustain them throughout the season. Apart from the impacts of crop losses, climate change had caused severe losses in livestock, adding pressure to food security among households. Also of significant impact in livestock production was the drying of surface water sources like dams, streams and rivers. Farmers were faced with challenges of increasing pests and diseases which had become a major issue due to the increasing temperatures and reduced rainfall. The late onset of rain had affected households in the sense that it delayed the availability of wild fruits and animals which households could use as food sources. The lack of proper adaptation in the area is mostly attributed to the lack of knowledge of the adaptation practices while some households' regards adaptation practices as laborious and time consuming. Some adaptation practices that have been adopted included contour ploughing, reducing the size of land planted land, practising crop rotation and use of organic fertilisers.

Acknowledgements

The authors are grateful to the Food, Agriculture and Natural Resources Policy Analysis Network (FANRPAN) for funding the finalization and publication of the research findings through the Strengthening Evidence-Based Climate Change Adaptation Policies (SECCAP) project. The authors are also grateful to the University of Swaziland for availing time and facilities to conduct the study.

References

[1] Chambwera, M. and Stage, J. (2010). Climate change adaptation in developing countries: issues and perspectives for economic analysis.

[2] Creative Research Systems, (2014). Sample size calculator. http:// www.surveysystem.com/sscalc.htm. 12/01/2014

[3] Deressa, T.T, Hassan, R.M., Ringler, C., Alemu, T., and Yesuf, M. (2011). Perception of and Adaptation to Climate Change by Farmers in the Nile Basin of Ethiopia. The Journal of Agricultural Science, Vol. 149, pp. 23-31

[4] FAO. (2007). Climate Change and Food Security: A Framework for Action. Report by an Interdepartmental Working Group on Climate Change. FAO, Rome, Italy.

[5] FANRPAN (2013). Assessemnt of Climate Change Impact and Adaptation. Questionnaire. Food, Agriculture and Natural Resources Policy Analysis Network. Pretoria, South Africa.

[6] Government of Swaziland (2007a). Swaziland Population and Housing Census. Swaziland Central Statistical office. http://unstats.un.org/unsd/demographic/sources/census/2010_ PHC/Swaziland/Swaziland_more.htm . 07/04/2014.

[7] Government of Swaziland, (2007b). Progress report on the achievement of the Millennium Development Goals. Ministry of Economic Planning and Development, Mbabane, Swaziland.

[8] Government of Swaziland (2010). Swaziland's second national communication to the United Nations framework convention on climate change. Ministry of Tourism and Environmental Affairs. Mbabane, Swaziland.

[9] Government of Swaziland (2011). HASSP community training workshop report for Swaziland. Seed quality control services. Ministry of Agriculture, Malkerns.

[10] Government of Swaziland (2013). Report on baseline study of the Swaziland seed sector. Ministry of Agriculture, Mbabane, Swaziland.

[11] Government of Swaziland (2014). Rainfall and Temperature data. Swaziland Meteorology Department. Mbabane, Swaziland.

[12] IPCC (2007). Climate Change Impacts, Adaptation and Vulnerability. IPCC WGII Fourth Assessment Report.

[13] Kandji, S.T., Verchot, L., and Mackensen, J. (2006). Climate Change and Variability in the Southern Africa: Impacts and Adaptation Strategies in the Agricultural Sector. United Nation Environmental Programme, Nairobi, Kenya.

[14] Manyatsi, A.M., Thomas, T.S., Masarirambi, M.T., Hachigonta, S., and Sibanda L.M. (2013). Swaziland.pp 213-253. In: Southern African Agriculture and Climate Change. Hachigota, S., Nelson, G.C., Thomas, T.S., and Sibanda, L.M. (Eds). International Food Policy Research Institute, Washington, USA.

[15] Manyatsi, A.M. (1998). Soil Erosion and Control Training Manual. Soil Conservation, Watershed and Dam Management Training Manual. Environmental Consulting Services, Mbabane, Swaziland.

[16] Manyatsi A.M. and Mhazo, N. (2014). Comprehensive Scoping Study of Climate Smart Agriculture Policies in Swaziland. Draft report submitted to Food, Agriculture and Natural Resources Policy Network (FANRPAN). Pretoria, South Africa.

[17] Manyatsi A.M., Mhazo N, and Masarirambi M.T. (2010). Climate variability and changes as perceived by rural communities in Swaziland. Res. J. Environ. Earth Sic., Vol 2: (3), 165-170.

[18] Mudzonga, E. (2011). Farmers' Adaptation to Climate Change in Chivi District of Zimbabwe.Trade and development studies centre. 3 Downie Avenue Belgravia, Harare, Zimbabwe.

[19] Ngigi, S.N. (2009). Water Resources Management Options for Smallholder Farming Systems in Sub-Saharan Africa. The MDG Centre for East and Southern Africa of the Earth Institute at Columbia University, New York, USA.

[20] Shongwe P., Masuku M.B. & Manyatsi A.M. (2013). Cost Benefit Analysis of Climate Change Adaptation Strategies on Crop Production System: A case of Mpolonjeni Area Development Programme (ADP) in Swaziland. Sustainable Agriculture Research: 3:1, 37-49.

[21] SPSS (Statistical Package for the Social Sciencies), (2008). SPSS for Windows. Release 17.00.SPSS Inc., Chicago, USA.

[22] UNAIDS (2002). The Status of the HIV/AIDS Epidemic in Sub-Saharan Africa. Report on the Global AIDS Pandemic.

[23] UNFCCC (2007). Climate Change Impacts, Vulnerabilities and Adaptation in Developing Countries. http://nufccc.in/resources/docs/publications, 12/11/2014

[24] Vilane B.R.T., Manyatsi A.M. and Shabangu K. (2015). Drought coping strategies at Lunhlupheko community, a semi-arid rural area in Swaziland. African Journal of Agricultural Research. Vol 10 (8), 783-788

[25] WFP, (2005). World Food Programme Annual Report 2005. www.wfp.org/../wfp-annual-report-2005. 7/10/2014

[26] World Bank, (2003). Climate change and agriculture. A review of impacts and adaptations. Agriculture and rural development department.

[27] Young, M. H. (2008). Global Climate Change: What Does it Mean for the World's Women? Population Action International (PAI) blog. www.populationaction.org/./global-climate-change-what-doe.html, 7/11/2014.

19

Biological Control of Insect Pests of Medicinal Plants - *Abelmoschus moschatus, Gloriosa superba* and *Withania somnifera* in Forest Nursery and Plantation in Madhya Pradesh, India

Premanand Balkrishna Meshram, Nahar Singh Mawai, Ramkumar Malviya

Forest Entomology Division, Tropical Forest Research Institute (ICFRE), Jabalpur, India

Email address:

meshrampb@icfre.org (P. B. Meshram)

Abstract: The present investigation was carried out to evaluate the parasitoids (*Trichogramma raoi, T. chilonis*), predator (*Chrysoperla cornea*) and biopesticides i.e. botanicals (neem based Gronim) / mocrobials (*Bacillus thuringensis* and *Beauveria bassiana*) against five major insect pests viz. *Polytela gloriosae, Anomis flava, Earias vitella, Dysdercus cingulatus* and *Aphis gossypi* of important target species of medicinal plants- A*belmoschus moschatus, Gloriosa superba* and *Withania somnifera* in forest nursery, Tropical Forest Research Institute, Jabalpur and Delakhari west Chhindwara forest division, Madhya Pradesh (India). The results revealed that *Bacillus thuringensis* 1% followed by neem based pesticide (Gronim) 1% was found to be most effective against defoliators *Polytela gloriosae* on *G. superba, Anomis flava* and shoot/fruit borer *Earias vitella* on *A. moschatus*. Neem based pesticide (Gronim) followed by *Bt* 1% was found to be most effective against red bug *Dysdercus cingulatus* on the fruits of *A. moschatus* and aphid, *Aphis gossypi* on *W. somnifera*. Predator, *Chrysoperla cornea* @ 500 per 100 sq m followed by parasitoid *Trichogramma chilonis* @1500 per 100 sq m was also found to be most effective for reduction of the larval population of defoliators *P. gloriosae* and *A. flava*.

Keywords: Insect Pests, Medicinal Plants, Parasitoids, Predator and Biopesticides

1. Introduction

Muskdana, *Abelmoschus moschatus* (Medic.) [syn. *Hibiscus abelmoschus* (Linn.)] is considered as an important medicinal plant in Indian system of medicine and perfumery. Kalihari, *Gloriosa superba* (Linn.) is used under threatened plant species and used as tonic, stomachic, anthelmintic and having the most important alkaloid colchicine. Cluster winter cherry or ashwagandha, *Withania somnifera* (L.) Dunal, is a high value medicinal plant, extensively exploited in Indian system of medicine and used as an adaptive with antistress, antioxidant, antiflammatory, mind boosting and rejuvenating properties [24]. Central India (Madhya Pradesh, Chhattisgarh and Maharashtra states) is known for their rich diversity of medicinal plants. State Forest Departments through Minor Forest Produce Federations are encouraging farmers to take up medicinal plants for additional income. During cultivation of medicinal plants. The quality and quantity of raw materials

obtained from the medicinal plants are adversely affected by the attack of number of insect pests in cultivated areas. Use of chemical pesticides is hazardous and causes environmental pollution. Therefore, environmentally safe biological pesticides are to be involved for the management of insect pests of important cultivated medicinal plants. These potential medicinal plants are always under serious threat of insect attack. The major insect pests of *A. moschatus* are defoliator, *Anomis flava* (Fab.), shoot / fruit borer, *Earias vitella* (Fab.), sap sucker, *Dysdercus cingulatus* (Fab.) and their incidence varies from 50-62.5, 30-70 and 70-90 per cent respectively [1,8,15]. While *W. somnifera* suffers attack by insect pests like white grub, *Holotrichia serrata* (Fab.), mealy bug, *Coccidohystrix insolitus* (Green.) and spotted leaf beetle, *Epilachna vigintioctopunctata* (Fab.) [14,22]. No specific information is available on the biological control of insect pests of target species of medicinal plants viz. *A. moschatus, G. superba* and *W. somnifera*. Hence, the present study was undertaken to investigate the biological control of major insect pests of target species of medicinal plants.

2. Methodology

Egg parasitoid, *Trichogramma raoi* was reared and multiplied in the laboratory of Forest Entomology Division, Tropical Forest Research Institute (ICFRE), Jabalpur, India. The cards of parasitoid, *T. chilonis,* and predator, *Chrysoperla carnea* were procured from the laboratory of Entomology, Agriculture College, Nagpur (Maharashtra, India) and released in the forest nursery (experimental plot) at TFRI, campus Jabalpur and Delakhari (West Forest Division, Chhindwara, Madhya Pradesh, India). The observations on the intensity of damage caused by two different insect pests i.e. defoliators, *Polytela gloriosae* on *G. superba* and *Anomis flava* on *A. moschatus* were taken in the parasitoid and predator released areas and compared with non released (control) plots.

The experiments on the efficacy of some biopesticides against the key insect pests were laid out in Randomized Block Design (RBD) with three replications. Biopesticides i.e. microbial agents- *Bacillus thuringiensis, Beauveria bassiana,* and botanicals- commercial neem based pesticides (Gronim) were tested for their efficacy against the key insect pests i.e. defoliator, *Polytela gloriosae* on *Gloriosa superba*; defoliator *Anomis flava,* shoot/fruit borer *Earias vitella,* fruit sucker/ red cotton bug, *Dysdercus cingulatus* on *A. moschatus* and leaf sap sucker / aphid, *Aphis gossypi* on *W. somnifera.* Data thus obtained were subjected to statistical analysis for better interpretations of results [4].

3. Result and Discussion

The results of the field experiments on the efficacy of different biopesticides against major insect pests are described separately.

3.1. Effect of Selected Biopesticides Against Defoliators, Polytela gloriosae (Fab.) on G. superba and Anomis flava (Fab.) on A. moschatus

The data pertaining to larval mortality are presented in Table 1. The data pertaining to the mortality of larvae after 7 days of treatment revealed that all the four treatments were significantly superior over control in respect of larval mortality. Observations showed that the mortality of larvae of *P. gloriosae* varied from 13.33 to 73.33 per cent. The highest mortality of larvae was observed in the treatment T4 *Bacillus thuringensis* (Dipel) 1% (73.33 per cent) followed by T1 Neem based (Gronim) 1% (60.00 per cent). All the biopesticides tested, caused larval mortality of *P. gloriosae.* Among the treatments, *Bacillus thuringensis* (Dipel) 1% followed by Neem based (Gronim) 1% was significantly superior. This trend was not recorded by [20], because the evaluation of insecticides against *P. gloriosae* was entirely different from those biopesticides tested in the present study. According to them chlorpyrifos 0.02 per cent and quinalphos 0.05% were proved most effective in rapidly killing *P. gloriosae* in laboratory conditions. Since the work was not done on the evaluation of biopesticides against *P. goriosae.* Use of biopesticides in medicinal plants for control of the insect pests [5]. The use of chemical insecticides be restricted in medicinal plants as they affect the quality of medicines [19].

In case of *A. flava,* Table 1 revealed that all the treatments were significantly superior when compared to untreated control in the protection of plants. The treatment T4 *B. thuringensis* 1% followed by T1 Neem based (Gronim) 1% ; T3 *B. bassiana* (Traps) 1×10^8 + Neem based (Gronim) 1% resulted in 72.22, 69.44, 69.44, per cent larval mortality respectively. Treatments T3 and T1 found equally effective. Fenvalerate 0.01% was found to be most effective against defoliator, *Anomis flava* on *A. moschatus* [8].

Table 1. Effect of selected biopesticides against defoliators, Polytela gloriosae and Anomis flava.

Treatment	Mean (%) larval mortality after 7 days of treatment	
	P. gloriosae	*A. flava*
T1. Neem based (Gronim) 1%	60.0	69.44
T2. *Beauvaria bassiana* (Traps) 1×10^8	13.33	68.33
T3. *B. bassiana* 1×10^8 + Neem based(Gronim)1%	26.66	69.44
T4. *Bacillus thuringensis* (Dipel) 1%	73.33	72.22
T5. Control (Untreated)	0.0	0.00
S E m ±	8.36	6.58
CD at 5%	19.29	14.35

3.2. Effect of Selected Parasitoid / Predator Against Defoliators, P. gloriosae and A. flava

The data pertaining to per cent damaged plants and reduction of population of defoliator, *P. gloriosae* on *G. superba* are presented in Table 2. The per cent incidence on damaged plants and reduction of larval population varied from 10.56 to 30.32 per cent and 4.66 to 79.44 per cent respectively. The lowest per cent incidence of damaged

plants and highest per cent reduction of larval population observed in plots released with T3 predator, *Chrysoperla carnea* (10.56 and 79.44 per cent) followed by T2 parasitoid *Trichogramma chilonis* (14.30 and 70.83 per cent). All the treatments were found to be significantly superior over untreated control. The lowest per cent incidence of damaged plants of *G. superba* and highest per cent reduction of larval population of *P. gloriosae* were observed in plots released with predator, *Chrysoperla carnea* @500 /100 sq m followed by parasitoid *Trichogramma chilonis* and *T. raoi* @1500/sq m. *T. chilonis* @1.5lakh per hectare reduced fruit borer, *Earias vitella* damage in okra [3]. Release of *T. chilonis* comprised with other treatments reduced fruit borer infestation in okra [7]. Release of *T. raoi* @ 1.25 lakh / ha to minimize the attack of teak skeletonizer *Eutectona machaeralis* in in natural teak forest areas in west Mandla forest division, Madhya Pradesh [16]. Release of parasitoid and predator against defoliator, *Polytela gloriosae* was not reported on *G. superba* so far.

In case of *A. flava*, the data pertaining to per cent damaged plants and reduction of larval population of *A. flava* presented in Table 2. The per cent incidence on damaged plants and reduction of larval population varied from 13.45 to 23.20 per cent and 58.89 to 82.22 per cent respectively after 30 days of release of parasites and predator. The lowest per cent incidence of damaged plants and highest per cent reduction of larval population were observed in plots released

with T3 predator, *Chrysoperla carnea* (13.45 and 82.22 per cent) followed by T2 parasitoid *Trichogramma chilonis* (18.25 and 71.03 per cent). All the treatments were found to be significantly superior over untreated control. The lowest per cent incidence of damaged plants of *A. moschatus* and highest per cent reduction of larval population of *A. flava* were observed in plots released with predator, *Chrysoperla carnea* @ 500 /100 sq m followed by parasitoid *Trichogramma chilonis* and *T. raoi* @1500/sq m. *T. chilonis* @1.5 lakh per hectare reduced fruit borer, *Earias vitella* of o damage in okra [3]. Release of *T. chilonis* comprised with other treatments reduced fruit borer infestation in okra [7]. Release *T. raoi* @ 1.25 lakh / ha to minimize the attack of teak skeletonizer *Eutectona machaeralis* in in natural teak forest areas in west Mandla forest division, Madhya Pradesh, India [16]. Release of parasitoid and predator against defoliator, *A. flava* was not reported on *A. moschatus* so far.

Table 2. *Effect of selected parasitoid/predator against defoliators, P. gloriosae and A. flava (after 30 days of treatment).*

Treatments	*Polytela gloriosae*		*Anomis flava*	
	% Damaged plants	% reduction of larval population	% Damaged plants	% reduction of larval population
T1. Parasitoid *Trichogrmma raoi* (1500 eggs per 100 sqm.)	18.37	46.66	23.20	58.89
T2. Parasitoid *Trichograma chilonis* (1500 eggs per 100 sqm.)	14.30	70.83	18.25	71.03
T3. Predator *Chrysoperla cornea* (500 eggs per 100 sqm)	10.56	79.44	13.45	82.22
T4. Control (Untreated)	30.72	0.00	55.98	0.00
SE m ±	1.96	3.76	5.00	7.74
CD at 5%	4.79	9.209	12.23	18.96

3.3. Effect of Selected Biopesticides Against Shoot/Fruit Borer, Earias vitella (Fab.) on A. moschatus

The data pertaining to shoot and fruit borer are presented in Table 3. All the treatments were found to be significantly superior over control in respect of infested shoots and fruits. The infestation of *E. vitella* on shoots and fruits varied from 7.90 to 25.38 and 10.28 to 28.55 per cent. The lowest infested muskdana shoots and fruits was observed in plots treated with treatment T1 *Bacillus thuringensis* 1% (7.69 and 10.28 per cent) followed by T3 Neem based (Gronim) 1% (8.30 and 12.90 per cent) as compared to control (untreated). Other treatments like release of predator *Chrysoperla carnea*, parasitoid *Trichogramma chilonis* and *T. raoi* also reduced

the infestation of shoot and fruit borer as compared with control. The lowest infested shoots and fruits of *A. moschatus* was observed in plots treated with treatment *Bacillus thuringensis* 1% followed by Neem based (Gronim) 1% as compared to control (untreated). Other treatments like release of predator *Chrysoperla carnea*, parasitoid *Trichogramma chilonis* and *T. raoi* also reduced the infestation of shoot and fruit borer as compared with control. Fenvalerate 0.01% was found to be most effective against shoot / fruit borer *E. vitella* in on *A. moschatus* [8]. Least bhendi fruit damage by *E. vitella* due to the application of *Bacillus thuringiensis* [10] . Release of *T. chilonis* comprised with other treatments reduced fruit borer infestation in okra [3,5].

Table 3. *Effect of selected biopesticides against shoot/fruit borer, Earias vitella.*

Treatment	Average % infested shoot/fruits after 30 days of treatment	
	Shoot	Fruits
T1. Bacillus thuringensis 1%	7.90	10.28
T2. Trichogramma chilonis	15.50	20.50
T3. Neem based (Gronim) 1%	8.30	12.90
T4. Trichogramma raoi	12.90	18.28
T5. Chrysoperla carnea	11.54	15.52
T6. Control (Untreated)	25.38	28.55
SE m ±	0.54	0.55
CD at 5%	1.58	1.63

3.4. Effect of Selected Biopesticides Against Red Bug, Dysdercus cingulatus (Fab.) on A. moschatus

Table 4. Effect of selected biopesticides against Dysdercus cingulatus on A. moschatus.

Treatment	Average % reduction in population after treatment	
	3 days	7 days
T1. *Bacillus thuringensis* 0.5%	15.05	30.15
T2. *B. thuringensis* 1%	34.04	49.64
T3. Neem based (Gronim) 0.5%	21.59	33.53
T4. Neem based (Gronim) 1%	66.00	83.01
T5. *Beauveria bassiana*(Traps) 1x10^8	21.66	21.66
T6. Control (Untreated)	0.00	0.00
SE m ±	7.724	6.764
CD at 5%	16.830	14.739

The data (Table 4) on field trial showed that all the treatments were significantly superior to those of the control in reducing the population of bug after 3 and 7 days of application. T4 Neem based (Gronim) 1% followed by T2 *B. thuringensis* 1% proved highly effective by reducing 66.00, 83.01, 34.04, 49.64 per cent population of bug compared with untreated control. Among others T6 *Beauveria bassiana* 1% T3 Neem based (Gronim) 0.5%), T5 *Beauveria bassiana* (Traps) 0.7% and *B. thuringensis* 0.5% gave minimum bug population up to 23.33, 36.94, 21.59, 33.53,21.66,15.05, 30.15 per cent after 3 and 7 days of application respectively. Neem based (Gronim) 1% followed by *B. thuringensis* 1% was found superior to all other treatments for reduction of bug population after 3 and 7 days of treatment. Synthetic pyrethroid fenvalerate 0.01% and was found to be most

effective against red cotton bug, *Dysdercus cingulatus* on *A. moschatus* [15]. The use of chemical insecticides should be restricted in medicinal plants as they deteriorate the quality of medicines [19]. Since the work was not done on the evaluation of biopesticides against *D. cingulatus* on *A. moschatus*.

3.5. Effect of Selected Biopesticides Against Aphid Aphis gossypii (Glover) on W. somnifera

The data presented in Table 5 revealed that the average reduction per cent of aphid population varied from 12.60 to 55.00 and 15.00 to 65.00 per cent per three leaves among the four biocontrol treatments which were significantly lower than the untreated (control). Highest population reduction of 55 and 65 per cent after 15 and 30 days of treatment was recorded in T2. Neem based (Gronim) 1% followed by T1. *Bacillus thuringensis* 1% (44.00 and 59.00 per cent) and T4. *Chrysoperla carnea* (20.30 and 35.00 per cent). All the treatments were found to be significantly superior over control in respect of per cent reduction of aphid population. Highest population reduction of aphid after 15 and 30 days of treatment was recorded in Neem based (Gronim) 1% followed by *Bacillus thuringensis* 1%. Neem, a phytopesticides obtained from *Azadirachta indica* is a safe pesticide. It proved good control against aphid. However, a suspension of neem seed was the best against *Aphis gossypii* in agriculture crops [2]. Similarly the highest mortality of aphid in *Verticillium lecanii* 0.3% compared to Neem seed kernel 4% and dimethoate on Gerbera aphid, *Myzus persicae* [9].

Table 5. Effect of selected biopesticides against Aphis gossypii on W. somnifera.

Treatment	Average % reduction in aphid population after treatment	
	15 days	30 days
T1. *Bacillus thuringensis* 1%	44.00	59.00
T2. Neem based (Gronim) 1%	55.00	65.00
T3. *Beauveria bassiana*(Traps) 1x10^8	12.60	15.00
T4. *Chrysoperla carnea* (1500 eggs)	20.30	35.00
T5. Control (Untreated)	0.00	0.00
SE m ±	5.20	5.70
CD at 5%	12.70	13.62

4. Conclusion

On the basis of field experiments it can be concluded that *Bacillus thuringensis* (Dipel) 1% followed by neem based pesticide (Gronim) 1% was found to be most effective against kalihari defoliator, *P. gloriosae,* muskdana defoliator, *Anomis flava* and shoot/fruit borer *Earias vitella*. Neem based pesticide (Gronim) followed by *Bacillus thuringensis* (Dipel) 1% was found to be most effective against red cotton bug/stainer, *Dysdercus cingulatus* on the fruits of *A. moschatus* and aphid, *Aphis gossypi* on Ashwagandha, *Withania somnifera*. Predator, *Chrysoperla cornea* @ 500 per 100 sq m followed by parasitoid *Trichogramma chilonis* @1500 per 100 sq m was also found to be most effective for reduction of the larval population of defoliator *P. gloriosae*

on *G. superba* and *A. flava* on *A. moschatus*.

Acknowledgement

Authors are grateful to Dr. U. Prakasham, IFS, Director, Shri P. Subramanyam, IFS Group Co-ordinator (Research) and Dr. N. Kulkarni, Scientist-G, Head, Forest Entomology Division Tropical Forest Research Institute (ICFRE), Jabalpur, Madhya Pradesh, India for providing the necessary facilities for carrying out this study and encouragement financial support from ICFRE, Dehradun, India is also acknowledged. Authors are also grateful to the Divisional Forest Officers, West Forest Division, Chhindwara and south forest divisions Balaghat, Madhya Pradesh, India for providing the necessary field facilities.

References

[1] Anon. (1985). "*The Wealth of India: A Dictionary of Indian Raw Materials and IndustrialProducts*". Raw Materials, Vol. 1 (Revised Edn.), 513 pp. Publication and Information Directorate. C.S.I.R., New Delhi, India .

[2] Asari, P.A.R.and M.R.G.K. Nair (2003). "On the control of brinjal pests using deterrents". *Agric. Res. J. Kerala* 10:133-135. M. Gupta. Methods of biopesticides in medicinal plants against insect pests. *Udhyamita* 11: 42-43.

[3] Ghatge, S.M.(2002)."*IPM of summer okra*". M.Sc.(Agri.) thesis submitted to Mahatma Phule Krishi Vidhyapeeth, Rahuri, M.S.

[4] Gomez, A.K. and Gomez, A.A.(1984).Statistical Procedures for Agricultural Research. JOHN WILEY and SONS, Singapore. pp.130-138.

[5] Gupta, M. (2003). Methods of biopesticides in medicinal plants against insect pests. *Udhyamita*, 11(9):42-43.

[6] Hanumanthaswamy, B.C., D.Rajgopal, A.A. Farooqui and A.K. Chakravarthy(1993). "Insect pests of *Costus speciosus* Linn. a medicinal plant". *My Forest*, 29: 158-160.

[7] Jadhav, S.S. (2003). "*Integrated pest management of okra Abelmoschus esculentus*". M.Sc.(Agri.) thesis submitted to Mahatma Phule Krishi Vidhyapeeth, Rahuri, M.S.

[8] Joshi, K.C., P.B. Meshram, S. Sambat, U. Kiran, S. Humne and G.K. Kharkwal (1992). "Insect pests of some medicinal plants in Madhya Pradesh". *Indian J. Forestry,* 15 : 17-26.

[9] Kadam, J.R., P. Mahajan and A.P. Chauhan (2008). "Studies on the potential of *Verticilium lecanii* against sucking pests of Gerbera". *J. Maharashtra Agri. Univ.*, 33 :214-217.

[10] Kharbade, S.B.,A.G.Chandele and M.D. Dethe (2003). "Management of *Earias vitella* Fab. on summer okra with *Bacillus thuringiensis* varietal products". State level seminar on pest management for sustainable agriculture, MAU, Parbhani 24 pp.

[11] Kulkarni, S.S., K.V. Naik, V.N. Jalgaonkar and A.V. Rege (2008)."Survey of pest infestations on the important medicinal plants of Konkan region of Maharashtra". *Pestology,* 32 : 31-33.

[12] Kumar, H.R. (2007). "Survey of pests of medicinal plants with special reference to biology and management of Epilachna beetle, *Henosepilachna vigintioctopunctata* Fabricius (Coleoptera: Coccinellidae) on Ashwagandha". *M.Sc. (Agri) Thesis,* Uni. Agric. Sci., Dharwad.

[13] Mathur, A.C. and J.B. Srivastava (1996). "A noted list of the insect pests of important medicinal plants of Jammu & Kashmir". Symposium on proearth Laboratory Jammu, 16 pp.

[14] Meshram, P.B.(2005)."New report of defoliator *Psilogramma menephron* on *Rauvolfia serpentina* and white grub, *Holotrichia serrata* on *Withaniasomnifera*". *Indian Forester* 131:969-970.

[15] Meshram, P.B and K.C. Choudhury (2000). "Effect of some insecticides against red cotton bug, *Dysdercus cingulatus* (Fabr.) on *Abelmoschus moschatus* Medic". *Indian Perfumer*, 44: 7-9.

[16] Meshram, P.B., N. Roychoudhury M. Yousuf and R.K. Malviya (2013). "Efficacy of indigenous egg parasitoid *Trichgramma raoi* against teak skeletonizer *Eutectona machaeralis* (Lepidoptera : Pyralidae)". *Indain J. Trop. Biodiv.*, 21(1&2):129-134.

[17] Muralibaskaran, R.K., D.S. Rajavel and K. Suresh (2009). "Yield loss by major insect pests in Ashwagandha". *Insect Environ* 14: 149-151.

[18] Oudhia, P. (2001). "First record on orange banded blister beetle, *Zonabris putulata* on safed musli". Nat. Res. Seminar on herbal conservation, cultivation, marketing and utilization with special emphasis on Chhattisgarh, 82 p, 13-24 December,

[19] Oudhia, P. (2006). "Traditional medicinal knowledge about *Polytela gloriosae* Fab. (Lepidoptera: Noctuidae) feeding on kalihari (*Gloriosa superb*a) in Chhattisgarh, India". http://www.ecoport.org

[20] Patel, V.C. and Patel, H.K.(1967). Chemical control of Polytela gloriosae(F.) on ornamental lilies. *Indian Journal of Entomology*, 29(4):397.

[21] Rai, S.N. and J. Singh. (1996). "Biology and chemical control of carmine spider mite, *Tetranychus cinnabarinus* (Boiso.) (Acarina: Tetranychidae) on Ashwagandha, *Withania somnifera*". *Pestology* 20 11: 23-27.

[22] Ravikumar, A. R. Rajendran, C. Chinnaih, S. Irulandi and R. Pandi (2008). "Evaluation of certain organic nutrient sources against mealy bug, *Coccidohystrix insolitus* (Green.) and the spotted leaf beetle, *Epilachna vigintioctopunctata* (Fab.) on Ashwagandha, *Withania somnifera*". *J. Biopesticides* 1: 28-31.

[23] Singh, A.K (2006). "*Flower Crops: Cultivation and Management*". Published by M/S New India, Publishing Agency, Delhi, 463 pp.

[24] Singh, R.H., S.K. Nath and P.B. Behere (1990)."Depressive illness a therapeutic evaluation with herbal drugs". *J. Research in Ayurvedha and Siddha*, 11: 1-6.

[25] Talukder, D., A.R. Khan and M. Hasan (1989)."Growth of *Diacrisia obliqua* (Lep : Arctiidae) with low doses of *Bacillus thuringiensis* var. kurstaki)". *BioControl* 34: 587-589.

20

Long-Term Impacts of Cultivation and Residue Burning Practices on Soil Carbon and Nitrogen Contents in Cambisols of Southwestern Ethiopia

Yacob Alemayehu Ademe[*]

Department of Plant Science, College Agriculture and Natural Resource, Dilla University, Dilla, Ethiopia

Email address:

jaxx602@yahoo.com, yacobalem@gmail.com

Abstract: Soil organic carbon (OC) and total nitrogen (N) are important indices for evaluating land management system, so that assessing the management effects on soil OC and total N dynamics is essential for addressing sustainable land productivity and environmental quality issues. This study was carried out to determine the impact of long-term agricultural practices on the distribution and contents of OC and total N in Cambisols of Abobo, southwestern Ethiopia. Three adjacent fields: Cultivated field with continuous residue burning (CB), Grassland with annual burning (GB) and the Virgin land with native vegetation (VL) were used in this study. The soil in VL was used as a reference to assess extent of changes in soil OC and total N contents. Composite soil samples were collected from four soil depths (0-15, 15-30, 30-45 and 45-60 cm) of each land units, in the triplicate sites. A one-way ANOVA and correlation coefficient analysis were used to test the mean differences of the soil OC and total N contents in each soil depth, and to determine their degree of association with other soil variables. The result revealed that the existing management practices significantly affected soil OC and total N contents in all the studied soil depths. The depletion of soil OC and total N from CB and GB fields were up to 83 and 66%, respectively, as compared to those in the VL. Changes in soil OC and total N were more pronounced in the top 30 cm depth of soil, although significant reduction observed in the 30- to 60 cm depth. The contents of deeper soil layers (45-60 cm) in burned and burned/cultivated sites were comparable, implying that immediate fire/tillage impacts were restricted to the near surface soil depth. The overall results suggest that the existing land management is not sustainable; hence, proper residue management is imperative in order to sustain the soil quality and maintain long-term productivity of the farmland.

Keywords: Prolonged Cultivation, Residue Burning, Organic Carbon, Total Nitrogen, Soil Depth

1. Introduction

Soil organic carbon (OC) is one of the largest carbon pools on the earth's surface, accounting for 2344 Pg (56%) of overall global carbon [1]. Soil OC contents play a vital role in sustaining soil fertility for crop production and environmental quality due to its effects on soil physical and chemical properties, as well as biological activities [2]. This implies that soil quality is highly linked with soil OC content, whose status depends on biomass input and management, mineralization, leaching and erosion of soil organic matter. A type of agricultural landuse and/or management practices has a significant factor that controls soil OC levels, since it affects the amount and quality of litter input, decomposition rates and the processes of OM (~58% C) balance in soils [3,

29]. For instance, removal or burning residues in the field causes considerable loss of organic C and N [8, 9] and other nutrients by volatilization [10, 11], which might adversely affect soil microorganisms [12].

Ethiopia has one of the oldest agrarian cultures in the sub-Saharan Africa with a large agriculture potential. Today, agriculture is not only the backbone of the economy, but also a major occupation for nearly 83% of Ethiopia's inhabitants [9]. This sector is, however, beset by anthropogenic factors (land use/management systems) that adversely affect its productivity [4, 5]. Improper land use and/or management practices mainly: continuous cultivation with low input, removal of crop residues (as animal feed or fuel wood) or burning plant residues as practiced under the traditional system of crop production are major contributors to the loss of soil OC and nutrients [6, 27] that aggravates the decline of quality and productive capacity of soils in various parts of the country [5, 6, 7, 13, 33]. Different rates of decline in OC and total N after cultivation of forest soils have been reported, but most contain a great reduction. In the western Alfisols of Ethiopia, Wakene and Heluf [13] reported losses in OC and total N of 79 and 76%, respectively, from virgin soils after 40 year of cultivation. Tsehaye and Mohammed [14] found that surface OC and total N concentrations in Mollisols declined by 68% and 56%, respectively, after cultivation as compared to that of natural forest counterpart. Also reported by Nega and Heluf [15], significant decreases in OC and total N up to a depth of 50cm, in fifty year continuously cropped field when compared with forestland. Elsewhere, Reeder et al. [16] reported a decline of 18 and 26% in surface soil OC and total N, respectively, after sixty year of intensive cropping in the Great Plains.

The present study targets Abobo area, southwestern Ethiopia, where agricultural pressure became higher over the past three decades, and recurrent residue burning (both cultivated and open field) practice is common traditional agriculture. In this area, fire is widely used by farmers to clear vegetation and crop residues (in every dry season), because it provides an easy and economical means of access to fields. In this regard, very little is known about the soil variables with long-term different management practices. Previous studies [17, 18] observed variations in soil properties (viz. OM content, soil depth) along landscapes of the area. However, the aforementioned researches merely compare variation soil properties as a function of landscape, and have tended to ignore the variation found due to differences in the existing land use and/or management practices. As land management effect varies depending on land use scenario (duration of land use, cropping systems, residue management) and climate of the area, it is impractical to take reported data from other areas to assess the situation in Abobo area. Hence, knowledge about the condition of soil quality indicators is vital for replenishing and maintaining soil fertility of the area.

In this work, it was hypothesized that soil OC and total N contents would vary due to the extended variation in land management practices in the site. Therefore, the objectives of this study were (i) to determine the quantity and the distribution of soil OC and total N at depths of fields affected by different land management practices (ii) to estimate the changes in OC and total N of soils under prolonged period of cultivation and burning practices that could contribute toward improved management of the agricultural land in the area.

2. Materials and Methods

2.1. Location and Site Characteristics

The study site is located in Gambella region, at the village of Abobo, southwest Ethiopia (Fig. 1). The agro-ecology of the area is typically hot to warm sub-humid lowlands [30] with mean annual rainfall of 1039.4 mm (uni-modal type),

and mean annual temperature 26.4 ^{0}C [31]. Subsistent and mixed crop-livestock agriculture characterizes the farming system. The study site is flat terrain with elevation about 550 m a.s.l, and covering about 210 ha. The soils at the site were classified as *Haplic Cambisols* and *Fluvi-Mollic Cambisols*, according to WRB soil classification system [19], with moderately deep to deep soil profile and dark reddish brown color [18], and slightly acidic pH. For more understanding of the soil condition, selected physicochemical properties of soils of the site are presented in Table 1. The area is nearly level to gently sloping, and no salinity and drainage problems existed.

2.2. Land Use/Management

The study area is known to have a three distinguishable land use and/or management practices: namely *long-term cultivated field, grassland and forest/shrub land*. Based on the land use and/or management history, all the sites were similar before, and changes in land use have been introduced since the last three decades. Information on the land use and/or management history and characteristics of the site is briefly presented as follows:

Long-term cultivated field with annual residue burning (CB): The field has been cultivated continuously for about 30 years with annual mechanized plow, and cropped predominantly to maize (*Zea mays* L.), cotton (*Gossypium hirsutum* L.) and haricot bean (*Phaseolus vulgaris* L.). Small vegetation and crop residues burning in the dry season (January-March) of every year, has been common practice of land preparation for cultivation. It covers about 154 ha of land.

Grassland with annual burning practice (GB): the site occupied by annual grasses, mainly Sudan grass (*Sorghum bicolor* subsp. *drummondii*), which is used as forage for livestock. Similar to cultivated land, this field has been undergoing annual burning practice, for rejuvenation and establishment of the grass for better forage. This site has never been cultivated for several years, and covering nearly 36 ha.

Virgin forestland (VL): located adjacent to the Grassland field, which is occupied by various shrubs and tree species such as: *Azadirachta indica, Acacia sp.* and other local tree species. In addition, plantations have been established in early 1980s.

2.3. The Research Approach and Soil Sampling

A necessary assumption made in this research approach was that soil conditions or parameters for all the sites should be similar before changes in the land use/management have been introduced. Accordingly, three treatments based on agricultural management practices: long-term cultivated field with annual residue burning (CB), grassland which had been received annual burning (GB), and virgin forestland (VL) were set up; with four sampling depths (0-15, 15-30, 30-45 and 45-60 cm) for each. Three replicate sampling locations approximately 200m apart were bordered within the specific land unit.

The replicate fields were sampled randomly over the whole area. At the time of soil sampling (in March, 2012), both under cultivation and grassland had been received residue burning, as usual as it was done before. A total of 36 composite soil samples (each composite sample made from a pool of 9 subsamples) were collected from each sampling

depth of soils of the three land management classes. The sites had comparable slope (gradient and aspect) and parent materials they developed [18], hence, similarity in slope and topographical conditions among sampling plots were maintained as much as possible in order to minimize extraneous errors.

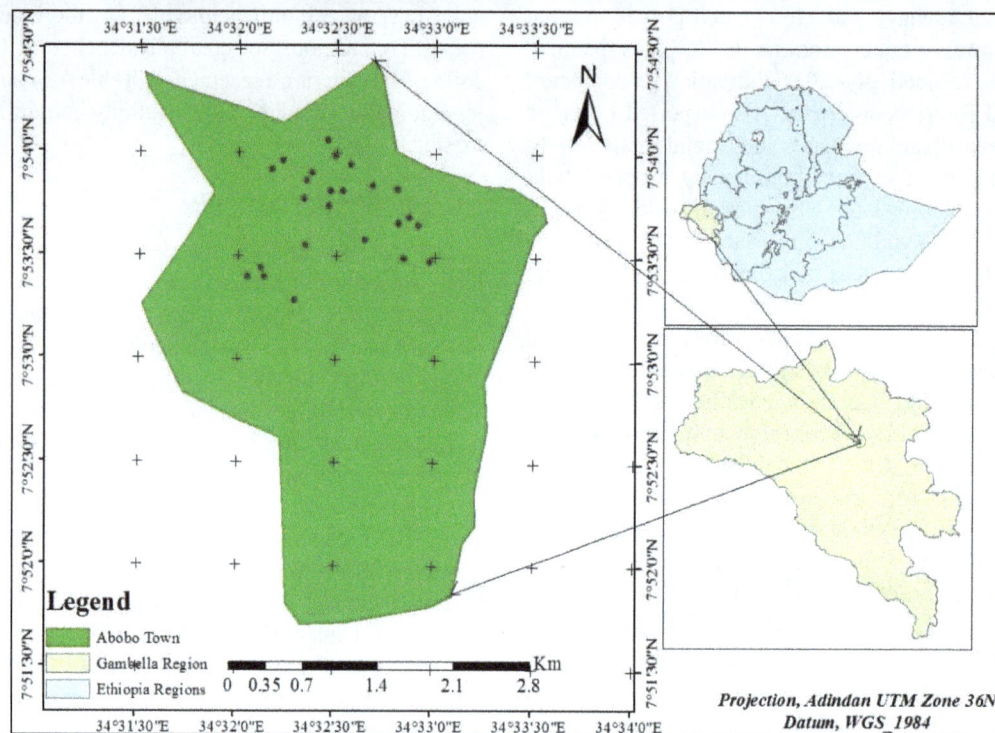

Figure 1. *Location map of the study area and the sampling points (Abobo area, southwestern Ethiopia).*

Table 1. *Mean values of the selected physicochemical properties of soils sampled from different lend use/management practices (data collected in 2012).*

Land unit #	Depth (cm)	sand	silt	clay	pH	P (mg kg⁻¹)	Na	K	Ca	Mg	CEC	PBS (%)
		--------------(%)---------------					--------------------(cmol$_c$ kg⁻¹)--------------------					
CB	0-15	15.60	37.23	47.80	6.06	9.64	0.57	1.00	9.11	9.51	31.51	64
	15-30	16.00	34.83	47.13	5.88	8.37	1.55	1.57	8.81	10.12	31.88	69
	30-45	20.16	39.00	40.80	6.06	8.41	1.97	2.13	8.10	9.51	32.31	67
	45-60	19.23	37.05	45.61	6.02	8.66	1.97	2.21	6.13	9.21	30.30	64
GB	0-15	16.40	40.10	42.63	6.37	10.73	1.59	1.59	8.67	9.32	34.17	62
	15-30	21.06	33.80	45.13	6.19	11.67	2.17	1.20	10.42	10.51	35.71	68
	30-45	16.40	40.63	43.03	6.01	9.72	2.66	2.26	8.66	10.62	35.61	68
	45-60	--	--	--	--	--	--	--	--	--	--	--
VL	0-15	13.20	32.43	54.37	6.60	27.61	0.33	1.73	10.14	15.95	38.91	72
	15-30	19.07	34.7	47.13	6.23	25.18	1.02	1.72	11.45	15.23	36.97	80
	30-45	29.8	29.83	40.36	6.33	22.82	1.57	1.22	10.93	13.18	35.67	75
	45-60	--	--	--	--	--	--	--	--	--	--	--

CB (long-term cultivated field with annual residue burning) and GB (grassland with annual burning) and VL (virgin forest soil).
-- = Not determined

2.4. Soil Sample Analysis

Composite samples collected from the respective depths of each land units were air-dried and crushed to pass a 2-mm sieve, and the selected soil physicochemical parameters were determined in the laboratory, using standard analytical methods. Soil OC was determined by the Walkley-Black wet digestion method [20]. Total N was determined using the micro-Kjeldahl digestion, distillation and titration procedure

[21]. Soil texture was determined by hydrometer method as described by Van Reeuwijk [22] after dispersion of clays with sodium hexametaphosphate. Available P quantified by Olsen method [23], as the method recommended for all types of Ethiopian soils [24]. The pH of soil samples were measured in 1:2.5 soil-water ratio. Cation exchange capacity (CEC) was determined after extracting the soil samples by ammonium acetate at pH 7.0 [25]. Exchangeable Ca and Mg in the extracts were determined using atomic absorption

spectrophotometer, whereas Na and K were quantified by flame photometer [22], and then percent base saturation (PBS) was computed as:

$$PBS = \frac{Mg^{2+} + Ca^{2+} + K^+ + Na^+)}{CEC} \times 100 \qquad (1)$$

2.5. Statistical Analysis

A one-way analysis of variance (ANOVA) was used to test differences in soil OC and total N contents and C:N ratio among land units for each soil depth, using the General Linear Model (GLM) procedure of SAS program [26]. All tests of significance were made at $p < 0.05$, and the least significant difference (LSD) test to separate means between treatments. Similarly, individual land units were evaluated for soil OC and total N contents in the four depths down to 60 cm. Pearson's correlation coefficient analyses were performed to determine the relationships between organic C and selected soil variables. In addition, the changes due to long-term cultivation and residue burning were determined by comparing the current values of C and N under CB and GB fields with those C and N content of undisturbed virgin soils (VL), assuming to have similar properties earlier to the present land uses.

3. Results and Discussion

3.1. Soil Organic Carbon Content

Analysis of variance (ANOVA) showed an overall significant effect ($p<0.01$) of land management practices on soil OC contents in all the considered soil depths (Table 2). Across the land units, the mean value of OC ranges between 12.2 g kg^{-1} and 72.7 g kg^{-1} of soil. The largest apparent variation in OC concentration was found in surface soils (0-

30 cm) of long-term cultivated field (Table 3). The reduction of OC from the topsoil (0-15 cm) under long-term cultivated field with annual burning (CB) and grassland with annual burning (GB) were 83 and 72%, respectively, as compared to the virgin forestland, VL (Fig. 2A). The amount of soil OC decline in CB, in this case, is about 32% greater than that of reported from the Great Plains soils subjected to intensive cultivation for 60 years [16]. The reduction occurred mostly in the top 30 cm of soil, although significant reductions were observed in the 30 to 60 cm depth. This is due to the fact that prolonged cultivation coupled with frequent burning of crop residues practices have been accelerated the rapid turnover rates of organic materials in both fields. Conversely, the virgin soil (VL) produced high mean value of OC (Table 3) which is mainly due to the continuous accumulation of decomposed plant and animal residues in the absence of disturbance of soil environment over a long time period.

Compared to GB field, long-term cultivated field with annual burning (CB) has produced significantly lower quantity of OC ($p < 0.01$) at the topsoil (0-15 cm depth) (Table 3), in which the difference in OC corresponds to about 38% lower than the long-term burned/grass field. However, the concentrations in the all the subsoil depths (15-60 cm) of both fields were remained statistically at par, that the contents of deeper soil layers (30-60 cm) in burned/cultivated and burned sites was similar, suggesting that immediate fire/tillage impacts were restricted to the surface soil (0-30 cm). Low quantity of OC under CB than GB field is due to the combined effect of prolonged cultivation and recurrent residue burning practices.

However, despite the absence of such soil disturbance by cultivation practices, periodic burning and removal of grasses in GB field has brought to a considerable decline in OC, compared with the neighboring virgin forest field.

Table 2. Mean square (MS) and results of one-way ANOVA for soil organic carbon (OC) and total nitrogen (TN) contents in different land management at different soil sampling depth.

Sampling depth	OC			TN			C:N		
	MS	F-value	p-value	MS	F-value	p-value	MS	F-value	p-value
0-15	3265.1	6.29	<0.0001	6.94	150.46	0.0002	69.42	114.54	0.0003
15-30	2631.3	424.1	0.001	4.48	28.14	0.004	93.54	33.57	0.032
30-45	324.9	41.64	0.002	0.69	18.21	0.009	32.80	578.88	<0.0001
45-60	29.04	16.56	0.012	0.074	26.80	0.0048	12.30	7.48	0.045

* Land units: long-term cultivated field with annual residue burning (CB), grassland with annual burning (GB) and virgin forestland (VL).

Table 3. Mean values soil organic carbon (OC), total N (TN) and C:N ratio under different land units are compared within each sampling depth.

land units	0-15 cm			15-30 cm			30-45 cm			45-60 cm		
	OC	TN	C:N	OC	TN	C:N	OC	TN	C/N	OC	TN	C:N
	---(g kg^{-1})---			---(g kg^{-1})---			----(g kg^{-1})----			---(g kg^{-1})---		
CB	12.2^c	1.7^b	7.9^c	10.9^b	1.5^b	6.8^b	10.3^b	1.1^b	9.1^c	9.4^b	0.9^b	10.1^{ab}
GB	19.6^b	1.9^b	10.8^b	14.9^b	1.6^b	9.9^b	9.7^b	1.3^b	7.9^b	9.3^b	1.1^a	7.9^b
VL	72.7^a	4.4^a	16.7^a	63.8^a	3.6^a	17.7^a	28.1^a	2.0^a	13.8^a	14.6^a	1.2^a	11.8^a
LSD(0.05)	6.12	0.49	1.76	5.64	0.48	3.78	6.36	0.44	0.54	3.01	0.11	2.91
CV(%)	7.74	8.22	6.68	8.36	9.66	14.6	17.5	13.1	2.35	11.9	4.74	12.8

Comparison is based on one-way ANOVA and LSD test; Means within column followed by the same letter are not significantly different; triplicate samples for each land management type and depth were used.

Considering the soil depths of each field, the ANOVA also indicated the existence of highly significant difference

($p<0.001$) in OC among the depths of VL, and significant difference ($p=0.05$) GB and CB fields (Table 4). Similarly, mean comparison test showed a consistent and significant decrease ($p=0.05$) with increasing depth of VL field. Compared to their respective surface soils, the amount of soil OC at 45-60 cm depth was declined by 23%, 53% and 79%, respectively, for burned/cultivated field, burned/grassland and virgin soils. Statistically higher values at 15-30 cm depth of virgin land than its deeper depths (Table 4), suggest that the layer is still the most biologically active part of the soil profile. In contrast, the decrease of OC was gradual with depth of CB as compared to other fields, this might be due to disturbances by tillage implements, which could mix different soil layers. This observation is consistent with David et al. [2] who reported the decline in OC in paired fields after 50 year of cultivation in central Illinois.

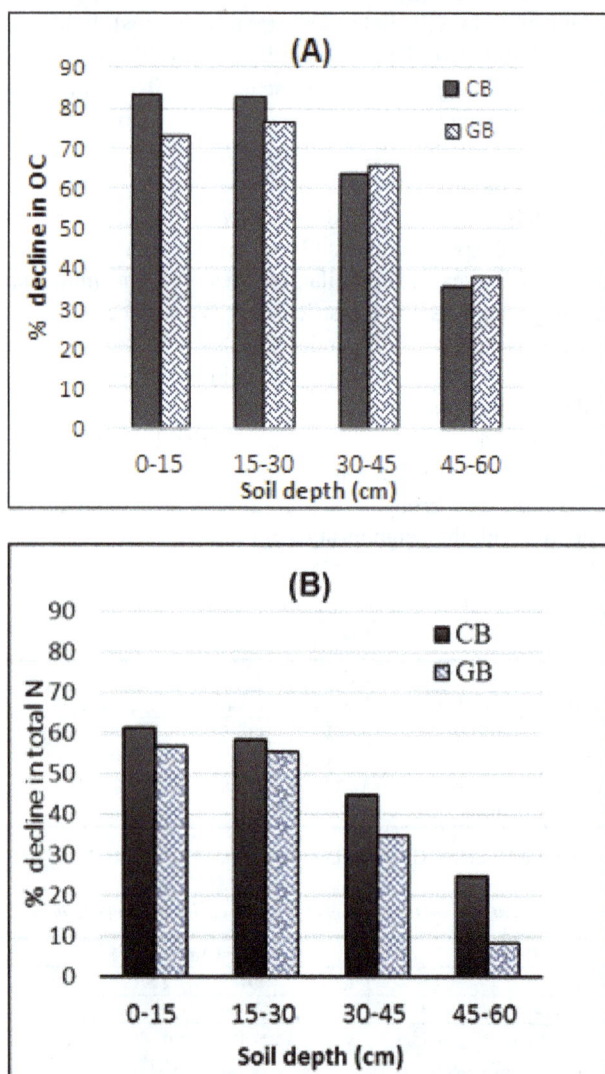

Figure 2. Percentage decline in (A) soil organic carbon contents and (B) total nitrogen contents for each soil depth of CB (long-term cultivated field with annual residue burning) and GB (grassland with annual burning), as compared to virgin forest soils.

Following the ratings for tropical soils [28], the OC concentration of topsoil of both CB and GB fields qualify for

'low' level; hence such condition could render the soil quality which directly relates to reduced agricultural production. The result supports the suggestions that most cultivated soils of Ethiopia are poor in OC content due to the effect of reduced soil OM (~ 58% C) inputs apparently complete removal of crop residues from cultivated fields [7, 15, 27]. The values obtained under VL are comparable to those reported in forest soils of southern Ethiopia [6, 32] and Alfisols of western highlands Ethiopia [13], however, the magnitude of decline in this case is higher than those reported by the authors mentioned. On the other hand, tremendously higher OC decline (up to 90%) from burned soils compared to unburned soils have also been reported [8, 11] within a short-term (< 10 year) duration.

3.2. Total Nitrogen Content

The result presented in Table 2 indicated that soil total N content was significantly (p ≤ 0.01) affected by land management in all the measured soil depths. Considering the topsoil (0-15 cm) of land units, the highest soil total N content (4.4 g kg-1) was found in VL, followed by CB (1.9 g kg-1) and GB (1.7 g kg-1). Mean comparison of soil total N also showed that the forest soil (VL) significantly differ (p < 0.05) in all the soil depths from the soil of other land units, while the values under GB and CB remain statistically at par except 30-45 cm (Table 3). Variations in soil total N were more pronounced in the top 30 cm of soil, in which the contents were normally 58 - 61% less in CB compared with VL, whereas 57% decline in total N content of GB fields (Fig. 2B). The value observed in surface of CB was about 27% greater than that of reported [17] from the nearby site that was subjected to burning and removal of crop residue for several years.

The quantity of total N was strongly associated with OC (r = +0.59; p ≤ 0.01) (Table 5), and decreased consistently with increasing soil depth of all land units. Compared to the respective soil depth of the virgin forestland, about 46 and 35% (30-45 cm) as well as 25 and 8% (45-60 cm) of total N decline were observed in the subsoil of CB and GB fields, separately. It can be suggested that the low content of total N in CB and GB fields, possibly due to the effects of continuous cultivation and subsequent burning of crop residues, which aggravated the oxidation of organic C. Comparable studies from northwest Ethiopia [33] and boreal interior of Alaska [8] also noticed that frequent burning crop residues had significantly reduced soil total N contents under cultivated land as compared to the uncultivated counterpart of the same site. Earlier study [14] also reported that surface total N contents declined by 56%, after cultivation as compared to that of natural forest counterpart.

Soil total N contents significantly (p ≤ 0.01) decreased with the increasing soil depth of VL and GB fields, but the decrease was non-significant (p > 0.05) in CB field and almost similar for soil depths (Table 4). Moreover, in all the land units, the soil depth below 45 cm was similar decline in total N concentrations, indicating that the depletion in OM occurred at the upper layers. In general, the amount of total N at the surface soils of CB and GB fields can be rated as 'very low' level [28], whereas the VL field exceeds the 'medium' range. Such decline might have significant consequences on crop production in the area, because OM supplies most of the nitrogen taken up by unfertilized crops [8].

3.3. Soil Carbon-Nitrogen Ratio

Soil C:N ratio is often considered as an indication of soil N mineralization capacity [29]. The mean value of C:N ratio of soils varied from 6:1 in CB to 18:1 in VL. In all the soil depths, the undisturbed virgin soil (VL) had significantly ($p \leq 0.01$) higher C:N ratio (17:1) than the other soils (Table 3). On the contrary, the C:N ratio of most of depths in CB field was found to be narrow, below the 'optimum' range (10 - 12:1) for arable soils [29]. In fact, natural lands usually have higher C:N ratio than prairies and cultivated areas, since cultivation/burning leads to losses of C and N, but the loss of C was much higher than the loss of N, the C:N ratio narrows. Hence, massive burning of crop residues are reasons for low C:N ratios in CB and GB fields, and this could accelerate the process of microbial decomposition of OM and N [10]. This observation is in consistent with earlier studies [14, 28] which reported greater C:N ratios in forest soils than

agricultural soils.

3.4. Association of OC with Other Soil Variables

The computed correlation coefficient indicated a strong and positive association of OC with some important soil variables: which was highly significant with CEC (r = +0.76; $p \leq 0.001$), P (r = +0.72; $p \leq 0.01$) and Mg (r = +0.61; $p \leq 0.01$), C:N (r = +0.88; $p \leq 0.001$), and positive but non-significant with pH (r = +0.46; $p > 0.05$), Ca (r = +0.37; $p > 0.05$), PBS (r = +0.22; $p > 0.05$) (Table 5). This implies that deprivation of OC by the existing management practices had also left the soils of CB and GB fields with a discernible decrease in some vital soil variables. On the other hand, increased soil variables in VL with the soil depth (Table 1) indicates vegetation restoration has implication for improvement of soil nutrients. Equally, as CEC depends largely on clay and OM contents of the soil [29], continuous burning of organic residues had depleted soil OM that attributed to reduction in CEC of soils in CB and GB compared to the undisturbed VL counterpart. The findings are in harmony with the earlier studies [12, 18], who reported higher reduction of CEC, PBS and P contents of soils under comparable management.

Table 4. Mean values for soil OC, total N and C:N ratio of each land units with the soil depth.

Sampling depth (cm)	CB			GB			VL		
	OC	TN	C:N	OC	TN	C:N	OC	TN	C/N
	-----(g kg⁻¹)----			-----(g kg⁻¹)----			-----(g kg⁻¹)----		
0-15	12.2ᵃ	1.7	7.2ᵇ	19.6ᵃ	1.9ᵃ	10.8ᵃ	72.7ᵃ	4.4ᵃ	16.7ᵃ
15-30	10.9ᵇᵃ	1.5	6.8ᵇ	14.9ᵇ	1.6ᵇ	9.9ᵃ	63.8ᵇ	3.6ᵇ	17.7ᵃ
30-45	10.3ᵇ	1.1	9.1ᵃ	9.7ᶜ	1.3ᶜ	7.9ᵇ	28.1ᶜ	2.0ᶜ	13.8ᵇ
45-60	9.4ᵇ	0.9	10.2ᵃ	9.1ᶜ	1.1ᶜ	7.9ᵇ	14.6ᵈ	1.2ᵈ	11.8ᵇ
Mean (60cm)	10.5	1.3	8.3	13.5	1.5	9.1	44.8	2.8	15.1
LSD	2.05	-	1.55	3.68	0.17	1.78	8.22	0.69	2.79
Significance	*	NS	**	*	**	**	***	***	**
CV	9.78	6.1	9.7	14.3	5.97	9.6	9.18	12.1	9.31

Soil OC, TN and C:N values of the sampling depths are compared within land unit; comparison is based on one-way ANOVA (LSD test; p<0.05); values followed by the same letter are not significantly different.
*, **, *** indicates that the ANOVA is significant at 0.05, 0.01 and 0.001 probability levels, respectively; NS = non-significant.

Table 5. Pearson's Correlation Coefficients for the selected soil chemical parameters.

Parameters	pH	TN	C/N	P	Na	K	Ca	Mg	CEC	PBS
OC	0.46	0.59**	0.88***	0.72**	-0.64***	-0.14	0.37	0.62**	0.76***	0.22
TN	0.51*	1.00	-0.18	-0.18	-0.51**	-0.42	0.08	0.02	-0.05	-0.20
C/N	0.31	0.02	1.00	0.88***	-0.49**	-0.02	0.36	0.56**	0.61**	0.25

*, **, *** indicates that the correlation is significant at 0.05, 0.01 and 0.001 significant levels, respectively.

4. Conclusions

Assessing long-term land management effects on soil OC and total N dynamics is essential for addressing sustainable land productivity issues. Results of the present study showed that the existing management practices significantly affected soil OC and total N contents and distribution in all the studied soil depths. Prolonged cultivation coupled with residue burning practices had severely depleted soil OC and total N contents of the cultivated field, as compared to the

uncultivated grassland site. Similarly, great declines in soil OC and total N content was found in the grass field that has been undergoing annual burning practice, compared with the adjacent virgin forest field. Variations in soil OC was more pronounced at topsoil, and the contents of deeper soil depths in burned and unburned/cultivated sites was more or less similar, suggesting that immediate fire/tillage impacts were restricted to the surface soil depth. On the other hand, positive and strong association of OC with some soil variables as well as increased soil variables in forest field with its soil depth indicates vegetation restoration has implication for improvement of soil nutrients. Since soil organic matter is a key resource of soil nutrients for plant growth, soil structural stability and carbon stock, it must be well managed if agricultural activities were to be sustainable. Thus, precautions should be taken with residue burning to avoid loss of plant nutrients around the root zone, and improvement in soil management (such as fallowing, biomass transfer) should be implemented to ensure sustainability of the farming system. Because, the adverse impacts of residue burning and intensive cultivation are not only dropping input of biomass OC, but also reduction in nutrient cycling, which in turn brought to decline in soil quality.

Acknowledgements

Funding was provided by the Rural Capacity Building Project (RCBP) for training and research in Ethiopia. My sincere thanks goes to the anonymous field and laboratory assistants for their unreserved collaboration in collecting samples and analyzing the soil parameters. Also, Mr. Ararsa Boki and two anonymous reviewers, who provided helpful suggestions on the manuscript are highly acknowledged.

References

[1]   Jobbágy, E.G., Jackson R.B. The vertical distribution of soil organic carbon and its relation to climate and vegetation. *Ecological Applications*, 10: 423-436, 2000.

[2]   David, M.B. Gregory F.M., Robert G.D., and Rex A.O., Long-Term Changes in Mollisol Organic Carbon and Nitrogen. *J. Environ. Qual.*, 38: 200-211, 2009.

[3]   Mikhailova, E.A., Bryant R.B., Vassenev, S.J and Post. C.J., Cultivation effects on soil carbon and nitrogen contents at depth in the Russian Chernozem. *Soil Sci. Soc. Am. J.*, 64: 738–745, 2000.

[4]   Girma T., Land Degradation: A Challenge to Ethiopia. *Environmental Management*, 27(6), pp. 815-824, 2001.

[5]   Girmay G., Singh B., Mitiku H., Borresen T. And Lal R., Carbon stocks in Ethiopian soils in relation to land use and soil management. *Land Degrad. Develop.*, 19: 351-367, 2008.

[6]   Lemenih M., Karltun E., Olsson M., Assessing soil chemical and physical property responses to deforestation and subsequent cultivation in smallholders farming system in Ethiopia. *Agriculture, Ecosystems and Environment*, 105: 373-386, 2005.

[7]   Aklilu A, Stroosnijder L., and Graaff J., Long-term dynamics in land resource use and the driving forces in the Beressa watershed, highlands of Ethiopia. *Journal of Environmental Management*, 83: 448-459, 2007.

[8]   Neff, J.C., Harden, J.W., and Gleixner G., Fire effects on soil organic matter content, composition, and nutrients in boreal interior Alaska. *Can. J. For. Res.*, 35: 2178-2187, 2005.

[9]   United Nations Country Team March. Ethiopia United Nations Development Assistance Framework 2012 to 2015, 2011

[10]   Zhao H, Daniel Q, Tong Q., Xianguo L., Guoping W., Effect of fires on soil organic carbon pool and mineralization in a Northeastern China wetland. *Geoderma*, 190: 532-539, 2012.

[11]   Stan V., Fîntîneru G., Mihalache M. Multicriteria analysis of the effects of field burning crop residues. *Not. Bot. Horti. Agrobo*, 42(1): 255-262, 2014.

[12]   Graham MH., Haynes RJ., Meyer JH. Soil organic matter content and quality: effects of fertilizer applications, burning and trash retention on a long-term sugarcane experiment in South Africa. *Soil biology and biochemistry*, 34: 93-102, 2002.

[13]   Wakene N., and Heluf G. Influence of land management on morphological, physical and chemical properties of some soils of Bako, Western Ethiopia. *Agropedology*, 13(2): 1-9, 2003.

[14]   Tsehaye G., Mohammed A., Effects of Land-Use/Cover Changes on Soil Properties in a Dryland Watershed of Hirmi and its Adjacent Agro Ecosystem: Northern Ethiopia. *International Journal of Geosciences Research*: 1(1): 45-57, 2013.

[15]   Nega E., and Heluf G., Effect of land use changes and soil depth on soil organic matter, total nitrogen and available phosphorus contents of soils in Senbat watershed, western Ethiopia. *ARPN Journal of Agricultural and Biological Science*. 8(3): 206-212, 2013.

[16]   Reeder, J.D., Schuman, G.E., and Bowman, R.A. Soil C and N changes on conservation reserve program lands in the Central Great Plains. *Soil Tillage Res.*, 47: 339-349, 1998.

[17]   Yacob A. Soil Characterization for Sustainable Land Management: Potentials and constraints of Ethiopian lowland soils. LAMBERT Academic Publishing, Germany. pp.65-96, 2012.

[18]   Yacob A., Heluf G., and Sheleme B. Pedological characteristics and classification of soils along landscapes at Abobo, southwestern lowlands of Ethiopia. *J. Soil Sci. Environ. Manage.* 5(6): 72-82, 2014.

[19]   IUSS Working Group. World Reference base for Soil Resources 2006. 2nd ed. World Soil Resources Reports No. 103. FAO, Rome, 2006.

[20]   Walkley, A. and Black, C.A. An examination of the Degtjareff method for determining soil organic matter and modifying the chromic acid method. *Soil Science*, 37: 29-38, 1934.

[21]   Bremner J.M. and Mulvaney C.S. Nitrogen total, pp 595-624. *In*: A.L. Page (ed). *Methods of Soil Analysis, Part II. Chemical and microbiological properties*. 2nd ed. American Society of Agronomy, Wisconsin, 1982.

[22]   Van Reeuwijk, L.P. (ed.). Procedure for soil analysis, 6th ed. International Soil Reference and Information Centre, Wageningen, the Netherlands, 2002.

[23]   Olsen, S.R., Cole, C.V., Watanabe, F.S. and Dean. L.A. Estimation of available phosphorus in soils by extraction with sodium bicarbonate. USDA circular 939. U.S. Govt. Printing, Washington D.C. pp.1-19, 1954.

[24]   Tekalign M., and Haque, I. Phosphorus status of some Ethiopian soils. III. Evaluation of soil test methods. *Tropical Agriculture*, 68: 51-56, 1991.

[25]   Chapman, H.D. Cation exchange capacity. *In*: Black, C.A., Ensminger, L.E. and Clark, F.E. (eds.). Methods of soil analysis. *Am. Soc. Agro.*, 9: 891-901, 1965.

[26] SAS Institute Inc. SAS Users Guide. Version 9. SAS Institute Inc., Cary, NC. USA, 2002.

[27] Gebeyaw T. Soil fertility status as influenced by different land uses in Maybar, north Ethiopia. MSc Thesis. Haramaya University, Ethiopia. 56p., 2007.

[28] Landon, J.R. (ed.). Booker tropical soil manual: A handbook for soil survey and agricultural land evaluation in the tropics and subtropics. Longman, 1991.

[29] Havlin, J.L., Tisdale, S.L., Nelson, W.L., and Beaton, J.D. Soil Fertility and Fertilizers, 6th ed. Macmillan Publishing. USA, pp.85-196, 1999.

[30] MoA (Ministry of Agriculture). Agro-ecological zones of Ethiopia. Natural Resource Management and Regulatory Department, Ministry of Agriculture, Addis Ababa, Ethiopia. 1998.

[31] NMSA (National Metrological Service Agency). Report on temperature and rainfall distribution for Abobo District. Abobo Metrological Office, GNRS, Ethiopia, 2012.

[32] Yifru A. and Taye B. Effects of landuse on soil organic carbon and nitrogen in soils of bale, Southeastern Ethiopia. *Tropical and Subtropical Agroecosystems*, 14: 229 – 235, 2011.

[33] Habtamu K, Husien O, Haimanote B, Tegenu E, Charles F, Amy S, Tammo S. The effect of land use on plant nutrient availability and carbon sequestration. pp 208-219. Proc. 10th conf., March 25-27, Addis Ababa, 2009.

21

Physiochemical Parameters Analysis to Get an Upgraded Composting System

Edson Edain González-Arredondo, Juan Carlos González-Hernández[*]**, Jorge Rodríguez-López, Christian Omar Martínez-Cámara**

Laboratory of Biochemistry, Department of Biochemical Engineering, Technological Institute of Morelia, Lomas de Santiaguito, Morelia, Mexico

Email address:

jcgh1974@yahoo.com (J. C. González-Hernández)

Abstract: Physicochemical determinants parameters were analyzed in different stages of organic matter decomposition, using various chemical and biological treatments: Manure, Legumes, Mineral solution, and Vermicompost. Specifically, we studied the importance of decisive physicochemical parameters for obtaining an improved composting system. To do this, were used different techniques, such a C/N ratio, pH, organic matter content, atomic absorption to determining concentration of several mineral, ashes quantification for moisture content and temperature measurement. The vermicompost was the most effective treatment for decomposition of matter, achieving speed up the composting process just 35 days, accounting for 5 months the minimum estimated time to have a complete degradation using a conventional composting system, this represents a decrease of 23.3%, values obtained mainly from C/N ratio were close to 25:1 (25%), a final pH of 8.2, a percent of organic matter lower than 48%, and a concentration of minerals and heavy metals within the norm.

Keywords: Organic Matter, Mineral Solution, C/N ratio, Vermicompost, Manure, Organic Fertilizer, Conventional System

1. Introduction

Municipal Solid Waste represent a potential economic development to be possible to generate monetary resources from its use and, above all, achieving lower environmental movement of pollutants and greenhouse gases, which convey disease vectors and the accumulation in the disposal sites [1]. The generation of municipal solid waste has increased by 66.86% during the decade of 2001-2012, going from 31488.51 to 42102.75 (thousand tons), with a production of 1.12 Kg per capita/day. Composting is a viable option for solving the problem of municipal solid waste alternative, whereas the organic fraction covers between 48 and 55% of the total, approximately 22,584.4 tons [1]. That's the reason they have implemented various measures to counter the increasing fraction of the municipal solid waste. Methodologies such as those used by Amador et al., 2011, who used TiO_2 as a chemical agent for the degradation of organic matter in the effluent water [2]. As a result, there has been discussion about the advantages and disadvantages that come with using chemical treatments.

Compost is a fertilizer obtained from the bacterial degradation of organic matter, it is an odorless, stable humus-like material that does not pose a health risk to the natural and social environment. Currently a small percentage is recovered to 50% of potentially recyclable waste [3].

Within conventional composting processes various treatments have been applied to try to optimize the decomposition of organic matter not succeeding in most occasions. However, there are chemical and biological treatments that can greatly accelerate the rate at which waste is broken down, with the use of Californian red worms (Eisenia foetida) and legumes, especially alfalfa and lentils because of their high binding capacity nitrogen when performing interactions synergism with nitrogen-fixing microorganisms [4].

The aim of this study was to design a treatment process for the organic fraction of municipal solid waste by composting, which would reduce the processing time compared to conventional methods.

2. Materials and Methods

The research was conducted at the facilities of the Technological Institute José María Morelos y Pavón, located in the city of Morelia, Michoacán. In a land with an area of 63 m², with an approximate slope of 6° west (Source: Google Maps - ©2013 Google), allowing direct this leachate that occurred towards the pits intended for storage.

2.1. Selection of Treatments

Manure, legumes, mineral solution and vermicomposting: 4 different treatments for composting systems were used. These were selected because of the various antecedents are as to their use for the decomposition of organic matter. All were compared against a conventional composting system (A), which consists of a stack of layers of waste soil and plant debris, without the intervention of any outside treatment or benefits affecting microbial activity present. Identifying treatments: manure (B), legume (C) mineral solution (D) and vermicompost (E) (Mexican standard NMX-FF-109-SCFI-2008, which establishes the procedure for conducting a vermicompost).

2.2. Establishment and Fitness Site

Size beds estimated one third of the volume recommended by the Municipal Composting Manual [3], measures 0.8 meters were established long, 1 meter wide and 0.2 meters high. It also considered the volume detracting from the beds the polyethylene layer and other components of fitness, increasing about 5 centimeter in each direction. Site cleanup because it had plenty of waste building materials and excess vegetation was conducted and used to establish perimeters and stuff like coffee, respectively. Having cleaned the soil were measured and marked the areas allocated to the beds and

lagoons, this according to a scaling in which took into account the volume of organic matter and the number of beds.

The pits were excavated in triplicate for treatments and white, for a total of 15 beds. Also the drainage system and leachate containment was installed, for which hoses introduced at the bottom of the pit and connected in series between the graves of the same treatment and directed towards their respective lagoon to contain the leachate were used there. Polyethylene impermeable layers are placed to prevent the leachate from leaking to the ground. This was done both in the pits for the beds to the leachate lagoons.

2.3. Collection, Sorting and Processing of Organic Matter

The organic fraction was used in this work was obtained from the municipal solid waste produced in the Supply Center in Morelia, Michoacán, Mexico. The first step was to characterize the waste according to the Mexican standard NMX-AA-022-1985, used for characterization of solid waste. Once the characterization is noted that only green waste containing organic matter, that is, food waste as tomatoes, lettuce, etc.

The residues were ground using a mill 5.5 HP® Central Machinery brand with certain amounts of organic coffee (garden waste) to achieve adequate homogeneity. A particle size of 10-50 mm suitable for the decomposition process was reached [5].

2.4. Armed Composting Beds

Plates were installed on the bed to facilitate aeration once the compostable material is placed. The beds were assembled in layers depending on the type of treatment at issue, differences exist only in the method of treatment when treated with earthworms and legumes (Table 2.1).

Table 2.1. Establishment and armed of composting beds.

Treatment	Layer					
Target	Ground	Organic matter*	Ground	Plastic layer		
Manure	Ground	Cow manure	Organic matter	Ground	Plastic layer	
Legumes	Ground	Organic matter*	Alfalfa and lentils	Ground	Plastic layer	
Mineral solution	Ground	Organic matter*	Ground	Mineral solution	Plastic layer	
Vermicomposting	Ground	Organic matter*	Earth worms	Organic matter*	Ground	Membrane

*Organic matter was composed by garden trash and wood, 50-50 proportion.

2.5. Determinations

Each determination was performed in triplicate to find significant statistical differences as well as their corresponding standard deviations. For matters relating to the analysis of the results obtained the statistical method of Tukey confidence intervals set at 95%, to thereby obtain statistically which of the 4 treatments the process of decomposition and mineralization are best performed organic matter to organic fertilizer. Parameters of temperature, pH, carbon-nitrogen ratio (C/N), and percentage of organic matter; mineral concentration and moisture content were measured at intervals of 4 days during the first 2 weeks after the samples were taken every 7 days up to a total of 5 samples over a

period of 35 days. The values of these measurements were obtained by Mexican standards and technical methodologies: pH (NMX-AA-025-1984, with a range between 7.5-9), C/N (NMX-AA-067-1985, stablishing a 25% as a allowed value), percentage of organic matter (NMX-AA-021-1985), and moisture content (NMX-AA-016-1984, having an allowed range between 50-60%), the temperature measurement was performed with the use of a bayonet thermometer, likewise mineral quantification was performed with the use of equipment atomic absorption Perkin Elmer, model 400A Analyst in the Instituto Tecnológico de Estudios Superiores de Tacámbaro.

3. Results and Discussion

3.1. pH Determination

The initial pH value was 6.45, increased as the passage of time due to the formation of ammonia from the decomposition of organic matter, which causes an increase of the pH to between 8.5 and 9 during the first few days remaining in this range about twelve days, then begin to decrease due to the formation of organic acids from aliphatic molecules that are metabolized by microorganisms, mainly bacteria and fungi [6]. The optimum pH for the decomposition sort is developed in a range of 6.5-8.0, as a reference the results obtained in the present study were within those values [7]. Shown graphically that a significant difference in behavior of the pH in each composting systems depending on the treatment that was used. Was observed between the results obtained for the treatment

with manure and those obtained for the mineral solution, in which case the pH values when they reached the stable value was between 8.5 and 9 had different changes (Fig. 1a). For the treatment with mineral solution pH drop was with a much stronger tendency than any of the other treatments including the "target" system. Statistical analysis showed that pH do not determinate which of 4 treatments is the best because there is no significant difference between theirs.

Vasquez Diaz reported in 2010 composting systems optimized by using native microbial consortia, for pH values between 8.6 and 9.6 for systems where minerals also were obtained solutions were added, these values reported as stable at a time of 40 days approximately, being similar to those obtained in this investigation for the system in which the mineral solution used values [8].

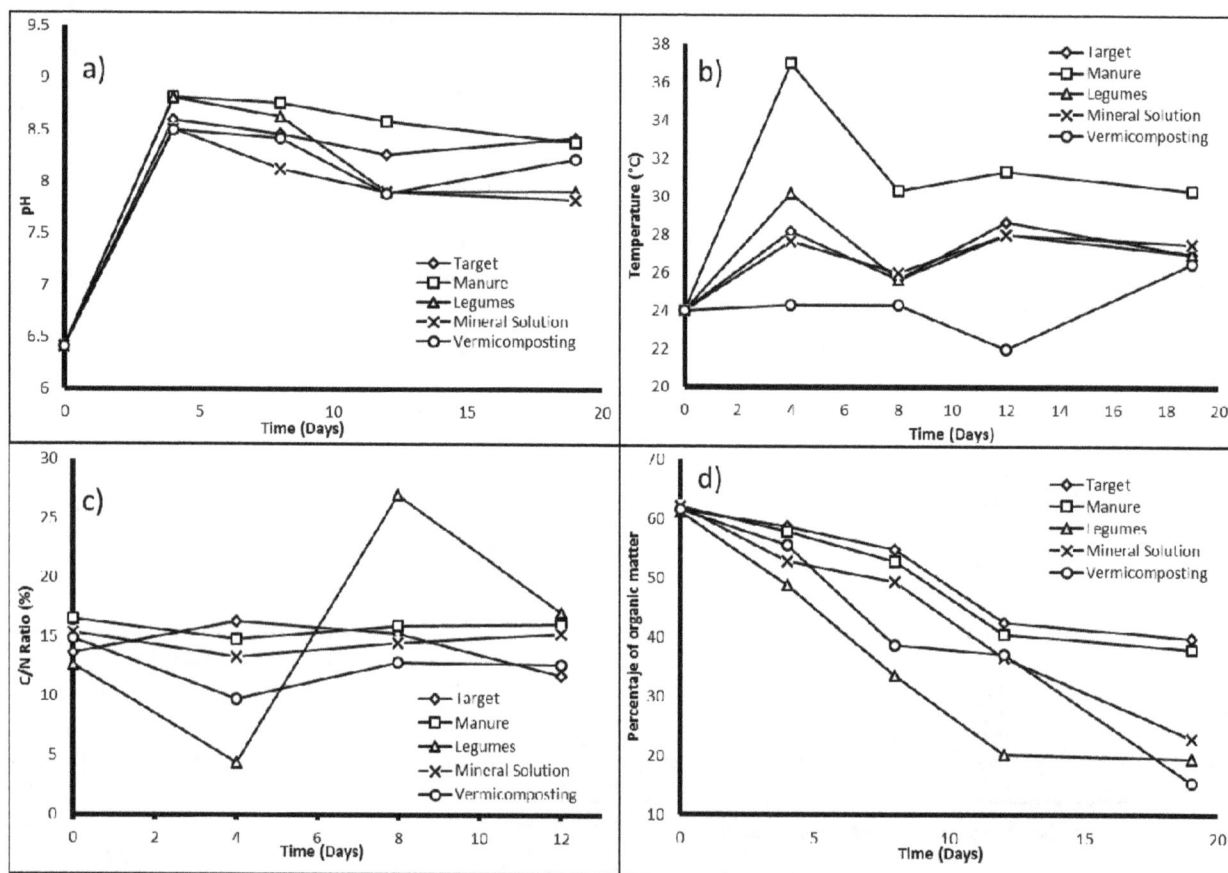

Figure 1. a) pH behavior during the composting period. Representative sampling every 4 days. Ex-situ measurements on a wet basis; b) Temperature variation. In- situ measurements using thermometers with bayonet; c) Study of the C/N ratio in each of the systems through decomposition time. Measurements from the fourth day for proper homogenization of nutrients; d) Determination of the reduction of organic matter. Acid digestions aerobic digester using phosphoric acid as the agent. White (◊), legumes (Δ), Vermicompost (○), manure (□), mineral solution (×). Data are representative of 3 experiments per sample.

3.2. Temperature

Temperature control was difficult, due to high ambient temperatures developed during the course of the experiment, reaching values of up to 30 °C. These temperatures affected the moisture retention in the soil. An initial value in the range of 24 to 30 °C for vermicomposting systems, legume and mineral solution was obtained, while for the system with manure the initial temperature was 38 °C (Fig. 1b). Rink

(1992) reported an optimal temperature range for proper decomposition of organic matter comprising values between 65-75 °C, also in giving a range which can still be considered as acceptable temperature, 55-75 °C [6]. An important factor to be considered is the use of appropriate thermometers, in this case two types, and one laboratory conventional bayonet used. Martinez et al., 2013, reported an analysis of the relationship between the size of the pits used for composting systems and temperatures that can be achieved, explained that the size of a compost pile will influence the duration and intensity thermophilic phase, and therefore the duration of the bioprocess. His research is the ideal dimensions were obtained to reach thermophilic temperatures within the range, these dimensions were only in relation to the height of the pits, 0.5m and 1m being the most suitable [9]. The decrease of the temperature range, compared with that shown in the Manual composting of Secretary of Environment and Natural Resources (SEMARNAT), is possible due to the scaling of composting systems and the concentration of organic matter was used.

3.3. Determination of the C/N Ratio

This is one of the most important factors in determining the degree of maturity of compost; Soto (2002) reported optimum values of this ratio, with the optimum range of 25-30% [7]. In this case determinations began the fourth day to allow it to begin to consume the carbon and nitrogen, for every four days to perform the sampling. After this time it was found that the systems had a lower ratio of 16%, this represented a loss of nitrogen we lack carbon. Espinoza (2013) reported a systems research reactors using aerobic composting diapers outpatient treatment, obtaining values of C/N in the range 10 to 12, with an average value of 20 as optimal [10], according to standard NMX-FF-109-SCFI-2008, which stablished optimal C/N ratio values. The only system that showed good ratio C/N in which alfalfa was used, with a value of over 25% (Fig. 1c), a value that is between the average required for the degradation of organic matter, this is thanks to that legumes have a symbiotic relationship with nitrogen-fixing microorganisms [4]. After 15 days, the C/N ratios stabilized up to values between 20 and 12% giving this results in a stable system and a partially complete decomposition, but statistically, there is no difference between the four treatments and all are grouped within the same range due to the proximity of their respective mean values (Table 3.1).

Table 3.1. C/N ratio, Statistical analysis.

Treatment	N	Media	Group
Legumes	3	16.211	A
Manure	3	15.660	A
Target	3	14.450	A
Mineral solution	3	14.416	A
Vermicomposting	3	11.777	A

Simultaneous confidence intervals of 95% Tukey
All comparisons paired-levels between treatments
Individual confidence level = 99.18%

3.4. Determination of the Percentage of Organic Matter

Percentage of composted organic matter refers to the amount of substrate that exists to be consumed by the microorganisms present. In this investigation the average

initial percentage organic matter, obtained by a random sampling of 3 of the 15 beds, was 61.7, almost starting to decrease steadily. Legume systems were those with a more rapid decline. However, vermicomposting treatment achieved the highest reduction in the percentage reaching a value of 15% (Fig. 1d), this representing a reduction of almost 47% over a period of 20 days. Statistical comparison shows a similarity between the four treatments (Table 3.2), however, according to the trend following treatments with vermicompost and manure system, these are considered the most effective treatments.

Table 3.2. Organic matter percentage, Statistical analysis.

Treatment	N	Media	Group
Manure	4	15.886	A
Legumes	4	15.340	A
Mineral solution	4	14.662	A
Target	4	14.255	A
Vermicomposting	4	12.560	A

Simultaneous confidence intervals of 95% Tukey
All comparisons paired-levels between treatments
Individual confidence level = 99.18%

3.5. Concentration of Minerals

During the process of decomposition and mineralization of organic matter in compost piles study the concentration of different minerals that are present during this process, phosphorus, sodium, calcium, and potassium, was performed, as well as other nutrients that rather than are considered pollutants, including zinc, iron, copper, and manganese were found. In the case of the concentration of phosphorus, which is one of the most representative nutrients found in the soil naturally allowing correct growth of vegetable organisms. Analyses were performed by atomic absorption using samples taken only at the end of the composting process to assess the quality of the end product and analyze the possibility of marketing it. Table 3.3 shows the concentrations obtained; was observed that fall within the ranges reported by several authors in similar investigations, e.g. Meneses (2012) reported values of concentrations for potassium (0.793 mg/L), sodium (0.217 mg/L), zinc (77.29 mg/Kg, with a concentration of 200 mg/Kg the maximum allowable) [11], being within the limits established in the Mexican standard NMX-FF-109-SCFI-2007. Paul and Clark (1996) report phosphorus values in a range of 0.15-1.6% [12]. While the concentrations reported in Table 5 within that range. Soto (2002) reports different concentrations in a study using composting systems with coffee pulp and mucilage, obtaining values of 0.27% and 0.25%, respectively [7]. In the same investigation concentrations of some heavy metals such as iron (3413.53 ppm) and manganese (155.17 ppm), close to those obtained for the 57260 ppm Fe and 500 ppm Mn values are reported.

Table 3.3. *Mineral concentrations.*

Treatments	Samples (g)	Zn*	Ca	Mg	Mn*	K	Na	Cu*	Fe*	P**
		Units: mg/g (dry basis)								
A1	0.50	0.06	1.74***	4.55	0.63	18.75	1.81	0.04	58.30	0.24
A2	0.50	0.05	5.18	4.32	0.42	16.52	1.96	0.04	56.11	0.36
A3	0.50	0.06	6.57	3.80	0.46	18.81	1.69	0.04	57.38	0.31
Media		0.06	5.88	4.22	0.50	18.03	1.82	0.04	57.26	0.30
B1	0.50	0.10	1.52	3.77	0.70	14.72	1.88	0.05	60.30	0.51
B2	0.50	0.16	6.84***	6.73	0.47	28.84	3.21	0.09	56.52	1.66
B3	0.50	0.14	2.47	6.30	0.71	24.40	3.00	0.08	58.72	1.63
Media		0.13	2.00	5.60	0.63	22.65	2.70	0.07	58.51	1.27
C1	0.50	0.06	1.73	5.05	0.62	19.83	1.41	0.04	60.36	0.35
C2	0.50	0.06	2.79	4.96	0.49	22.33	1.79	0.04	57.55	0.48
C3	0.50	0.08	8.72***	7.06	0.49	25.28	1.93	0.04	55.73	0.58
Media		0.067	2.260	5.690	0.534	22.480	1.710	0.040	57.880	0.471
D1	0.50	0.06	1.32	5.21	0.59	19.78	1.71	0.04	61.27	0.98
D2	0.50	0.09	2.60	5.82	0.57	21.62	1.72	0.06	59.10	1.18
D3	0.50	0.09	2.59	4.67	0.61	19.70	1.47	0.06	60.63	1.04
Media		0.08	2.17	5.23	0.59	20.37	1.63	0.05	60.33	1.07
E1	0.50	0.09	8.97***	4.68	0.47	10.47	2.20	0.04	58.44	0.29
E2	0.50	0.08	1.37	4.49	0.77	9.40	1.92	0.04	61.52	0.45
E3	0.50	0.09	0.80	4.41	0.69	7.89	1.90	0.04	63.19	0.32
Media		0.09	1.09	4.53	0.64	9.25	2.01	0.04	61.05	0.36

* Heavy metals

** P is in weight percentage already

*** Values untaken because they are out of ratio

3.6. *Moisture*

The initial values of moisture that had been over 50% (Fig. 2) because this type of waste that were used were mainly tomato, lettuce. Which present water concentrations of up to 95% (FAO, 2008). However, after 4 days the values decreased approximately 15% to values below the optimum values of 50-60% [6], remaining all the time how hard the decomposition process, reaching values above 40% for the system with legumes, but not to exceed that range. The system in which manure is used as a treatment provided a humidity value below 35% due to the high temperatures developed, which were the highest of the 5 systems, this combined with high ambient temperatures provided the percentages of moisture less stable. Soto et al., 2001 obtained values greater humidity of 85%, however, showed that a system of volts programmed it is possible to reduce these optimal values of 60%, approximately, being this close to the value obtained from the first show for all treatments [13].

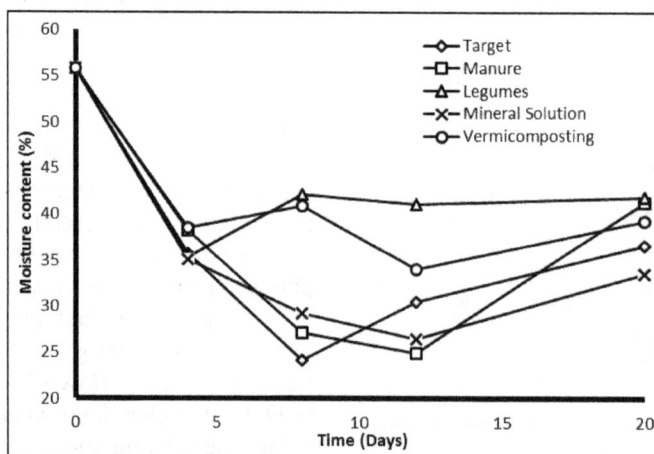

Figure 2. *Comparison of changes in moisture between individual treatments. Method for determination of ash porcelain capsules. White (◊), legumes (Δ), Vermicompost (○), manure (□), mineral solution (×). Data are representative of 3 experiments per sample.*

4. Conclusions

Applying conventional treatments composting systems results in decomposition of the organic matter and efficient desirable physicochemical characteristics. Characterization of composting systems identifies registered common compliance requirements in various investigations that allow control of product quality is sought offer. In this paper conducted a series of test methods recommended by different regulations, locating parameters that are outside the recommended range, as in the case of the presence of moisture above the regulated rate and temperature. From which we conclude that aeration and moisture are critical for proper decomposition of organic matter in a compost system factors, however, are not the only ones that determine the speed with which this process takes place. It is therefore not considered when making the decision on what was the most effective treatment.

The atomic absorption method allowed to study the concentration of minerals and heavy metals. The coexistence of these was more dependent on the treatment used to natural soil conditions. High concentrations of cations such as manganese (Mn) and iron (Fe) suggest that the quality of the compost obtained depends on the conditions established in composting systems.

The pH of composting systems was studied by use of potentiometers. The shape of the curves of each treatment pH of said ammonia formation is accelerated and is maintained constant for a period of time, then triggering a decrease in the values due to the mineralization of the nitrogen fraction. Therefore, the design of conventional composting systems using physicochemical treatments can reduce the time of mineralization of nitrogen fractions and suitable carbon, but without reaching pH values higher than those reported.

We conclude that the establishment of the beds composting conditions size and organic matter content allows a correct decomposition, yielding values of physicochemical parameters ideal for the process, also reducing composting time by 24%. These conditions were achieved first by vermicompost systems, can be say that this was the most effective treatment.

Acknowledgements

Special thanks to IBQ. Paleo Francisco Servin, head of the Laboratory of Analytical Instrumentation of the Instituto Tecnológico de Estudios Superiores de Tacámbaro and its partner the IIA. Arcelia Vargas Elizabeth Vargas and IIA. Gardenia Yanet Maldonado Arévalo for their support in conducting the atomic absorption technique for quantification of minerals.

References

[1] Secretaria del Medio Ambiente y Recursos Naturales. Base de datos estadísticos del SNIARN (Badesniarn) 2011.

[2] Amador-Hernández J, Enríquez A, Velázquez-Manzanares M & Anaya G E, Seguimiento en tiempo real de la degradación de compuestos orgánicos mediante procesos fotocatalíticos heterogéneos con TiO_2, Revista Mexicana de Ingeniería Química, 10 (2011) 471-486.

[3] Rodríguez M & Córdova A, Manual de Compostaje municpal: Gestión Ambiental y Manejo Sustentable de los Recursos Naturales 2008, 14-16.

[4] Ariz I, Asencio A, Zamarreño A, García-Mina J, Aparicio-Tejo P & Moran J, Changes in the C/N balance caused by increasing external ammonium concentrations are driven by carbon and energy availabilities during ammonium nutrition in pea plants: The key roles of asparagine synthetase and anaplerotic enzymes, Physiologia Plantarum, 10 (2012) 12-16.

[5] Mustin M, Le Compost, Gestion de la Matière organique (Editions Francois DUBUSC, Paris) 1987, 954.

[6] Rykn R, On-farm composting handbook, Northeast Regional Agricultural Engineering Service (New York) 1992, 184-186.

[7] Soto G & Muñoz C, Consideraciones teóricas y prácticas sobre el compost, y su empleo en la agricultura., Revista de Manejo Integral de Plagas y Agroecología, 65 (2002) 123-129.

[8] Vásquez de Díaz M C, Prada P P A & Mondragón M A, Optimización del proceso de compostaje de productos post-cosecha (cereza) del café con la aplicación de microorganismos nativos, Publicación Científica en Ciencias Biomédicas, 14 (2010) 216-218.

[9] Robles-Martínez F, Nieto-Monteros D A, Picasso-Muñoz D, Macías-Hernández M & Osorio-Mirón A, Efecto de las Dimensiones de las Pilas en el Desarrollo de la Fase Termofílica en un Proceso de Composteo, in 5° Congreso Interamericano de Residuos Sólidos, 2013.

[10] Espinoza-Valdemar R M, Sotelo-Navarro P X, Beltrán-Villavicencio M & Vázquez-Morillas A, Evaluación de la Degradación de Pañales Desechables usados Mediante Composteo en Biorreactores Aerobios, in 5° Congreso Interamericano de Residuos Sólidos, 2013.

[11] Meneses G, De La Rosa P & Monroy J L, Caracterización de humus de lombriz. Disertación por parte del Departamento de Ingeniería y Ciencias Químicas a través del Laboratorio de Ingeniería Ambiental, Universidad Iberoamericana, 2012.

[12] Paul E A & Clark, F E, Soil Microbiology and Biochemistry, 2nd ed. (Academic Press, New York) 1996, 340-349.

[13] Soto G, Abonos orgánicos: producción y uso de compost. Disertación por parte de la Universidad de Costa Rica, San José, Costa Rica, 2001.

22

Technical Efficiency of Boro Rice Production in Meherpur District of Bangladesh: A Stochastic Frontier Approach

Md. Nehal Hasnain, Md. Elias Hossain, Md. Khairul Islam

Department of Economics, University of Rajshahi, Rajshahi-6205, Bangladesh

Email address:

nehalhasnainshoron@ymail.com (M. N. Hasnain), khairul06eco@gmail.com (M. K. Islam), eliaseco@ru.ac.bd (M. E. Hossain)

Abstract: Although rice is the main crop in Bangladesh and the country is ranked as the sixth largest rice producer in the world, researchers observe that rice is not produced with full efficiency in the country. It is also observed that owing to the application of high yielding variety seeds, chemical fertilizer, pesticide, and irrigation, productivity of rice in Bangladesh has increased in the recent years though it is still lower compared to other Asian countries. A review of existing literature reveals that so far little attention has been given by the researchers in investigating the efficiency of rice production in Bangladesh. Thus, the objective of the present study is to analyze the technical efficiency of rice production in Bangladesh using data from boro rice farmers. Required data are collected from 115 boro rice producing farmers of Meherpur district selected using multistage random sampling procedure. In analyzing the data, farm specific technical efficiency scores are estimated using the Translog Stochastic Frontier Production function approach. The study found that technical efficiency of boro rice farms in Meherpur district is 89.5%. It is also found that 'labor', 'fertilizer and pesticide', 'seed' and 'irrigation' are the significant factors that affect the level of technical efficiency while 'farm size' and 'ploughing cost' are found insignificant in affecting technical efficiency of boro rice production in the study area. The results indicate that boro rice farms in the study area have been operating below the maximum level of production frontier and given the available technology, farmers can increase their production by 10.5% through increasing the use of labor, seed and irrigation inputs and also by using proper doses of fertilizer and pesticide inputs.

Keywords: Boro Rice, Technical Efficiency, Stochastic Frontier Approach, Bangladesh

1. Introduction

Agriculture is one of the most important sectors of Bangladesh economy (Nargis and Lee, 2013). The sector contributes around 16.77% to the gross domestic product (GDP) of the country and employs around 47.5% of the total labor force (GoB, 2014). Moreover, the sector feeds up around 160 million people of the country and provides survival and nutrition for the farm households of rural areas (GoB, 2014). In addition, this sector provides raw materials to agro-based and other industries operating in the country.

The main agricultural commodities in Bangladesh are rice, wheat, maize, jute, sugarcane, potato, vegetables, oilseeds, pulses, tea, etc. Among these crops, rice is widely cultivated all over the country throughout the year. In Bangladesh, rice is grown in three distinct seasons: boro (post-monsoon rice) from January to June, aus (pre-monsoon rice) from April to August and aman (monsoon rice) from August to December (Nargis and Lee, 2013). Of these three types of rice, boro alone comprises of about 55% of total food grain production. According to the report of the Bureau of Statistics of Bangladesh (BBS, 2010) productivity of *boro* rice per unit of land is higher (3.84 MT per hectare) compared to *aus* (1.76 MT per hectare) and *aman* (2.16 MT per hectare).

Rice yield in Bangladesh has increased at a significant rate as a result of introducing the use of high yielding seed varieties, fertilizers, pesticides, irrigation and mechanized cultivation system. Rice production has increased by 23.78% in fiscal year 2012-13 compared to fiscal year 2006-07 (BBS, 2013). However, until now the rate of growth of rice production in the country is lower than the rate of growth of demand for rice in the country. To meet additional demand, the country has to import rice almost every year in the previous decades (Nargis and Lee, 2013). For example, Bangladesh imported 18.72 lakh metric tons of food grains in fiscal year 2012-13 (GoB, 2014). To the opinion of some researchers, Bangladesh would not have to import rice if productivity could be enhanced through increasing technical efficiency in production (Hossain and Rahman, 2012).

However, Bangladesh agriculture has already been operating at its land frontier and there is little or no scope to expand cultivable land to meet increasing demand for food for its ever-increasing population (Hossain and Rahman, 2012). Moreover, average farm size in Bangladesh is relatively smaller compared to other countries due to existing ownership and inheritance system in the country and in many cases it becomes very difficult to adopt modern agricultural technologies. Again, there is lack of information and awareness among marginal farmers about the efficient use of inputs and proper cultivation methods of rice production. Given these situations, measurement of technical efficiency of rice production is an important issue in Bangladesh agriculture from the standpoint of agricultural development. This would provide pertinent information about the existing inefficiencies and facilitate to make sound policies related to management decision, resource allocation and institutional development toward enhancing efficiency of rice production in the country (Nargis and Lee, 2013).

Thus, the objectives of this study are to determine the level of technical efficiency of *boro* rice production in Meherpur district of Bangladesh and to assess the effects of the key inputs on technical efficiency of boro rice production in the study area.

2. Literature Review

A comprehensive review of literature regarding different aspects of technical efficiency of agricultural production in the context of Bangladesh as well as in other countries has been done.

Khan *et al.* (2010) examined the technical efficiency of rice production and its determinants in Jamalpur district of Bangladesh. Through using random sampling technique a total of 150 rice farmers were selected for the study and a stochastic production frontier approach was employed to estimate technical efficiency and determinants of efficiency in boro and aman rice production. The study found that the mean technical efficiency of boro rice production in the study area is 95% and in the case of aman rice production it is 91%. The study also found that younger farmers are more efficient than elderly farmers. Moreover, the study found that education and experience of the farmers substantially reduce farm inefficiencies in the study area. Hasan (2008) conducted a study in sadar upazila of Dinajpur and Panchagarh districts of Bangladesh to estimate costs, returns and economic efficiency of boro rice farming using the Cobb-Douglas stochastic frontier production function taking 100 farm households from each district. The study found that average technical efficiency of boro rice farming is 0.84 in Dinajpur district and 0.80 Panchagarh district.

Mohapatra (2013) estimated the technical efficiency scores and the factors of inefficiency of paddy production in Odisha. Using both Cobb-Douglas and Translog production function approach this study found mean technical efficiency as 97.04%. It also found that farming experience and high

school as well as college education have significant and positive contribution in improving the technical efficiency.

Hossain and Rahman (2012) estimated technical efficiency of rice farmers in Naogaon district of Bangladesh using stochastic frontier approach. The study found that the mean technical efficiency of rice farming is 79.58% in the study area. The study observed that appropriate use of labor, seed, fertilizer, insecticide and irrigation may increase the level of technical efficiency of rice production in the study area. Shantha *et al.* (2013) investigated the technical efficiency of rice farming in a major irrigation scheme in Sri Lanka. A total of 357 paddy farmers under Nagadeepa reservoir were selected randomly for collecting relevant information. Applying the translog stochastic frontier production function, the study found that average technical efficiency of selected farmers is 72.80%.

Tijani (2006) estimated technical efficiency of rice farms in Osun State of Nigeria and identified some socioeconomic factors which influence productive efficiency. In this study, stochastic frontier production function approach was applied to estimate the level of production efficiency. The findings of the study revealed that the level of technical efficiency in the study area ranges from 29.4% to 98.2% with a mean efficiency of 86.6%. It is also found that the level of efficiencies is significantly and positively correlated with the application of traditional land preparation methods and off-farm income.

Chirwa (2007) explored technical efficiency among smallholder maize farmers in Malawi using Cobb-Douglas stochastic frontier production function. The study identified the sources of inefficiency using farm level data. It is found that smallholder maize farmers in Malawi are inefficient and the average efficiency score is only 46.23% in the study area. It is also found that 79% of the plots have efficiency scores below 70%.

3. Methodology

3.1. Study Area and Data Collection

The present study is mainly based on primary data. Using multistage random sampling technique, required data are collected from a total of 115 boro rice producing farmers. The study is carried out in four villages, namely, Chandpur and Shibpur of Kutubpur Union, and Rajnagar and Borshibaria of Pirojpur Union under the Sadar Upazila of Meherpur district in Bangladesh.

3.2. Concept of Technical Efficiency

Technical efficiency of a farm is the ratio of farm's actual output to the technically maximum possible output, at given level of resources (Battese and Coelli, 1988; Adedeji1 *et al.*, 2013). It can be classified broadly into three categories, namely, deterministic parametric estimation, stochastic parametric estimation and non-parametric mathematical programming (Udo and Akintola, 2001). Parametric frontier approach assumes functional form on the production function and makes assumptions about the data. The most common functional forms include the Cobb-Douglas, Constant Elasticity of Substitution (CES) and Translog production functions. Deterministic frontiers assume that all the deviations from the frontier are a result of firms' inefficiency, while stochastic frontiers assume that part of the deviation from the frontier is due to random events (reflecting measurement errors and statistical noise) and part is due to farm specific inefficiency (Forsund *et al.* 1980; Battese, 1992 and Coelli *et al.*, 1998). On the other hand, nonparametric frontier assumes no functional form on the production frontiers and does not make assumptions about the error term. It uses linear programming approaches. The most popular nonparametric approach is the Data Envelopment Analysis (DEA).

The concept of efficiency of farms has widely been studied by a number of researchers. Most of the studies estimated technical efficiency using the stochastic frontier production function approach (Bravo-Ureta and Pinheiro, 1993; Parikh and Shah, 1994; Ajibefun and Abdulkadri, 1999; Ajibefun and Daramola, 1999; Sharma *et al.*, 1999 and Ajibefun *et al.*, 2002). Following these studies the stochastic parametric model is formulated in this study to analyze the technical efficiency of boro rice production in the study area.

3.3. Stochastic Frontier Production Function

The stochastic frontier model decomposes the error term into a two-sided random error that captures the random effects outside the control of the firm (the decision making unit) and a one-sided efficiency component. The model was first proposed by Meeusen and van den Broeck (1977) and Aigner *et al.* (1977). Assuming a suitable production function, the general functional form of the model is as follows.

$$Y_i = f(X_i, \alpha) + \varepsilon_i \qquad (1)$$

Where, Y_i is the level of output of the ith sample farm, X_i is the value of input of the ith sample farm, α is unknown parameters to be estimated and ε_i is the error term that is composed of two independent elements V_i and U_i, such that $\varepsilon_i = V_i - U_i$. The composite error term V_i is the two-sided error term, and U_i is the one-sided error term. The components of the composed error term are governed by different assumptions about their distribution. The random (symmetric) component V_i is assumed to be identically and independently distributed as $N(0, \sigma_v^2)$ and is also independent of U_i. This random error represents random variations in output due to factors outside the control of the farmers reflecting luck, weather, natural disaster, machine breakdown and variable input quality as well as the effects of measurement errors in the output variable, statistical noise and omitted variables from the functional form (Aigner *et al.* 1977). Following Battese and Coelli (1995) the U_i is nonnegative random variable that represents the stochastic shortfall of outputs from the most efficient production. Therefore, U_i is associated with the technical inefficiency of the farmers and are assumed to be independently and identically distributed truncations of the half normal distribution as $N(0, \sigma_u^2)$ and

also independently distributed of V_is.

The parameters of the stochastic frontier model can be consistently estimated by the maximum-likelihood estimation method. The variance of the parameters of the likelihood function are estimated as

$$\sigma_s^2 = \sigma_v^2 + \sigma_u^2$$

and

$$\gamma = \sigma_u^2 \Big/ \sigma_s^2 = \sigma_u^2 \Big/ \left(\sigma_v^2 + \sigma_u^2\right)$$

Some empirical studies have attempted to analyze production risk and technical efficiency in a single framework. Kumbhakar, (1993) demonstrated a method to estimate production risk and technical efficiency using a flexible production function to represent the production technology. Battese *et al.* (1997) specified a stochastic frontier production function with an additive heteroskedastic error structure that is adopted in the present study. The model of Kumbhakar, (1993) permits negative or positive marginal effects of inputs on production risk which is consistent with the Just and Pope (1978) framework. Following their studies, the error specification in equation (1) is

$$\varepsilon_i = g(X_i, \beta)[V_i - U_i] \qquad (2)$$

Thus, from equation (1) and (2) we have

$$Y_i = f(X_i, \alpha) + g(X_i, \beta)[V_i - U_i] \qquad (3)$$

Equation (3) is the specification of the stochastic frontier production function with flexible risk properties (Battese *et al.*, 1997). The mean and variance (risk function) of output of the ith farmer given the values of inputs and technical inefficiency effect can be estimated as

$$E(Y_i \setminus X_i, U_i) = f(X_i, \alpha) - g(X_i, \beta)U_i \qquad (4)$$

and

$$Var(Y_i \setminus X_i, U_i) = g^2(X_i, \beta) \qquad (5)$$

Using this variance (risk function), the marginal production risk can be obtained by partial derivative of variance of production with respect to inputs which can be either positive or negative. That is

$$\frac{\partial Var(Y_i \setminus X_i, U_i)}{\partial X_{ij}} > 0, or < 0 \qquad (6)$$

Accordingly, the technical efficiency of the ith farmer (TE_i) is defined by the ratio of the mean production for the ith farmer (given the values of the inputs, X_i, and its technical inefficiency effect, U_i) to the corresponding mean maximum possible production (production with no technical inefficiency) can be specified as

$$TE_i = \frac{E(Y_i \setminus X_i, U_i)}{E(Y_i \setminus X_i, U_i = 0)} = 1 - TI_i \qquad (7)$$

Where TI_i is technical inefficiency defined as potential output loss and represented as

$$TI_i = \frac{U_i \cdot g(X_i, \beta)}{E(Y_i \setminus X_i, U_i = 0)} = \frac{U_i \cdot g(X_i, \beta)}{f(X_i, \alpha)} \quad (8)$$

If the parameters of the stochastic frontier production function are known, the best predictor of U_i would be the conditional expectation of TE_i, given the realized value of the random variable $E_i = V_i - U_i$ (Jondrow et al. 1982). It can be shown that $U_i \setminus (V_i - U_i)$ is distributed as $N(\mu_i^*, \sigma_*^2)$, where μ_i^* and σ_*^2 are defined by

$$\mu_i^* = -\frac{(V_i - U_i)\sigma_u^2}{(1 + \sigma_u^2)} \quad (9)$$

$$\sigma_*^2 = \frac{\sigma_u^2}{(1 + \sigma_u^2)} \quad (10)$$

It can also be shown that $E[U_i \setminus (V_i - U_i)]$, denoted by $\hat{U}_i$ is

$$\hat{U}_i = \mu_i^* + \sigma_* \left[\frac{\varphi\left(\mu_i^* / \sigma_*\right)}{\Phi\left(\mu_i^* / \sigma_*\right)} \right] \quad (11)$$

Where, $\varphi(.)$ and $\Phi(.)$ represent the density and distribution functions of the standard normal random variable. Equation (11) can be estimated using the corresponding predictors for the random variable, E_i, given by

$$\hat{E}_i = \frac{Y_i - f(X_i, \hat{\alpha})}{g(X_i, \hat{\beta})} \quad (12)$$

After estimating equation (11), equation (8) can be estimated as

$$TI_i = \frac{\hat{U}_i \cdot g(X_i, \hat{\beta})}{f(X_i, \hat{\alpha})} \quad (13)$$

The technical efficiency of the ith farmer is predicted by $\hat{TE}_i = 1 - \hat{TI}_i$. Technical efficiency of the ith farmer can also be calculated as $TE_i = \exp(-U_i) \ast 100$ (TE is converted into percentage through multiplying this equation by 100). It is calculated using the conditional expectation of the above equation, conditioned on the composite error ($\varepsilon_i = V_i - U_i$).

3.4. Empirical Specification of the Translog Production Function Model

Empirically, Translog stochastic production frontier model is employed in this study to estimate the level of technical efficiency of rice producing farms in the study area. For this purpose, total amount of boro rice production of farmers are taken as dependent variable and inputs of boro rice production used by farmers are incorporated as independent variables. Thus, following Villano and Fleming (2004), the empirical model for the present study is specified as

$$\ln Y_i = \beta_0 + \sum_{j=1}^{6} \beta_j \ln X_{ji} + 0.5 \sum_{j=1}^{6} \sum_{k=1}^{6} \beta_{jk} \ln X_{ji} \ln X_{ki} + V_i - U_i \quad (14)$$

Where, Y = total production of boro rice (mound), X_1 = farm size (bigha), X_2 = cost of labor used in boro rice cultivation (Tk., Bangladeshi currency), X_3= cost of fertilizer and pesticide used to produce rice (Tk.), X_4 = cost of seed planted to produce rice (Tk.), X_5 = irrigation cost (Tk.), X_6 = ploughing cost (Tk.), and β_js are unknown parameters to be estimated.

The Translog production function is the most frequently used flexible functional form in efficiency analysis in recent years. It is considered as more general function due to its flexible functional form. It permits the partial elasticities of substitution between inputs to vary, i.e. the elasticity of scale can vary with output and factor proportions. On the other hand, Cobb-Douglas functional form imposes severe restrictions on the use of technology by restricting the production elasticities to be constant and the elasticities of inputs substitution to be unity implying that capital and labor are substitutable in both the short and the long run. Another commonly used production function is Constant Elasticity of Substitution (CES) production function. Unlike Cobb-Douglas production function, the CES production function permits one to vary the elasticity of substitution (Villano and Fleming, 2004).

4. Description of Data

In the present study, data on output and inputs are used to estimate farm level technical efficiency of rice production. Before estimation, some properties of data such as mean, minimum and maximum are calculated. The properties of data are shown in Table 1.

From Table 1 it is seen that the mean farm size of sample farmers is 5.77 bighas with minimum of 1.50 bighas and maximum of 20 bighas in the study area. Again, the average labor cost, fertilizer cost, pesticide cost, seed cost, irrigation cost and ploughing cost are Tk.3240, Tk.1920, Tk.302.26, Tk.491.22, Tk.2700 and Tk.1078.32, respectively, of sample farmers.

Table 1. *Description of Collected Data.*

Subject	Mean	Minimum	Maximum
Farm size (bigha)	5.77	1.50	20.00
Labour cost (Tk.)	3240	3000.00	4200
Fertilizer cost (Tk.)	1920	1220.00	2800.00
Pesticide cost (Tk.)	302.26	130.00	500.00
Seed cost (Tk.)	491.22	240.00	780.00
Irrigation cost (Tk.)	2700	2500.00	2800.00
Ploughing cost (Tk.)	1078.32	800.00	1500.00
Per bigha production (Mound)	28.02	20.00	35.00

Source: Authors own calculation

Table 1 also reveals that the average production of boro rice per bigha in the study area is 28.02 mounds with minimum of 20 mounds and maximum of 35 mounds in the study area.

5. Discussion of Results

The estimated results of the stochastic frontier production function are discussed in this section. Maximum Likelihood (ML) method is applied to estimate the coefficients of Translog production function. The findings of the present study are compared with those of the earlier studies to check the variation of the results found in the present study.

5.1. Technical Efficiency of Boro Rice Production

Table 2. Technical Efficiency (TE) of Boro Rice Production in the Study Area.

Technical efficiency (%)	No. of farm	% of farm
51-60	0	0
61-70	1	0.87
71-80	9	7.83
81-90	52	45.22
91-100	53	46.09
Total	115	100.0
Mean TE (%)	89.5	
Minimum TE (%)	69.8	
Maximum TE (%)	99	

Source: Authors own calculation

Technical efficiency of all sample farms is estimated using the Translog production function and it is classified into five categories on the basis of efficiency level. The estimated results of technical efficiency of boro rice farms are summarized in Table 2. From the table it is observed that the average level of technical efficiency of sample boro rice farms is 89.6% with minimum efficiency of 69.8% and maximum efficiency of 99%.

From the above table it is also found that among all sample farms there are only 0.87% farms within the efficiency level between 61%-70% and 7.83% farms within the efficiency level between 71%-80%. It is interesting that most of the farms, around 91.31%, have been operating between the efficiency level 81%-100%.

However, the average level of technical efficiency of the sample farms indicates that there is a certain level of technical inefficiency in boro rice production in the study area. This result suggests that in the short run it is possible to increase the amount of boro rice production in the study area by increasing the efficiency level. Farmers may increase the level of efficiency of their farms by controlling the use of inputs of production that have significant contribution to influence efficiency level of production. For this purpose, it is essential to identify the significant inputs of production which have positive or negative contribution to production. In section 5.2, significant factors of production are estimated using Translog production function.

5.2. Results of Translog Production Function

Table 3. Maximum Likelihood Estimation for Boro Rice Farms.

Variables	Parameters	Coefficients	Standard error	t-ratio
Constant	β_0	10.51*	1.72	6.13
ln farm size	β_1	1.05	0.57	1.83
ln labor	β_2	0.002*	0.0004	4.66
ln fertilizer & pesticide	β_3	-0.01*	0.004	- 2.97
ln seed	β_4	0.19*	0.03	5.73
ln irrigation	β_5	0.001**	0.001	2.29
ln ploughing	β_6	0.0001	0.0003	0.44
$(\ln \text{ farm size})^2$	β_7	0.01	0.08	0.07
$(\ln \text{ labor})^2$	β_8	0.37	0.43	0.85
$(\ln \text{ fertilizer & pesticide})^2$	β_9	0.26	0.70	0.37
$(\ln \text{ seed})^2$	β_{10}	0.14	0.36	0.40
$(\ln \text{ irrigation})^2$	β_{11}	1.00***	0.49	2.03
$(\ln \text{ ploughing})^2$	β_{12}	0.17	0.17	1.01
ln farm size*ln labour	β_{13}	-0.25	0.13	-1.90
ln farm size*ln fertilizer & pesticide	β_{14}	0.03	0.16	0.19
ln farm size*ln seed	β_{15}	0.15	0.14	1.09
ln farm size*ln irrigation	β_{16}	0.18*	0.04	4.81
ln farm size*ln ploughing	β_{17}	-0.01	0.16	-0.04
ln labour*ln fertilizer & pesticide	β_{18}	1.15*	0.20	5.83
ln labour*ln seed	β_{19}	-0.15	0.43	-0.36
ln labour*ln irrigation	β_{20}	-0.50	0.58	-0.86
ln labour*ln ploughing	β_{21}	0.11	0.38	0.29
ln fertilizer & pesticide *ln seed	β_{22}	-0.12	0.58	-0.21
ln fertilizer & pesticide *ln irrigation	β_{23}	-0.20	0.47	-0.42
ln fertilizer & pesticide *ln ploughing	β_{24}	0.53	0.27	1.93
ln seed*ln irrigation	β_{25}	0.20	0.40	0.50
ln seed*ln ploughing	β_{26}	0.00002	0.48	0.0001
ln irrigation*ln ploughing	β_{27}	0.18	0.43	0.43
Sigma-squared	σ^2	0.01*	0.001	5.93
Gamma	γ	0.97*	0.35	2.82
Log likelihood function: 152.21; LR: 25.76				

Source: Authors own calculation; Note: *, **, *** indicates 1%, 5% and 10% level of significance

Production of boro rice is generally affected by some factors such as farm size, labor cost, fertilizer and pesticide cost, seed cost, irrigation cost, ploughing cost etc. The estimated effects of these factors are shown in Table 3.

From Table 3 it is found that the estimated coefficients of labor cost, fertilizer and pesticide cost, seed cost and irrigation cost are statistically significant. This indicates that these factors of production are the major determinants that affect boro rice production in the study area. The sign of coefficients of all these variables is positive except fertilizer and pesticide cost. On the other hand, the coefficients of farm size and ploughing cost are statistically insignificant. So, farm size and ploughing cost do not bear any significant meaning to affect boro rice production.

In Table 3 coefficient of labor is found 0.002 indicating that a 1% increase in labor cost may increase output by 0.002%. Similarly, coefficient of seed cost and irrigation costs are 0.19 and 0.001, respectively, indicating that a 1% increase in seed cost and irrigation cost may increase output by 0.19% and 0.001%, respectively. Again, coefficient of fertilizer and pesticide cost is -0.01 indicating that a 1% increase in fertilizer and pesticide cost may decrease output by 0.01%. The negative coefficient of fertilizer and pesticide cost may seem interesting but it is also similar to the findings of Islam et al. (2004), Backman et al. (2010), Khan et al. (2010) in case of Bangladeshi farms.

Among six square parameters only the coefficient of irrigation square is statistically significant and has positive sign. It means that an increase in irrigation cost will increase boro rice production at an increasing rate. Other square parameters are statistically insignificant. From Table 3 it is also found that among the interactive variables 'farm size*irrigation' and 'labor*fertilizer and pesticide' are significant at 1% level and have also positive sign. Besides, 'farm size*labor' and 'fertilizer and 'pesticide*ploughing' are significant at 10% significance level and have negative and positive sign, respectively. Moreover, coefficients of other interactive variables are statistically insignificant indicating no significant meaning in explaining boro rice production. The estimated value of γ is found as 0.97, which means that 97% of the total variation in rice output is due to technical inefficiency. It means that about 97% of the discrepancies between observed output and the frontier output are due to technical inefficiency.

6. Conclusion

Estimation of production efficiency is important to increase the productivity of agriculture sector in a developing country like Bangladesh. The present study finds that the average level of technical efficiency of boro rice farms in the study area is 89.5%. This result means that the boro rice farms in the study area have been operating below the maximum level of production frontier. Given the available technology, farmers can increase their production by 10.5%. The estimated results of Translog production function shows that labor cost, seed cost and irrigation cost have positive and significant effect on the level of technical efficiency of boro rice production in the study area. Thus, farmers may increase

production level by increasing the use of these inputs. On the other hand, fertilizer and pesticide cost is found as negative contributor to the level of production efficiency. This result might indicate that fertilizer and pesticides are being used at high doses by the farmers in the study area and therefore, they should use these inputs with appropriate doses. Thus, on the basis of the findings of the present study, it can be suggested that the government and non-government organizations operating in the study area should make the farmers aware about proper use of inputs of boro rice production to increase the level of technical efficiency.

References

[1] Adedeji, I. A., Kazeem .O., Adelalu, S.I., Ogunjimi, A.O. and Otekunrin (2013). Application of stochastic production frontier in the estimation of technical efficiency of poultry egg production in Ogbomoso metropolis of Oyo state, Nigeria. World Journal of Agricultural Research. 1(6), 119-123.

[2] Aigner, D. J., Lovell, C. A. K., and Schmidt, P. (1977). Formulation and estimation of stochastic frontier production function models. Journal of Econometrics, 6(1), 21-37.

[3] Ajibefun, I. A., and Abdulkadri, A. O. (1999). An investigation of technical inefficiency and production of farmers under the national directorate of employment in Ondo State, Nigeria. Applied Economics Letter, 6, 111–114.

[4] Ajibefun, I. A., and Daramola, A. G. (1999). Measurement and source of technical inefficiency in poultry egg production in Ondo State, Nigeria. Journal of Economics and Rural Development, 13, 85–94.

[5] Ajibefun, I. A., Battese, G.E., and Daramola, A. G. (2002). Determinants of technical efficiency in smallholder food crop farming: application of stochastic frontier production function. Q. J. Int. Agric., 41(3), 225-240.

[6] Anonymous (2009). Annual Report for (2008). IRRI, Los Banos, Philippines.

[7] Backman, S., Islam, K. M. Z., and Sumelius, J. (2010). Determinants of technical efficiency of rice farms in North-Central and North-Western Regions in Bangladesh. An unpublished paper.

[8] Battese, G. E., and Coelli, T. J. (1988). Prediction of farm-level technical efficiencies with a generalized frontier production functions and panel data. Journal of Econometrics, 38, 387-399.

[9] Battese, G. E., Rambaldi, A. N., and Wan, G. H. (1997). Stochastic frontier productions function with flexible risk properties. Journal of Productivity Analysis, 8, 269-280.

[10] BBS (2010). Statistical Yearbook of Bangladesh. Bangladesh Bureau of Statistics, Statistics Division, Ministry of Planning, Government of the People's Republic of Bangladesh, Dhaka, Bangladesh.

[11] BBS (2013). Statistical Yearbook of Bangladesh. Bangladesh Bureau of Statistics, Statistics Division, Ministry of Planning, Government of the People's Republic of Bangladesh, Dhaka, Bangladesh.

[12] Bravo-Ureta, B. E., and Pinheiro, A. E. (1993). Efficiency analysis of developing country agriculture: A review of the frontier function literature. *Agricultural and Resource Economics Review*, 22(1), 88-101.

[13] BRRI, (2000). Annual report, Bangladesh Rice Research Institute, Gaziupur, Bangladesh.

[14] Chirwa, W. E. (2007). *Sources of technical efficiency among smallholder maize farmers in Southern Malawi*. African Economic Research Consortium, Nairobi, AERC Research Paper 172.

[15] Coelli, T., Rao, P. D. S., and Battese, G. E. (1998). *An introduction to efficiency and productivity analysis*. London: Kluwer Academic Publisher.

[16] FAOSTAT (2012). *Food and Agricultural Commodities Production 2010. Food and Agriculture Organization of the United Nations*. Available at http://faostat. fao.org/site/339/default.aspx.

[17] FAPRI (2009). *The Agricultural Outlook 2009. World Rice. Food and Agricultural Policy Research Institute*. Available at http://www.fapri. iastate. Edu/outlook/2009/.

[18] Forsund, F. R., Lovell, C. A. K., and Schmidt, P. (1980). A survey of frontier production functions and of their relationship to efficiency measurement. *Journal of Econometrics,* 13, 5-25.

[19] GoB (2014). *Bangladesh Economic Review 2008*. Ministry of Finance, Government of the People's Republic of Bangladesh, Dhaka.

[20] Hasan, F. M. (2008). Economic efficiency and constraints of maize production in the Northern Region of Bangladesh. *j. innov.dev.strategy.* 2(1), 18-32.

[21] Hossain, M. E., and Rahman, Z. (2012). Technical efficiency analysis of rice farmers in Naogaon district: An application of the stochastic frontier approach. *Journal of Economics and Development Studies,* 1(1), 1-20.

[22] Islam, R. M., Hossain, M., and Jaim, H. M. W. (2004). Technical efficiency of farm producing transplanted aman rice in Bangladesh: A comparative study of aromatic, fine and coarse varieties. *Bangladesh Journal of Agricultural Economics*, XXVII (2), 1-24.

[23] Jondrow, J., Lovell, C. A. K., Materov, I. S., and Schmidt, P. (1982). On the estimation of technical inefficiency in the stochastic frontier production function model. *Journal of Econometrics,* 19, 233-238.

[24] Just, R. E., and Pope, R. D. (1978). Stochastic specification of production functions and economic implications. *Journal of Econometrics,* 7, 67-86.

[25] Khan, A., Huda, A. F., and Alam, A. (2010). Farm household technical efficiency: A study on rice producers in selected areas of Jamalpur District in Bangladesh. *European Journal of Social Sciences*, 14(2), 262-271.

[26] Kothari, C. R. (2003). *Research methodology method and technique*. New Delhi. Wishwa Prakashan.

[27] Kumbhakar, S. C. (1993). Production risk, technical efficiency, and panel data. *Economics Letters,* 41, 11-16.

[28] Meeusen, W., and van den Broeck, J. (1977). Efficiency estimation from Cobb–Douglas production function with composed error. *International Economic Review,* 18, 435-444.

[29] Mohapatra, R. (2013). Farm level technical efficiency in paddy production: A translog frontier production function approach. *International Journal of Advanced Research,* 1(3), 300-307.

[30] Nargis, F., and Lee, S. H. (2013). Efficiency analysis of boro rice production in North-central region of Bangladesh. *The Journal of Animal & Plant Sciences*, 23(2), 527-533.

[31] Parikh, A., and Shah, M. K. (1994). Measurement of technical efficiency in the northwest frontier province of Pakistan. *Journal of Agricultural Economics*, 45, 132-138.

[32] Sattar, A. (2000). *Bridging the rice yield gap in Bangladesh. In bridging the rice yield gap in Asia and the Pacific*. RAP Publication, 2000/16.

[33] Shantha, A. A., Ali, A. B. G. H., and Bandara, R. A. G. (2013). Technical efficiency of paddy farming under major irrigation conditions in the dry-zone of Sri Lanka: A parametric approach. *Australian Journal of Basic and Applied Sciences*, 7(6), 104-112.

[34] Sharma, K. R., Leung, P., and Zalleski, H. M. (1999). Technical, allocative, and economic efficiencies in swine production in Hawaii: A comparison of parametric and non-parametric approaches. *Agricultural Economics*, 20(1), 23–35.

[35] Tijani, A. A. (2006). Analysis of the technical efficiency of rice farms in Ijesha Land of Osun State, Nigeria. *Agrekon,* 45(2).

[36] Udoh, E. J., and Akintola, J. O. (2001). *Land management and resource use efficiency among farmers in South Eastern Nigeria". Award winning paper presented to the African Real Estate Society and the Rics foundation*. Elshadai Global Ventures Ltd., 20-32.

[37] Villano, R., and Fleming, E. (2004). *Analysis of technical efficiency in a rainfed low land rice environment in Central Luzon, Philippines using a stochastic frontier production function with a heteroskedastic error structure*. Working Paper series in agricultural and resource Economics, University of New England.

23

Evaluation of Satellite Rainfall Estimates for Swaziland

Absalom Mganu Manyatsi[*], Ntobeko Zwane, Musa Dlamini

University of Swaziland, Department of Agricultural and Biosystems Engineering, Manzini, Swaziland

Email address:

Manyatsi@uniswa.sz (A. M. Manyatsi)

Abstract: Swaziland is generally an arid country, with most rains falling during the period of October to March. The long term average annual rainfall ranges from 400 mm in the lowveld to 1,200 mm in the mountainous highveld. Raingauges have been used as reliable source of rainfall data, but the density of these ground based instruments is too low, offering poor spatial coverage. The use of satellite products to estimate rainfall can fill the gap created by poor spatial coverage of ground based instruments. The African Monitoring of the Environment for Sustainable Development (AMESD) project that was launched in 2007 aims to provide African Nations with resources for climate monitoring application through the use of Meteosat Second Generation (MSG) satellite data. In Swaziland, the satellite receiving station was installed in 2012. The satellite rainfall product has not been evaluated in the country. The objective of this paper was to evaluate the rainfall product by comparing it with rainfall data sourced from raingauge. Daily rainfall data were obtained for five weather stations (Big Bend, Malkerns, Matsapha, Mhlume and Nhlangano) for 1998 to 2006. These daily rainfall data were organized in 10-day (dekadal) totals. Dekadal satellite rainfall data were obtained from the local AMESD receiving station for the respective period. The data were exported to Statistical Package for Social Sciences (SPSS) computer software for analysis. Person correlation and linear regression tests were performed for average dekadals and yearly data for the five weather stations to compared gauged rainfall and satellite rainfall estimates. The correlation for average dekadal rainfall data was significant at 99% level of confidence for all weather stations. Correlation coefficients and R^2 were higher for weather stations in the middleveld (Malkerns and Matsapha). The magnitude of underestimation of rainfall by satellite products was higher during the wet season for weather stations receiving relatively higher rainfall. The correlation between yearly gauged rainfall and yearly rainfall estimates from satellite product was significant at 99% level of confidence for Big Bend, Mhlume and Matsapha. It was significant at 95% level of confidence for Malkerns, and not significant for Nhlangano weather station. The regression models that were developed could be used to adjust rainfall estimates from satellite products to ground (gauged) rainfall for an area or community.

Keywords: Correlation, Dekadal, Gauged, Regression, Satellite Data, Validation

1. Introduction

Swaziland is a landlocked country located on the southern part of the African continent along the geographic coordinates 31°30'E and 26°30'S and has an area of 17,363 km² (Figure 1). The country is divided into four ecological zones, namely (from west to east); highveld, middleveld, lowveld and Lubombo plateau. The highveld is the wettest, receiving an average annual rainfall of between 900 and 1,200 mm. The middleveld is drier with average annual rainfall of 850 to 1,000 mm. The lowveld is the hottest and driest region with average annual rainfall ranging between 400 and 600 mm. The Lubombo plateau has rainfall distribution more or less similar to the middleveld. Most rainfall is received in the form of convectional showers in summer, between October and March (Davis, 2011).

Swaziland does not have adequate ground-based weather stations to provide reliable gauged rainfall data. There are twelve weather stations that provide reliable data in the form of daily rainfall, and, daily minimum and maximum temperature. The spatial coverage of ground-based stations falls way below the World Meteorological Organisation (WMO) standards of one raingauge per 100 km² (WMO, 2010; Ochieng and Kimaro, undated). Each raingauge covers an average of 1,447 km². The weather stations are not evenly distributed and they are found in urban areas and plantations (agriculture and forest plantations). This problem is common not only in Swaziland but also in most developing countries in Africa because of poor topography, for socio-economic and political reasons, and also due to lack of funding for hydro-meteorological gauging stations (Hughes et al., 2006).

Rainfall monitoring is crucial for the country for agricultural and humanitarian purposes. Satellite based precipitation provide greater spatial coverage with higher temporal frequency than raingauge networks (Nicholson et al., 2003; Abiola et al., 2013). Satellite estimates are essentially averages over the area of a pixel, whereas raingauges provide measurements made at that point, applicable to a radius of a few kilometres (Grimes et al., 1999). The advantage of using satellites for monitoring the weather, rainfall in particular, include the fact that forecasts can be made well in advance, and the information can then be relayed to end-user within a short time delay (Hughes, 2006). A number of algorithms have been developed to estimate rainfall from satellite products. These include the TAMORA algorithm (Chadwick et al., 2010), the Regional Atmospheric Modeling System (RAMS) model (Orlandi et al., 2004) the Tropical Applications of Meteorological Satellite (TAMSAT) model (Teo and Grimes, 2007), and the Pitman model (Hughes at al., 2006).

Satellite based precipitation estimates provide greater spatial coverage with higher temporal frequency than many of the current rain gauge networks (Abiola *et al.,* 2013). Several systems have been developed to provide this kind of information, such as the National Oceanic and Atmospheric Administration (NOAA), the Global Precipitation Climatology Center (GPCC) and the National Aeronautics and Space Administration (NASA), which operate at a global scale (Nicholson et al., 2003). Satellite estimates are better during the wetter months (between July and September) in North Africa (Nicholson et al., 2003). High monthly precipitation can however result to a high number of false detection (Yilmaz et al., 2005). With gauges, errors may arise from sitting and gauge type, but these are small when compared with the bias in satellite estimates (Xie and Arkin, 1995).

Swaziland obtains weather data from weather stations situated in the country`s major towns and strategic points (Figure 1). Daily national newspapers provide forecasts for the subsequent day or two. Radio and television news broadcasts also provide weather forecasts, at least on 3-hourly basis (Manyatsi et al., 2010). Weather information finds utmost use in various disciplines such as for use by farmers in agriculture, for safety and emergency measures and for the general public as it influences day-to-day activities.

The African Monitoring of the Environment for Sustainable Development (AMESD) project was launched in 2007 to provide all African nations with resources through extending the operational use of Meteosat Second Generation (MSG) satellite data for environmental and climate monitoring application (European Space Agency, 2015). The programme is managed by the African Union Commission in Addis Ababa. Satellite data and products are supplied through the EUMETcast dissemination system (EUMESTSAT, 2015).

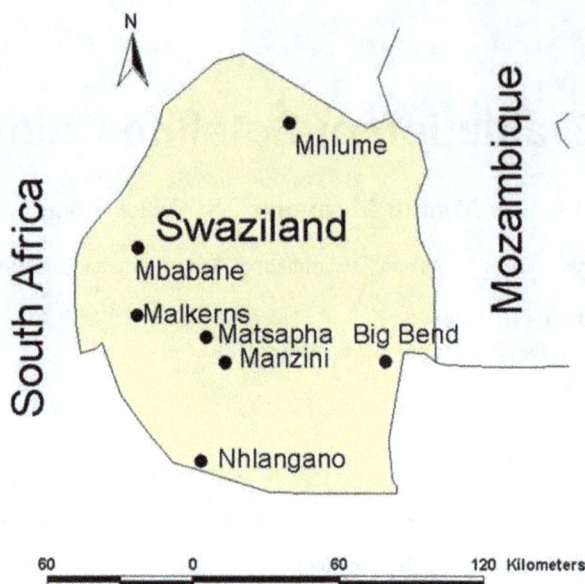

Figure 1. *Location of weather stations used in research.*

In Swaziland the satellite receiving station was installed in 2012, and it is based at the Luyengo campus of the University of Swaziland. Among other information provided by the system are historical rainfall amounts for different areas within the country. Such information could be useful in closing the gap created by sparse raingauge coverage. Meteorological satellites tend to perform poorly over Africa and there is a need to calibrate them so that they fit local use (Teo and Grimes, 2007). This paper aimed to evaluate the rainfall data sourced from satellites by comparing it with data sourced from raingauge.

2. Methodology

Five weather stations in Swaziland that were considered to have reliable rainfall records were selected for the research. These were Nhlangano, Malkerns, Matsapha, Big Bend and Mhlume (Table 1). Two of the weather stations were situated in the lowveld, a semi-arid region of the country. Another two were situated in the middleveld. One station was situated in the highveld of the country.

Daily rainfall data for the years 1998 to 2006 were obtained from Swaziland Meteorological Services in the form of an Excel spreadsheet. These were organized into 10-day (dekadal) totals. Dekadal satellite rainfall data for the selected weather stations were obtained from the local AMESD receiving system for the respective period (1998 to 2006). The data were exported to the Statistical Package for Social Sciences (SPSS, 2008) software package for analysis. Pearson correlation and linear regression tests were performed for average dekadals and yearly data for the five weather stations to compare gauged rainfall and satellite rainfall estimates. Regression models that could be used to estimate actual (gauged) rainfall from the satellite estimates were developed.

Table 1. Details of the five weather stations used in the study.

Weather Station	Ecological zone	Coordinates		Mean elevation (m)	Mean annual rainfall (mm)
Big Bend	Lowveld	S 26.8163	E 31.1985	105	480
Malkerns	Middleveld	S 26.5312	E 31.1875	758	968
Matsapha	Middleveld	S 26.5243	E 31.3093	627	898
Mhlume	Lowveld	S 26.0431	E 31.8132	278	576
Nhlangano	Highveld	S 27.1178	E 31.1985	1,045	767

Source: Government of Swaziland, 2014.

3. Results and Discussions

3.1. Analysis of Average Dekadal Data

The correlation for average dekadal rainfall data was significant at 99% level of confidence for all stations (Table 2). The correlation coefficient for combination of all the five stations was the highest at 0.901 with R^2 of 0.813. For individual stations, the correlation coefficients and R^2 were higher for weather stations in the middleveld (Malkerns and Matsapha) and lower for weather stations located in the lowveld (Big Bend and Mhlume). The lowveld is semi-arid and receives the lowest rainfall in the country (Table 1). The relationship between satellite estimates and gauged rainfall is more pronounced in areas with high rainfall as opposed to areas with low rainfall (Nicholson et al., 2003; Asadullah et al., 2010). There was a variation in biasness of satellite products as compared to gauged rainfall over the dekadals. However on overall the satellite product reported lower rainfall than the gauged rainfall (Figure 2). The magnitude of underestimation by the satellite product was higher during the wet seasons for weather stations that received higher rainfall (Malkerns and Matsapha).This could be a result of satellite failing to pick orographic enhancement of rainfall (Ebert et al., 2007). Overall satellite products overestimate low and underestimate high dekadal rainfall values (Tote et al., 2015).

Table 2. Dekadal correlation and regression.

Weather Station	Correlation	R^2	Significance	Regression Model (GAG=)
All	0.901	0.813	**	1.938 + (1.025 x SAT)
Big Bend	0.558	0.312	**	4.197 + (0.487 x SAT)
Mhlume	0.790	0.623	**	6.612 + (0.900 x SAT)
Matsapha	0.803	0.645	**	0.172 + (0.991 x SAT)
Malkerns	0.855	0.731	**	-1.225 + (1.490 x SAT)
Nhlangano	0.794	0.630	**	6.247 + (0.831 x SAT)

**- Correlation is significant at 0.01 level.

Average all weather stations.

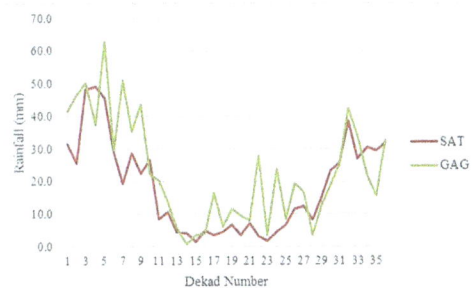

Mhlume weather station

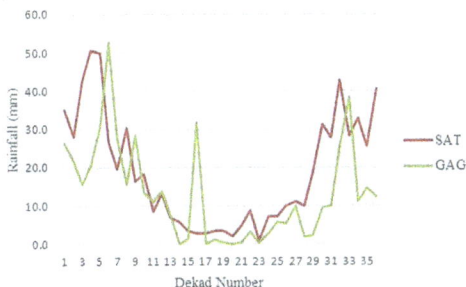

Big Bend weather Station

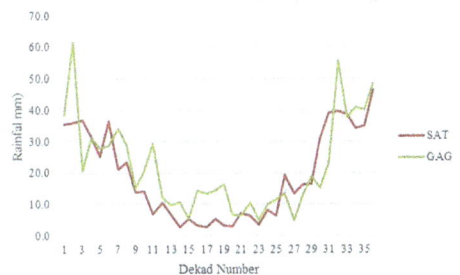

Nhlangano weather station

Malkerns weather station

Matsapha weather station

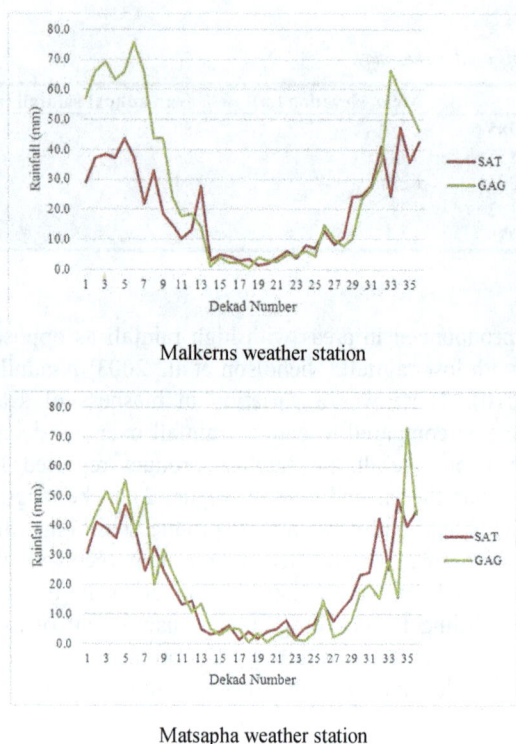

Figure 2. *Long term satellite and gauged rainfall for dekads (SAT = Satellite estimates; GAG = Gauged rainfall).*

Swaziland has four seasons; spring, summer, autumn and winter. Spring is when the offset of rainfall is expected and it occurs from the last week of September to the last week of December (dekadals 27 to 34). Summer occurs from the last week of December to the last week of March the following year (dekadals 35 to 9 of the following year). Autumn starts from the last week of March to the last week of June

(dekadals 10 to 18). Winter starts on the last week of June to the last week of September (dekadals 19 to 27). Winter is the driest season of the year and thus the both the gauged and the rainfall estimates from satellite products are very low for all the weather stations. Over the years there has been shift on the period of onset of rainfall (spring), as effective rainfall now starts in the last week of October or first week of November (dekadal 30 or 31). The shift in seasons is attributed to climate change (Manyatsi, 2011).

3.2. Analysis of Yearly Data

The correlation between yearly gauged rainfall and rainfall estimates from satellite product was significant at 99% level of confidence for Big Bend weather station, Mhlume weather station and Matsapha weather station (Table 3). It was significant at 95% level of confidence for Malkerns weather station, and not significant for Nhlangano weather station. The correlation coefficients were lower than those observed in Uganda where five satellite derived rainfall products were assessed. The mean annual rainfall that ranged from 1,100 mm to 1,300 mm for the Uganda conditions were higher than those for Swaziland condition (Asadullah et al., 2010). The annual rainfall for Swaziland weather stations was low, as they ranged from 480 mm for Big Bend to 968 mm for Malkerns weather station. The satellite product underestimated the annual rainfall in most years for all the weather stations, except for Big Bend, where the estimated rainfall from satellite products were higher that the gauged rainfall for all the years. The R^2 ranged from 0.328 (for Nhlangano) to 0.776 (for average all weather stations). The correlation was greater than 0.5 for all stations, with Nhlangano having the lowest of 0.573 and Big Bend having the highest of 0.817 (Table 3).

Table 3. *arly correlation and regression.*

Weather Station	Correlation	R²	Significance	Regression Model (GAG=)
ALL	0.881	0.776	**	30.582 + (1.083 x SAT)
Big Bend	0.817	0.668	**	-105.985 + (0.855 x SAT)
Mhlume	0.694	0.482	**	140.730 + (1.049 x SAT)
Matsapha	0.815	0.664	**	130.160 + (0.818 x SAT)
Malkerns	0.758	0.575	*	169.492 + (1.284 x SAT)
Nhlangano	0.573	0.328		133.601 + (0.964 x SAT)

**- Correlation is significant at 0.01 level.
*- Correlation significant at 0.05 level.

The year 2000 was the wettest for all the three weather stations. The country experienced floods that resulted in destruction of infrastructure, property and loss of lives (Manyatsi et al., 2010; IFRC, 2000).The year 2000 was followed by a period of rainfall that was below the long term average for all the stations. Drought was experienced in the country during 2001 to 2005. The magnitude of underestimating or overestimating of rainfall by the satellite product varied with the years and weather stations. For Big Bend the satellite product overestimated the rainfall during

the periods when rainfall was above the long term average of 480 mm/annum, and it underestimated the rainfall during the period when rainfall was below the long term annual average (Figure 3). The annual rainfall was underestimated for all the weather stations in the middleveld (Malkerns and Matsapha), due to failure to pick orographic enhancement of rainfall (Ebert et al., 2007).

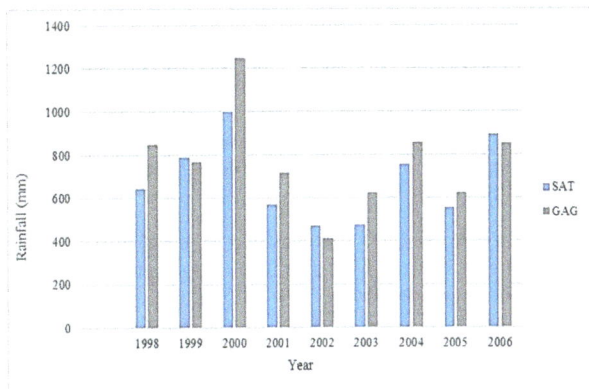

Average all weather stations

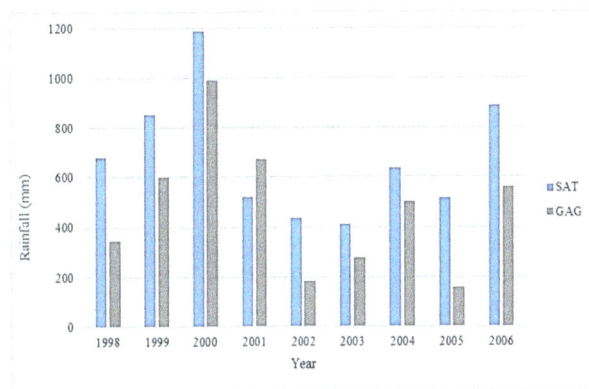

Big Bend weather Station

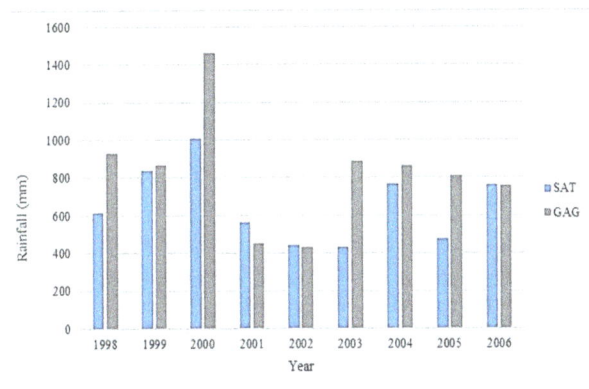

Mhlume weather station

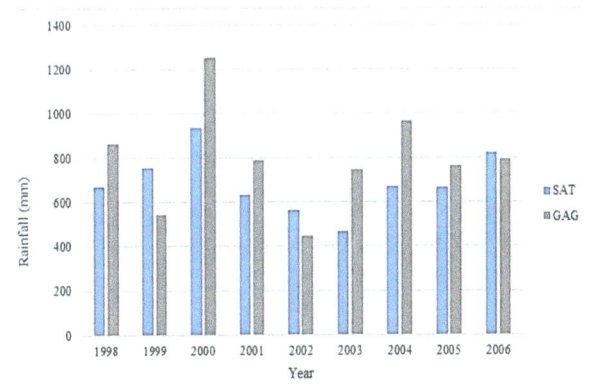

Nhlangano weather station

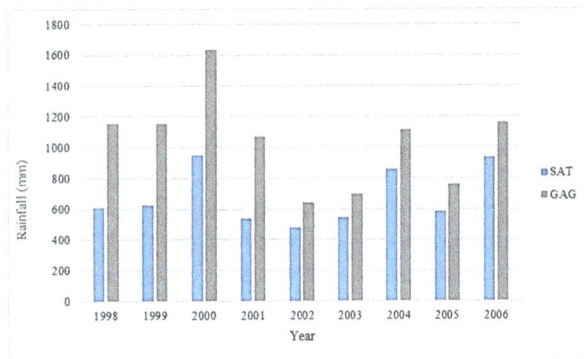

Malkerns weather station

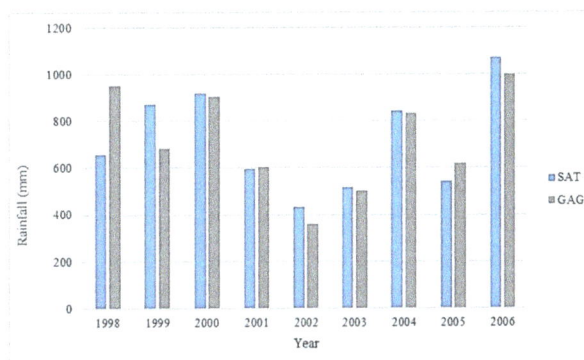

Matsapha weather station

Figure 3. *Yearly satellite and gauged rainfall for dekads (SAT = Satellite estimates; GAG = Gauged rainfall.*

4. Conclusions

Swaziland does not have adequate ground-based weather stations that provide reliable gauged rainfall data. The ground-based stations provide an average coverage of 1,447 km^2 per station. The weather stations are not evenly distributed and are found mostly in urban places. There are no weather stations in the rural community, where traditional farming is practiced, and where weather information is needed most. The AMESD project that is managed by the African Union Commission provide satellite data for environmental and climate monitoring in the continent. The satellite weather products need to be validated before they can be used in each of the country. The validation exercise that was carried for rainfall satellite products for Swaziland showed that the correlation was significant for all weather stations for dekadal (10 days) data. This was an indication that the satellite rainfall products could be used to estimate gauged rainfall. On the other hand the correlation for yearly data was not significant for Big Bend data, and yet it was significant for all the other weather stations. The regression coefficient and R^2 were low for Big Bend weather station for both dekadal data and yearly data. Using rainfall estimates from satellite products for estimating ground rainfall for Big Bend would be less reliable, compared to the other weather stations. Regression models were developed to covert rainfall estimates from satellite products for dekadal and yearly rainfall. The regression models could be used to adjust rainfall estimates from satellite product for a community, in order to effectively estimate the gauged rainfall for that community.

Acknowledgements

The authors are grateful to the Food, Agriculture and Natural Resources Policy Analysis Network (FANRPAN) for funding the finalization and publication of the research findings through the Strengthening Evidence-Based Climate Change Adaptation Policies (SECCAP) project. The authors are also grateful to the University of Swaziland for availing time and facilities to conduct the study.

References

[1]   Abiola S.F., Mohd-Mokhtar R., Ismail W., Mohamad N., and Mandeep J.S. (2013). Categorical statistics approach to satellite retreieved rainfall data analysis in Nigeria. Scienctific Research and Essays. 8(43), 2123-2137.

[2]   Asadullah A., N. McIntyre and Kigobe M. (2010). Evaluation of five satellite products for estiation of rainfall over Uganda. Hydrological Sciences Journal, 53-6, 1137-1150.

[3]   Chadwick R.S., Grimes D.I.F., saunders R.W., Francis P.N and Blackmore T.A. (2010). The TAMORA algorithm: satellite rainfall over West Africa using multi-spectral SEVIRI data. Adv. Geosci., 25, 3-9.

[4]   Davis, C.L. (2011). Climate Risk and Vulnerability: A Handbook for Southern Africa. Council for Scientific and Industrial Research, Pretoria, South Africa, pp. 92.

[5]   Ebert, E. E., Janowiak, J. E. and Kidd, C. (2007) Comparison of near-real-time precipitation estimates from satellite observations and numerical models. Bull. Am. Met. Soc. 88, 47–64.

[6]   EUMESTSAT, (2015). Monitoring weather and climate from space. http://www.eumetsat.int/website/home/index.html. 20/02/2015.

[7]   European Space Agency, (2015). Meteosat Second Generation. http://www.esa.int/Our_Activities/Observing_the_Earth/Mete osat_Second_Generation/MSGoverview2. 20/02/2015

[8]   Government of Swaziland (2014). Database of meteorological data. Ministry of Tourism and Environmental Affairs, Mbabane, Swaziland.

[9]   Grimes,D.I.F, Pardo-Iguzquiza, E. and Bonifacio, R (1999). Optimal areal estimation using rain gauges and satellite data. Elsevier Journal of Hydrology. 346: 33-50.

[10]   Hughes, D. A. (2006). Comparison of satellite rainfall data with observations from gauging stations. http://eprints.ru.ac.za/470/1/Hughes_Comparison_of_satellite _rainfall.pdf. 09/09/2013.

[11]   Hughes, D. A., Andersson, L., Wilk, J., Hubert and Savenije, H.H.G (2006). Regional calibration of the Pitman model for the Okavango River. Elsevier Journal of Hydrology 331: 30–42.

[12]   IFRC, (2000). Mozambique, Botswana, Swaziland, Zimbabwe: Floods: http://www.ifrc.org/docs/apeals/oo/o40004.pdf. 26/08/09.

[13]   Manyatsi A.M. (2011). Application of indigenous knowledge systems in hydrological disaster management in Swaziland. Current Research Journal of Social Sciences, 3:4, 353-357.

[14]   Manyatsi A.M., Mhazo N, and Masarirambi M.T. (2010). Climate variability and changes as perceived by rural communities in Swaziland. Res. J. Environ. Earth Sic., 2: 3, 165-170.

[15]   Nicholson, S.E. Some, B., Mccollum, J., Nelkin, E., Klotter, D., Berte, Y., Diallo, B.M., Gaye, I., Kpabeba, G., Ndiaye, O., Noukpozounkou, J.N., Tanu, M.M., Thiam, A And Toure, A.A. and A. K. Traore. (2003). Validation of TRMM and Other Rainfall Estimates with a High-Density Gauge Dataset for West Africa. Part I: Validation of GPCC Rainfall Product and Pre- TRMM Satellite and Blended Products. Journal of Applied Meteorology. 42: 1337-1354.

[16]   Ochieng W.O and T.A. Kimaro) undated). Coparative study of performance of satellite derived rainfall estimates: A case study of Mara River basin. http://www.waternetonline.org/Symposium/10/full%20papers /Water%20and%20Environment/Ochieng%20W.O..pdf. 20/03/2015.

[17]   Orlandi, A., Ortolania, A., Meneguzzoa, F., Levizzanic V., Torricellac, F. and Turkd, F. J. (2004). Rainfall assimilation in RAMS by means of the Kuo parameterisation inversion. Elsevier Journal of Hydrology. 288: 20–35.

[18]   SPSS (2008).Statistical Package for Social Science (SPSS). Polar Engineering and Consultancy. USA.

[19]   Teo C.K. and Grimes D.I.F. (2007). Stochastic modelling of rainfall from satellite data. Journal of Hydrology. 346 (1-2), 33-50.

[20]   Tote C., Patricio D., Boogaard H, van der Wijngaart R., Tarnavsky E. and Funk C. (2015). Evaluation of satellite rainfall estimates for drought and flood monitoring in Mozambique. Remote Sens. 7(2), 1758-1776.

[21]   WMO (2010). Guide to meteorological instruments and methods of observation. WMO-No 8. Geneva, Switzerland.

[22]   Xie. P. and Arkin, P. A. (1995): An intercomparison of gauge observations and satellite estimates of monthly precipitation. Journal of Meteorology. 34: 1134-1160.

[23]   Yilmaz, K. K., Houge, T. S., Hsu, K., Sorooshian, S., Gupta, V. H. and Wagener, T. (2005). Intercomparison of Rain Gauge, Radar, and Satellite-Based Precipitation Estimateswith Emphasis on Hydrologic Forecasting. Journal of Hydrometeorology. 6: 497- 517.

24

Grapevine Farming and its Contribution to Household income and Welfare among Smallholder Farmers in Dodoma Urban District, Tanzania

James Lwelamira, John Safari[*], Patrick Wambura

Institute of Rural Development Planning (IRDP), Dodoma, Tanzania

Email address:

jsafari@irdp.ac.tz (J. Safari)

Abstract: High incidence of poverty in semi-arid region of central Tanzania is one of the major development challenges in the area. This has mainly been caused by failures of major crops due recurrent drought. Production of high value horticultural crops under irrigation such as grapes could be one of the strategies to reduce the severity of poverty levels and food shortages in the area. A cross-sectional study was carried out in Dodoma urban district to (i) assess the role of grapevine farming on household income and welfare of small scale farmers, (ii) examine the factors that affect grapevine farming, and (iii) identify strategies for improving grapevine farming. Household food security status and consumption expenditure were assessed and used as proxy indicators of the household welfare. The study involved a total of 252 respondents (126 grape farmers and 126 non-grape farmers). Data were collected through interviews using semi-structured questionnaire and analyzed using Statistical Package for Social Sciences (SPSS) program version 20. Results show that grape farming contributes to more than one third (35.6%) of total household income and plays an important role in household welfare. Average household consumption expenditure for grape farmers was twofold higher than that of non-grape farmers (173,833 vs. 84,485 TZS; t = 13.3, $p < 0.001$). Score on household food insecurity index for grape farmers was 8.51 being lower than 11.9 for non-grape farmers (t = -5.7, $p < 0.05$). Nevertheless, there are a number of challenges in grape farming. These include low price of grapes, high costs of inputs, limited access to market, prevalence of pests and diseases, inadequate storage facilities and limited access to quality seedlings. This study gives insights into grape farming as a mitigation strategy of food shortage and the overall household welfare under the changing environmental and socio-economic circumstances.

Keywords: Food Security, High Value Crop, Semi-Arid Areas

1. Introduction

High incidence of poverty especially in the rural areas is a major development challenge in Tanzania. Dodoma region in central Tanzania is semi-arid and among the regions with wide spread poverty [18, 26] and high prevalence of food insecurity [1, 28]. The region experiences recurrent food shortages due to failure of major food crops [24, 27, 41]. The single most important trigger for crop failure has been drought [17, 39]. Rainfall is generally low and erratic which results in withered plants and greatly reduced yields [34, 37]. This condition is likely to be a manifestation of climate change which is often characterized by increased seasonal variability and frequency of extreme events. The changing conditions associated with climate change, generate more desire to build resilience into

agricultural systems [22]. Indeed, this needs to be seen in the context of farming systems in particular and agricultural growth and development in general. Although farming systems located in hot and dry areas are expected to be most severely affected by these changes, there may be considerable differences in adaptive capacity between cropping systems.

In this regard, choice of cropping system as an adaptation and mitigation measure is critical in helping farmers achieve their food, income and livelihood security objectives. A rational method to enhance productivity may be achieved through increased crop diversification and adoption of drought tolerant crops. Cultivation of such crops can lead to increased intensity, food security, commercialization and employment [3, 40, 7, 43]. It is argued that there has been no example of mass reduction of poverty in modern history that did not start with sharp rises in employment and self-employment income due to increased productivity among small family farms [23]

Evidence suggests that promoting high value and drought tolerant crops is a viable coping strategy of furthering poverty reduction and economic growth [5, 6, 15, 22, 38]. Grapevine (*Vitisvinifera*) is among such crops grown in Dodoma Urban district. Records show that during the past six years, the amount of grapes produced in the district increased by 73.8 % i.e. 3930 tons in 2008 to 6831 tons in 2014 [8]. During this period, the number of grape growers rose from 768 to 1012 (31.8 % increase) indicating that grapevine is a crop of growing significance in the rural livelihoods in Dodoma areas. Nevertheless, grape farming in Tanzania is a subject of very limited information. To our knowledge, this is the first empirical study that examines in greater detail the role of grape farming in rural livelihoods in Tanzania. Food security (food availability, access and adequacy) and household consumption expenditure are known to be key indicators of household welfare [19, 31, 32]. In this study, we identify the links between grape farming, household income, food security and consumption expenditure. Analysis is required to improve understanding of potentials, challenges and strategies needed to improve grape sub-sector especially as the scale of production is growing. The objectives of the study were (i) to assess the role of grape farming on household income and welfare of small scale farmers (ii) to examine the factors that affect grape farming and (iii) to identify strategies for improving grapevine farming

2. Methodology

2.1. Study Area

This study was carried out in six villages (Mbabala A, Mbabala B, Mpunguzi, Matumbulu, Veyula, and Mchemwa) which were randomly selected from three wards (Mbabala, Mpunguzi and Makutupora). These wards are among the five wards in Dodoma Urban district with high proportion (3%) of grape farmers [8]. This district has a total of 41 wards, 18 villages and 170 *mitaa*[1]. The district has 410,956 inhabitants of whom 211,469 (51.5%) are females and 199,487 (48.5%) are males with the average household size of 4.4 [42]. The district lies between latitude -6° 9' 35.028"N and longitude 35° 47' 52.8"E with a size of 2,969 km^2 (276,900 ha). A total of 196,000 ha are suitable for agriculture but only 107,007 ha are under cultivation [8]. In 2013/2014, an estimated 1242 ha (0.1% of the total area suitable for cultivation) were under grape cultivation [8]. The study area is semi-arid characterized by a long dry season starting late April to early December, and a short single wet season starting December to mid April. The average rainfall is 500mm annually, and about 85% of this falls in the four months between December and March. The study area cultivates a number of grape varieties including

Makutupora red, Cheninblanc, Regina, Syrah, Ugniblanc, Black rose, Alphoncelavalle, Tajitrozavij, Beauty seedless, Kismiscreveni and *Halelibelyji*. There are two harvesting seasons –one in the rainy season (March, April and May) and another in the dry season (August, September and October). High grape production occurs during the dry season.

2.2. Study Design

A cross-sectional design was applied to enroll a random sample of 126 grape farmers (adopters) and 126 non-grape farmers (non-adopters) from six villages in September, 2013. A sample size for grape farmers (n) was estimated from $\frac{(Z_{\alpha/2})^2 pq}{\lambda^2}$ [10]. Where, $Z_{\alpha/2}$ = 1.96, p=1-q=0.5, and λ=maximum error=10%. Further, 95% confidence interval and non-response rate of 10% were assumed. An equal number of non-grape farmers were included in the study for comparative analysis with respect to household income and welfare attributes. The study, therefore, involved a total of 252 respondents.

2.3. Data Collection

Respondents involved in this study were interviewed using a semi-structured questionnaire. The main aspects covered in the questionnaire were: Socio-demographic characteristics, farm size under grape, grape yield and household income. Others were household consumption expenditure, household food security status and constraints to grape production. Household consumption expenditure and household food security were used as proxy indicators of household welfare. Household consumption expenditure was derived from consumption expenditure on non- durable goods per adult equivalent per year as described in previous studies [4, 21, 2, 20]. Non-durable goods consisted of both food and non-food items. Food items consisted of dairy, grain, fruits/vegetables, eggs, meat/poultry/fish, legumes/nuts, oil/fats and beverages (coffee/tea/soft drinks). Non-food items included water bills and house rent, clothing (clothes, shoes and make-up), energy/fuel (firewood, charcoal, kerosene and electricity) and social activities (contribution to churches, mosques, local organizations, education and medical services). High total consumption expenditure on the above items indicated high level of the household welfare.

On the other hand, household food security status was assessed based on respondent's experience on household food status in the past 12 months prior to the survey. Respondents were therefore asked to rate the following items: worry about food, unable to eat preferred foods, go to bed hungry, go the whole day and night without food, eat food that really do not want to eat, eat small meal, eat fewer meals a day, no food of any kind in the household [4,30]. Rating of each of these items was based on a 3-point Likert scale: 0 = no, 1 = rarely, 2 = sometimes and 3 = often. Scores were then summed to obtain food insecurity index. High score on the index is an indication of high level of household food insecurity. In addition to household survey, data were collected from the

---

1 The *mtaa* (plural *mitaa*) is the lowest unit of government in urban areas in Tanzania. Each urban ward is divided into mitaa or neighbourhoods consisting of a number of households, which the urban council may determine.

department of agriculture at the district and key informants.

2.4. Data Analysis

Quantitative data were analyzed using SPSS version 20. The analysis involved descriptive statistics to determine distributions of means, standard deviations, frequencies and percentages. Inferential statistical analysis was performed to compare grape farmers with non-grape farmers with respect to socio-demographic characteristics, income levels and household welfare attributes. In addition, t-test was used to analyze continuous data while Chi-square test was used to analyze categorical data.

3. Results and Discussion

3.1. Characteristics of Respondents

Socio-demographic characteristics of respondents are presented in Table 1. Distribution of respondents varied among age groups and sex ($p< 0.05$) between grape farmers and non- grape farmers. The majority of grape farmers (80.2%) were aged 35-64 and that grape farming was a predominantly male activity (75.4%, male; 24.6% female). Married individuals constituted the highest group engaged in farming activities compared to other categories. Large family size (6 members and above) was a characteristic observed from 31% of grape farmers compared to 24% of non-grape farmers ($p<0.05$). Evidence from previous studies show large household size is associated with increased chances of adoption of new interventions [11, 16]. This scenario could be attributed to increased labour force in large families. In this study, adoption of grape farming was independent of the education level ($p>0.05$) which is probably due to the fact that majority of the respondents had similar education background. In other studies, however, increased education has been found to enhance adoption of agricultural technology [29, 33].

Table 1. Social demographic characteristics of the respondents.

Variable	Grape farmers (n=126)	Non-grape farmers (n=126)	All (n=252)	χ^2-value
Age				
< 35	18(14.3)	36(28.6)	54(21.4)	8.45*
35 -64	101(80.2)	81(64.3)	182(72.2)	
64+	7(5.6)	9(7.1)	16(6.3)	
Sex				
Male	95(75.4)	85(67.5)	180(71.4)	2.16*
Female	31(24.6)	41(32.5)	72(28.6)	
Marital status				
Married	93(73.8)	76(60.3)	169(67.1)	6.42*
Single	17(13.5)	32(25.4)	49(19.4)	
Others	16 (12.7)	18 (14.3)	34(13.5)	
Education level				
None	19(15.1)	30(23.8)	49(19.4)	4.25ns
Primary	89(70.6)	85(67.5)	174(69.0)	
Secondary+	18(14.3)	11(8.7)	29(11.5)	
Household size				
Less than 6	87(69.0)	98(77.8)	185(73.4)	2.54*
6 and above	39(31.0)	28(22.2)	67(26.6)	

Figures in parentheses are percentages; ns = not significant, * significant at $p< 0.05$

3.2. Farm Size and Grape Yield

Table 2. Period in grape farming, farm size and grape yield (n = 126).

Variable	Frequency	Percent
Number of years engaged in grape farming		
< 5	12	9.5
5 - 10	76	60.3
> 10	38	30.2
Farm size under grape production (acre)		
<2	94	74.6
2-4	22	17.5
>4	10	7.9
Total yield last year (ton)		
<1	74	58.7
1-2	45	35.7
>2	7	5.6

Results from Table 2 indicate that majority of respondents (90.5%) were engaged in grape farming for at least 5 years. About seven in ten grape farmers (74.6%) had less than two acres of land under grape production with average acreage of 2.1. More than half of grape growers (58.7%) harvested below 1ton of grapes per year. The average yield was 1.6 ton per acre per year which is somewhat lower than the district average of 2 tons per acre per year [8]. With improved technology and management practices, however, one acre can produce more than 8tons of grapes [25].

3.3. Contribution of Grape Farming to Household Income

To determine contribution of grape farming to household income, grape farmers were asked to indicate their annual income from grape farming and other sources. Income from farm produce was derived by multiplying the amount of a given crop harvested with its market price (TZS 800 for the case of grape; 1USD≈TZS1800). It was found that 73% of grape farmers earned < 1million TZS per year while 21.4% and 5.6% earned 1-2 millions and >2millions TZS per year, respectively. The average annual income per household from grape farming was 823, 151 TZS which is equivalent to 35.6% of the total income. This was followed by 620,079 (26.8%) from other crops, 546,111 (23.6%) from small scale business, 131468 (5.7%) from livestock and livestock products and 191,904 (8.3%) from other sources. Grape farming had the highest contribution to household income notwithstanding the observed low productivity and low price of grapes. This indicates high potential of reducing poverty levels through grape farming especially when grape productivity is improved.

3.4. Contribution of Grape Farming to Household Welfare

To determine the contribution of grape farming to the household welfare, household consumption expenditure on non- durable goods per adult equivalent and scores on household food insecurity index for grape farmers were compared with those for non-grape farmers. As shown in

Table 3, grape farmers had higher average household consumption expenditure per adult equivalent per year than non-grape farmers (t = 13.30, $p< 0.001$). On the other hand, low food insecurity index was recorded for grape farmers compared to non-grape farmers (t = -5.71, $p< 0.05$) meaning that grape farmers were more food secure than non-grape farmers. These findings are supported by results in Table 4 which present respondents' judgment on the status of food security in their households. Only 11% of the grape farmers reported having had problems in meeting their food needs a year before the survey compared to 56.4% of non-grape farmers ($\chi 2=57.68$, $p<0.001$). Overall, 33.7% of all respondents were less food secure. Unlike grape farmers who generate income from grapes, non-grape farmers in the area depend entirely on selling food crops for income. This practice partly explains the observed higher level of food insecurity among non-grape farmers.

Table 3. Household consumption expenditure and score on household food insecurity index.

Variable	Grape farmers(n = 126)		Non- grape farmers(n = 126)		t-value
	Mean	SD	Mean	SD	
Household consumption expenditure per adult equivalent per year (TZS)	173,833	62,562	84,485	45,281	13.30***
Score on household food insecurity index	8.51	4.17	11.98	5.41	-5.71*

SD = Standard deviation; *Significant at $p< 0.05$; *** = Significant at $p<0.001$

Generally, hunger is commonly experienced for five months in the area, and the most critical months of food shortage are November, December, January, February and March. Analysis also showed that at least for a number of months, 72.9% of the respondents met their daily food needs through purchase, own production or by adopting various coping strategies. These included sale of liquid or productive assets, eating less often, borrowing money to buy food and doing casual labour. Because grape-growers were less

subjected to the transient threat of food insecurity, most of them (96.8%) were able to meet basic necessities throughout the year without selling productive or household assets as compared to 82.5% of non-grape farmers. It is intriguing to note that although grape productivity was low, grape farming improved the status of food security. These findings show the importance of high-value crops in household welfare as reported elsewhere [14, 36].

Table 4. Opinions of respondents on household food security status.

Variable	Grape farmers(n=126)	Non-grape farmers (n=126)	All(n=252)	χ^2 -value
Food secure	112(88.9)	55(43.6)	167(66.3)	
Less food secure	14(11.1)	71(56.4)	85(33.7)	57.68***

Figures in parentheses are percentages, *** = Significant at $p<0.001$

3.5. Constraints to Grape Farming

Table 5 shows constraints of grape farming as experienced by farmers. These constraints are ranked and total weighted scores for each constraint is presented. Low price relative to input cost was the most important constraint. This is linked to limited access of grape market. The market is mainly domestic and grapes are sold as fresh fruits as there are no value-adding activities. However, there are few grape processing industries in Dodoma and their capacity is rather low. The market linkages for grapes are therefore weak and this is likely to undermine the overall growth of grape-subsector in the area.

During discussion with the key informants, it was revealed that except for a few cases in Mpunguzi, Mbabala A and Mbabala B where there are farmer organizations, most of them work independently. However, farmer organization is a critical factor in making markets work for the poor particularly in high

value products [13]. Where farmers operate on independently, access to farm inputs/or market outputs is invariably more limiting and products are usually low paid [35, 36]. Beyond prices, farmer groups function as important catalysts for adoption of innovation through promoting efficient information flows [9]. Thus, creating or strengthening farmer organizations needs to be considered as an important strategy for growth of grape- subsector.

Pests and diseases present another constraint. Important pests are grasshoppers, caterpillar, beetles and sucking insects while major diseases are powdery mildew and downry mildew. These diseases have significant effects on growth, yield and quality of grapes particularly sugar levels, juice colour and acidity [12]. Although data for grape losses due pests and diseases are not available, discussion with farmers clearly indicated that pests and diseases cause considerable damage to grapevine.

Water shortage is among the dominant environmental constraint for grape production in this area. Farmers showed concern over increased price of water for irrigation which resulted in limited water use. Thus, grapevine often faced some degree of drought stress especially in the dry season. To cope with this limitation, some farmers particularly in Mpunguzi, Mbabala A and Mbabala B villages have developed irrigation system from ground water sources. Significant increase in grape production was reported due to this scheme. Nevertheless, the capital cost for developing a

ground water supply is high. This is even more challenging as none of the farmers we interviewed had access to any credit scheme. As with any other agricultural activity, institutions in the commercial financial sector do not reach small scale farmers because of perceived risks. The farmers are, therefore, deprived of important tool for development. Grape farming is also constrained by the lack of appropriate storage facilities such that significant losses often occurred before the produce reached consumers.

Table 5. Constraints to grape farming.

Constraint[a]	Rank				Total frequency	Total weighted score[a]
	1st	2nd	3rd	4th		
Low price of grape	70(280)	34(102)	10(20)	3(3)	117	405
High costs of inputs[b]	20(80)	40(120)	10(20)	7(7)	77	227
Limited access to market	30(120)	10(30)	6(12)	7(7)	53	169
Pests and diseases	15(60)	4(12)	2(4)	5(5)	26	81
Water shortage	12(48)	5(15)	1(2)	3(3)	21	68
Limited access to financial services	8(32)	3(9)	7(14)	5(5)	23	60
Lack of storage facilities	5(20)	6(18)	10(20)	1(1)	22	59
Limited access to quality seedlings	2(8)	4(12)	6(12)	3(3)	15	35

Figures outside parentheses are frequencies while those in parentheses are weighted scores. Weighted scores were obtained by multiplying frequencies by a weight of a respective rank with the 1st, 2nd, 3rd and 4th ranks taking a weight of 4, 3, 2 and 1, respectively. [a]Total weighted score for each constraint was obtained by summing up individual weighted scores from different ranks (indicated in brackets) in a respective constraint. [b]Includes cost for seedlings, pesticides and labour

4. Conclusion and Recommendations

This study has shown that grape farming is a significant driver for growth and a major source of income for the small scale farmers. Indeed, this activity plays an important role in household food security and the overall welfare of small scale farmers. However, limited access to market is a major constraint to full exploitation of grape production potential and jeopardizes efforts to improving productivity. In this regard, government intervention is needed to attract investment in grape processing industries which would broaden the range of products from grapes and enhance grape market. Improved linkage to existing markets along with provision of technical support is critical to the development of the grape sub-sector, and this would subsequently reduce rural poverty and food insecurity. As part of the endeavour to address these constraints, efforts are also needed to establish groups of producers and processors or strengthen the existing ones as a strategy to facilitate access to technical support, farm inputs, output market and financial services.

References

[1]   Arndt, C., Farmer, W., Strzepek, K and Thurlow, J. 2012. Climate change, agriculture and food security in Tanzania. *Review of Development Economics*, 16(3), 378-393

[2]   Asfaw, S., Shiferaw, B., Simtowe, F and Lipper, L. 2012. Impact of modern agriculture technologies on small holder farmer's welfare. Evidence from Tanzania and Ethiopia. *Food policy*, 37, 283-295.

[3]   Brussaard, L., Caron, P., Campbell, B., Lipper, L., Mainka, S., Rabbinge, R and Pulleman, M. 2010. Reconciling biodiversity conservation and food security: scientific challenges for a new agriculture. *Current Opinion in Environmental Sustainability,* 2(1), 34-42.

[4]   Coates, J., Swindale, A and Bilinsky, P. 2007. Household food insecurity access scale (HFIAS) for measurement of food access. Indicator Guide. Food and Nutrition technical assistance, USAID.

[5]   Cooper, P., Dimes, J., Rao, K., Shapiro, B., Shiferaw, B and Twomlow, S. 2008. Coping better with current climatic variability in the rain-fed farming systems of sub-Saharan Africa: An essential first step in adapting to future climate change? Agriculture, Ecosystems and Environment, 126(1), 24-35.

[6]   Devereux, S and Edwards, J. 2004. Climate change and food security. *IDS Bulletin, 35*(3), 22-30.

[7]   Di Falco, S and Chavas, J.P. 2009. On crop biodiversity, risk exposure, and food security in the highlands of Ethiopia. *American Journal of Agricultural Economics,* 91(3), 599-611.

[8]   Dodoma Urban District Report, 2014. Annual Report.

[9]   Fischer, E and Qaim, M. 2012. Linking smallholders to markets: determinants and impacts of farmer collective action in Kenya. *World Development*, 40(6), 1255-1268.

[10] Fisher, A.A., Laing, J.E., Townsend and J.W. 1991. Handbook for family planning operations research and design. *Operations research.* Population council, USA

[11] Franzel, S. 1999. Socio-economic factors affecting the adoption potential of improved tree fallows in Africa. *Agro forestry Systems* 47,305- 321.

[12] Gadoury, D. M., Seem, R. C., Pearson, R. C., Wilcox, W. F and Dunst, R. M. 2001. Effects of powdery mildew on vine growth, yield, and quality of concord grapes. Plant Disease, 85(2), 137-140.

[13] Hellin, J., Lundy, M and Meijer, M. 2009. Farmer organization, collective action and market access in Meso-America. Food Policy, 34(1), 16-22.

[14] Hossain, M and Jaim, W. M. H. 2011. Empowering Women to Become Farmer Entrepreneur. In Paper presented at the IFAD Conference on New Directions for Smallholder Agriculture, Vol. 24, p. 25.

[15] IFAD.2008. International Fund for Agricultural Development. The role of high value crops in rural poverty reduction in the near east and North Africa.

[16] Junge, B., Deji, O., Abaidoo, R., Chikoye, D and Stahr, K. 2009. Farmers' Adoption of Soil Conservation Technologies: A Case Study from Osun State, Nigeria. The Journal of Agricultural education and extension.15 (3), 257- 274.

[17] Kahimba, F.C., Mutabazi, K.D., Tumbo, S.D., Masuki, K.F and Mbungu, W.B. 2014. Scaling –up of conservation agriculture in Tanzania: A case of Arusha and Dodoma Regions. Natural Resources, 2014

[18] Kisusu, R. W.,Mdoe, N. S. Y.,Turuka, F. M and Mlambiti, M. E. 2000. Contribution of smallholder dairy production to food security, household income and poverty alleviation: the case of Mvumi dairy development project, Dodoma. In: Proceedings of the 18th Scientific Conference of the Tanzania Veterinary Association held at AICC, Arusha, Tanzania, 5-7 December 2000. Vol. 20 pp 89- 99

[19] Kulig, A., Kolfoort, H and Hoekstra, R. 2010. The case for the hybrid capital approach for the measurement of the welfare and sustainability. Ecological Indicators, 10(2), 118-128.

[20] Kuwornu, J.K.M and Eric S. O. 2012. Irrigation access and per capita consumption expenditure in farm households: Evidence from Ghana. Journal of Development and Agricultural Economics, 4 (3), 78-92.

[21] Langat, J.K., Mutai, B.K., Maina, M.C and Bett, H.K. 2011. Effect of Credit on Household Welfare: The Case of "Village Bank" Credit in Bomet County, Kenya. Asian Journal of Agricultural Sciences 3(3), 162-170.

[22] Lin, B. B. 2011. Resilience in agriculture through crop diversification: adaptive management for environmental change. BioScience, 61(3), 183-193.

[23] Lipton, M. 2005. 'The family farm in a globalizing world: The role of crop science in alleviating poverty'. Policy Brief 74. Washington, DC: IFPRI.

[24] Liwenga, E. 2003. Food Insecurity and Coping Strategies in Semiarid Areas. The Case of Mvumi in Central Tanzania. PhD thesis, Stockholm University, Stockholm, Sweden.

[25] Mgwasa, D. 2012. Tanzania Distillers Limited (TDL) contribution on Dodoma vineyards development in Tanzania. Regional Workshop held on June 16- 17, 2012, Dar-es-salaam, Tanzania.

[26] Mkenda, A.F., Luvanda, E.G., Rutasitara, L and Naho, A. 2004. Poverty in Tanzania. Comparisons across administrative regions.

[27] Msaki, M.M., Mwenda, M.I and Regnald, I.J. 2013. Cereal bank as a necessary rural livelihood institute in Arid land, Makoje Village, Dodoma village, Dodoma – Tanzania. Asian Economic and Financial Review, 3(2), 259-269.

[28] Mukhebi, A., Mbogoh, S and Matungulu, K. 2012. Climate change, agriculture and food security in Tanzania. Review of Development Economics, 16(3), 378-393

[29] Odoemenen, I. U and Obinne, C.P.O. 2010. Assessing the factors influencing the utilization of improved cereal crop production technologies by small-scale farmers in Nigeria. Indian Journal of Science and Technology. 3(1), 180-183.

[30] Omidvar, N., Ghazi-Tabatabie, M., Sadeghi, R., Mohammadi, F and Abbasi-Shavazi, M.J. 2013. Food insecurity and its socio demographic correlates among Afghsn Immigrants in Iran. Journal of Health Population and Nutrition, 3, 356- 366.

[31] Rakodi, C. 2002. A livelihoods approach: conceptual issues and definitions. Urban livelihoods: A people-centred approach to reducing poverty, 3-22.

[32] Sahn, D. E and Stifel, D. 2003. Exploring alternative measures of welfare in the absence of expenditure data. Review of income and wealth, 49(4), 463-489.

[33] Saka, J.O., Okoruwa, V.O., Lawal, B. O and Ajijola, S. 2005. Adoption of Improved Rice Varieties among Small-Holder Farmers in South-Western Nigeria. World Journal of Agricultural Sciences 1 (1), 42-49.

[34] Schmied, D. 1993. Managing food shortages in Central Tanzania. GeoJournal, 30(2), 153-158.

[35] Shiferaw, B., Obare, G and Muricho, G. 2006. Rural institutions and producer organizations in imperfect markets: Experiences from producer marketing groups in semi-arid eastern Kenya. Journal of SAT Agricultural Research, 2(1), 1-41.

[36] Singh, S. 2013. Governance and upgrading in export grape global production networks in India: BWPI, The University of Manchester.

[37] Steduto, P., Hsiao, T. C., Raes, D and Fereres, E. 2012. Crop yield response to water: Food and Agriculture Organization of the United Nations.

[38] Stringer, L. C., Dyer, J. C., Reed, M. S., Dougill, A. J., Twyman, C and Mkwambisi, D. 2009. Adaptations to climate change, drought and desertification: local insights to enhance policy in southern Africa. Environmental Science and Policy, 12(7), 748-765

[39] Swai, O. W., Mbwambo, J. S and Magayane, F. T. 2012. Gender and perception on climate change in Bahi and Kondoa Districts, Dodoma Region, Tanzania. Journal of African Studies and Development, 4(9), 218-231.

[40] Toledo, Á and Burlingame, B. 2006. Biodiversity and nutrition: A common path toward global food security and sustainable development. Journal of Food Composition and Analysis, 19(6), 477-483.

[41] United Republic of Tanzania. 2007. National Adaptation Programme of Action (NAPA), Vice President's Office, Division of Environment, p. 52.

[42] United Republic of Tanzania. 2013. Population and Housing Census. National Bureau of Statistics (Tanzania).

[43] Weinberger, K and Lumpkin, T.A. 2007. 'Diversification into horticulture and poverty reduction: A research agenda'. *World Development* 35(8), 1464-1480

25

Women Economic Empowerment Through Non Timber Forest Products in Gimbo District, South West Ethiopia

Getahun Kassa[*], **Eskinder Yigezu**

Department of Agricultural Economics, Mizan-Tepi University, Mizan Teferi, Ethiopia

Email address:

zgetah@gmail.com (G. Kassa), eskobar_y@yahoo.com (E. Yigezu)

Abstract: In many communities there are limited income-generating opportunities for women. Fortunately, NTFP activities are one of the few cash-generating opportunities for women in marginalized rural communities. Nevertheless, increasing pressure on forests and biodiversity has increased the pressure on rural women. We investigated the factors influencing the income that women drive from NTFPs collection. In addition, our research also investigated the contribution of NTFPs income for the total annual income of the women and for reducing income inequality in Gimbo District. Two kebeles were selected based on NTFPs availability, level of forest exploitation activities of the people and nearness to road. Data were collected from 120 selected women through structured interviews, focus group discussions, market assessments as well as field observation. Though NTFPs accounted for 53.76% of the annual income for women, its contribution is affected by different factors. A multiple linear regression model was used to identify factors that affect the income from NTFPs. Out of eight variables included in the regression, four variables such as non NTFPs (other) income, time spent in NTFPs collection, proximity to the forest and distance to market significantly affected the income women derive from NTFP activities. The contribution of NTFPs income in reducing income inequality was analyzed by using Gini Coefficient. Comparing the Gini index with and without NTFPs production (income), the income disparity lowered from 0.40 to 0.27 in the inclusion of NTFPs. Therefore, empowering women through NTFP activities can create significant opportunity for women in terms of income and in reducing income inequality. Thus, policy programs should due attention in improving alternative sources of income for women in the study area.

Keywords: NTFPs, Determinants of Women Income, Gini Coefficient

1. Introduction

Women often face many disadvantages that hamper their ability to engage in economic activity [1]. A rural woman spends most of her time taking care of her husband, children, elders and other members of her family. Consequently, they become economically dependent and vulnerable, educationally backward as well as politically and socially disadvantaged. These in turn trigger off huge social, economic, and environmental costs on society as a whole and rural development in particular [2]. As women are generally the poorest of the poor, eliminating social, cultural, political and economic discrimination against them is a prerequisite of eradicating poverty [3]. Therefore, attention has been given to poor women's economic empowerment universally to catalyze development efforts.

Economic empowerment increases women's access to economic resources and opportunities including jobs, financial services, property and other productive assets, skills development and market information [4]. Women's economic empowerment is fundamental to enhancing and strengthening their economic and social status. According to Hill [2] rural women's economic empowerment can have a positive impact on their social and political empowerment through their increased respect, status, and self-confidence and increased decision-making power in households, communities, and institutions. It can also shift power relations between women and men by increasing women's control over household budgets [5]. Thus, if given responsibilities to managing critical household assets and natural resources, women are potential agents of change [6].

Around the world, resilient and resourceful rural women contribute in a multitude of ways through different livelihood strategies to lift their families and communities out of poverty

[2]. Women's enterprise activities have key importance in providing food and income that enable their families to exist [7]. If included in national accounts, this income represents between 15 to 50 percent of national Gross Domestic Products [8]. Considering women's employment, statistics shows that many women are employed in the informal sector and in the non-skilled areas [9]. Fortunately, NTFPs offer greater opportunities for women producers in the informal economy [10].

NTFPs include the product benefits or services that come from a forest except timber [11]. These include plants and plant materials used for food, fuel, storage and fodder, medicine, cottage and wrapping materials, biochemical, as well as animals, birds, reptiles and fishes, for food and feather [12]. It can also be referred as all the resources/products that are extracted from forest ecosystem and utilized within the household or marketed or meant for social, cultural or religious significance [13].

NTFP activities are one of the few cash-generating opportunities for women in marginalized rural communities [14]. Women find NTFP activities attractive because of the low technical and financial entry requirements, freely available resource base and instant cash in times of need [7]. By collecting NTFPs from the forest, women contribute significantly in meeting subsistence requirements and increasing family income [15]. To put it differently, income generated from NTFP activities by women can add significantly to their households' purchasing power [16]. Therefore, empowering women in the forest sector can create significant development opportunities for them and generate important spill over benefits for their households and communities [17].

NTFPs are playing a key role in generating income and improving the livelihood of rural women in Ethiopia. The money earned from collecting, selling or processing forest products has an indispensable contribution to household income [18]. However, the actual contribution of NTFPs to specific rural households especially for women is worth mentioning. Therefore, our research aimed at using a gender vantage point in revealing the role of women in collecting NTFPs; the constraints discouraging their role in NTFP activities; and estimating the contribution of NTFPs for total annual income and for mitigating income inequality in Gimbo District.

2. Research Methodology

2.1. Description of the Study Area

The study was conducted in Gimbo district, Kaffa zone, Southern Nations Nationalities and People's Region (SNNPR). It is found within the southwestern plateau of Ethiopia and 450km and 725km far from Addis Ababa and Hawassa respectively. The area lies within 07°00'- 7°25'North latitude and 35°55'-36°37'East longitude. The altitude of the study sites ranges from 1600 to 1900 m.a.s.l. The topography is characterized by slopping and rugged areas with very little plain land [19].

2.2. Data Collection

Gimbo district was selected purposively based on forest accessibility and availability of NTFPs. Two kebeles were purposively selected from the study area as a result of NTFPs availability, level of forest exploitation activities of the people and nearness to road. A total of 120 women respondents were selected randomly using proportional to size techniques based on the number of female respondents in Kebeles using NTFPs, so that, all sample units had equal chances of being selected. We used a method developed as in [20] to select the total sample size from the total households. Based on rules-of-thumb, "Ref [20]" suggested that the minimum number of respondents for each explanatory variable in a regression analysis should be 5 respondents to 1 explanatory variable. Therefore, in this study there are 10 explanatory variables and the minimum sample size should be 50. This is the minimum recommended sample size, but it is possible to select the total sample size above the recommended minimum sample size.

For the purpose of this study both primary data and secondary data were collected. Primary data were collected through focus group discussion and interview using semi structured questionnaire. For focus group discussions, women who have good experience in NTFP activities were selected to discuss specific issues related to the purpose of the study by forming small groups (members of 7-8) with homogenous composition. Secondary data were also collected from relevant zonal and district offices in the study area.

2.3. Data Analysis

To meet the objectives of the study descriptive statistics, Gini Coefficient and econometric model were used based on their importance for analyzing the quantitative data that have been collected from primary sources. Quantitative data were processed using Statistical Package for Social Sciences SPSS 21 and Microsoft Excel 2007.

Income inequality was measured with the use of the Gini Coefficient. The Gini coefficient is defined as a ratio with values between 0 and 1. Here, 0 corresponds to perfect income equality (i.e. everyone has the same income) and 1 corresponds to perfect income inequality (i.e. one person has all the income, while everyone else has zero income).

To answer the question of factors influencing the rural women income from NTFP production, econometric model was used. Factors influencing rural women income from NTFPs were estimated using multiple linear regressions model ([21]; [22]). The total monetary value of the various types of NTFPs collected by each woman per year is found to be the most appropriate unit of measurement to identify the factors that influence the rural women income from NTFPs. In this case, income (monetary value) derived from NTFPs by each woman was considered as dependent variable. The following econometric model was employed:

$$Y = \beta_o + \beta_i X_i + \varepsilon$$

Where: Y = income derived from NTFPs;

β_o = intercept (the value of Y when X=0)

β_i = a vector of estimated coefficient of the explanatory variables and;

X_i = explanatory variables

ε = the stochastic disturbance term

3. Result and Discussion

3.1. Household Characteristics

Of the total respondents, 95 % were married, 2.5% of them were single and the remaining 2.5% of them were divorced. Having 95% married respondents has a good advantage in providing financial support for husband and children. Regarding their religion, 77.5% of them were the followers of the Ethiopian Orthodox Thewahido, 16.67% were Protestant and the remaining 5.83% of them were followers of other religions in the country. With regard to ethnic composition of the respondents, Kaffa was the dominant group accounted for 79.17% of the total respondents, about 16.67% belonged to the Menja ethnic group and 4.16% were the remaining ethnic groups of the country. The uniqueness in the study area is that the Menja ethnic minorities consists the second largest percentage of the forest users in the study area. The Menja minorities are highly dependent on the forest as a means of livelihood.

Table 1. Socioeconomic Characteristics of the Respondents.

Variables	Minimum	Maximum	Mean	St. Deviation
Age	20	60	33.53	9.34
Family size	2	9	5.24	1.90
Education	0	10	1.69	2.70

Source: Household Survey (2015)

The mean age of the respondents was 33.53 years. The largest proportion of the respondents (94.17%) lay in the age range of 20-50 years. This implies the majority of the respondents were active work forces. The mean family size of the respondents was 5.25. With regard to educational attainment of the respondents, the mean educational attainment was 1.69 grades. The majority of the respondents (63.33%) were unable to read and write.

3.2. Major NTFPs Collected by Women

The women in the study area exploited various NTFPs like forest coffee, honey, feral honey, spices (Cardamom and long pepper), fuel wood, charcoal, bamboo and medicinal plants from the forest. The NTFPs were classified based on their importance for commercial as well as house use only. According to the information gathered through focus group discussion, women classified NTFPs like forest coffee, forest honey and spices as important NTFPs that generate them significant contribution of cash income. On the other hand, the women extracted fuel wood, charcoal, bamboo and medicinal plants for house-use only.

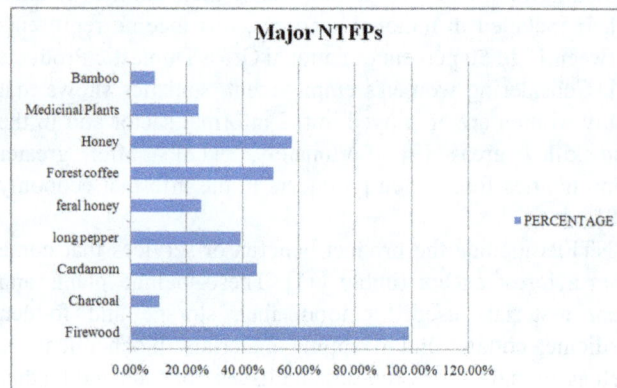

Source: Household Survey (2014)

Figure 1. Major NTFPs.

The figure (Figure 1) shows the majority of the women harvested fuel wood from near communal or state forests every day to meet the daily energy demand in the household. Charcoal is another source of energy that the women extracted from the forest. Our finding is consistent with the findings of those of [23] who reported wood fuel as the most common source of NTFPs income for 68% of households. Another finding as in "Ref. [24]" revealed that firewood constitutes the largest proportion (59%) of forest income.

In our finding, it is found that forest coffee, honey and spices were the most important cash generating NTFPs. In line with this, "Ref. [25]" revealed that 33% of households generate cash income by selling forest coffee. "Ref. [18]" also revealed that the benefits received from beekeeping encourage the preservation of the forest in South West Ethiopia.

3.3. Contribution of NTFPs for Women Income

Women in the study area earned income from limited sources such as agriculture (mainly crop production) and forests (NTFPs). There are limited economic opportunities in the study area which hindered the women from diversifying their sources of income. As a result, the women living at the proximity of the natural forest depend highly on the forest to extract many NTFPs.

Table 2. Sources and Proportion of Women Income.

No	Sources	Income	Percentage
1.	Agriculture (non NTFPs)	1,340.50 ETB	46.24%
2.	NTFPs	1,558.78 ETB	53.76%
	Total	2,899.28 ETB	100%

Source: household survey (2014)

NTFPs production was the main source of income (53.76%) for the women in the study area. Consistent with our findings, a study conducted in Nigeria revealed that the major source of income to the household is the income from NTFPs [26]. As in "Ref. [27]" in Southern Ethiopia forest products constitute an important part of the household income portfolio contributing 34% of total per capita income. "Ref. [23]" also supported that NTFPs account for about one third of total household income.

3.4. Role of NTFPs Income in Reducing Income Inequality

The figures (Figure 2 & 3) show the impact of NTFPs income on income inequality. Lorenz curves with the data for households' income including or excluding show that addition of NTFPs income to total income reduces the departure of the curve from the line of equal distribution.

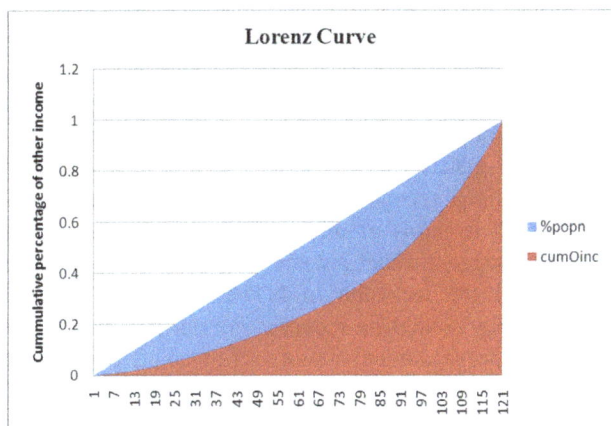

Figure 2. *Lorenz Curve without NTFPs Income.*

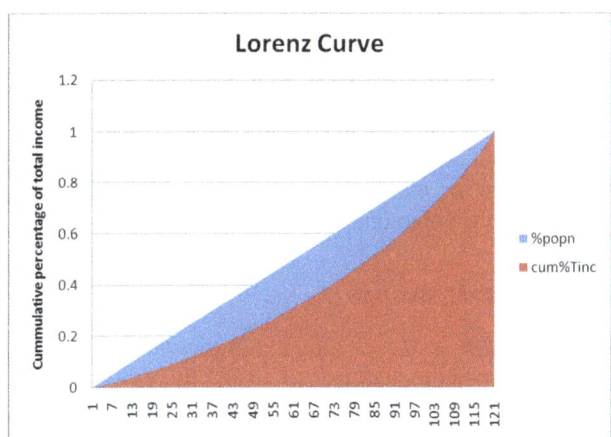

Figure 3. *Lorenz Curve with the inclusion of NTFPs Income.*

The Gini index for other incomes (non NTFPs income) is 0.40. This index is higher than the national value of Gini index (0.34) set by World Bank for the year 2014 [28]. However, when income from NTFPs was included in the total annual income for the women, the Gini index was calculated at 0.27. Comparing the Gini index with and without NTFPs production (income), the disparity was lowered by a coefficient of 0.13 in the inclusion of NTFPs. The result shows NTFPs play greater role in reducing income inequality among women in the study area.

Table 3. *Gini Coefficient for the two income sources.*

No	CASES	GINI INDEX (PERCENTAGE)	GINI COEFFICIENT
1.	Other income	39.55%	0.40
2.	Total income	27.05%	0.27
	Change in Gini index	12.5%	0.13

Ethiopian National Gini in 2014 is 33.6% (BTI, 2014)

In line with our finding, in Southern Ethiopia NTFPs income from gum and resin contributed to the reduction of the measured income inequality from 0.47 to 0.22 [29]. Another finding in Burkina Faso revealed that if forest (TFPs and NTFPs) sources of income are excluded from analysis, the estimated Gini coefficient increases from 0.47 to 0.89 which shows that addition of forest income reduces measured income inequality of 42% [30]. In "Ref. [27]" it is indicated that the increase in Gini coefficient becomes highest when forest income is excluded (0.45). Our study also concurs with the finding in "Ref. [24]" which indicated that income inequality is 0.28 when forest incomes are included, but the coefficient rises to 0.41 when forest incomes are excluded.

3.5. Determinants of Income from NTFPs

The multiple linear regression analysis was carried out to identify the factors influencing rural women income from NTFPs. The F test shows that the model is significant at less or equal to 0.000% probability level. The R^2 value of 0.607 indicates the explanatory power of the model. Thus, 60.7% of the variation in the dependent variable was explained by the regression. Explanatory variables such as educational attainment of the respondents (educ), the experience of rural women in collecting the NTFPs (experie), family size (Fsize) and time spent in the forest (Tspent) exhibited positive relation to income derived from NTFPs. But, age of the respondents (AGE), other income of the respondents (Oincome), proximity to the forest (DISFOREST) and distance to the market (DISMARKET) showed negative relation to income derived from NTFPs. However, out of the total variables, four were statistically significant at 10, 5 and 1% probability level. These were other (non NTFPs) income, time spent in the forest, proximity to the forest and distance to the market from the residence of respondents.

Table 4. *Factors Influencing Women Income from NTFPs.*

Variables	Coefficients	Std. Error	t statistics	significance
Constant	2370.79	577.74	4.10	.00
Age	-11.67	16.09	-.73	.47
Education	38.15	36.05	1.06	.29
Oincome	-.14	.06	-2.26	.03**
Famsize	13.09	52.13	.25	.80
Experience	9.98	31.59	.32	.75
Tspent	405.37	62.32	6.50	.00***
Pforest	-2347.98	372.92	-6.30	.00***
Dmarket	-215.98	116.55	-1.85	.07*

***, **, and * refer to significance at less than 1%, 5%, and 10% probability levels, Dependent Variable: income from NTFPs, N=120, R-Square=0.634, Adjusted R2=0.607, F=24.012

As predicted, proximity to the forest negatively affected the income from NTFPs at less than 1% significance level. Therefore, those residing far away from the forest are less interested in NTFP activities because of higher transaction cost. In line with our finding, a study conducted in Bonga Forest revealed that the negative and significant effect of distance to forest on income from NTFPs was due to the increase in cost of production [25]. Another study conducted in Nigeria noted that the negative association between income from NTFPs and distance to the forest may results from large transportation cost, if large quantities of NTFPs are gathered [26]. Another finding also noted that the negative association between distances of house from forest with share of forest income gives an idea that people nearby forest had more access to forest, easy to reach there and were more benefited [31]. A research conducted in Nigeria concluded that as one gets farther away from the forest area, she/he tends to use less of the forestry resource [32].

In our study, time spent in the forest positively affected the income from NTFPs at less than 1% significant level. This means those individuals who spent more time in the forest collect more quantity of NTFPs. Women devote their time in NTFPs collection if the return they earn from it is attractive. Therefore, the more time spent in the forest, the higher is the income women drive from NTFPs.

Other income (non NTFPs income) was another variable found to be negatively associated with NTFPs income. The negative association between other income and NTFPs income indicates if women receive more income from other activities, they will be less reliant on NTFPs income. Therefore, the higher income women drive from other sources; they will be less attracted to NTFPs. In line with our finding, "Ref. [33]" revealed that negative association between cash income per capita and the share of income generated by NTFPs. This happens because of low income from other sources, people have to collect commercial NTFPs to increase their household income [34].

Another important variable that significantly affected women income from NTFPs is distance to market. We found a negative association between distance to market from the residence of respondents and NTFPs income at less than 10% significance level. This means longer distance discourages women from being engaged in NTFP activities due to high transaction costs. Consistent with our finding, "Ref. [25]" and "Ref. [26]" revealed that distance from forest to point of sale has a negative and significant effect on NTFPs income which may results from large transportation cost. In "Ref. [23]" it is concluded that the farther away a household is from the market, the lower the likelihood to participate in NTFPs and also the less dependent a household is on income from NTFPs.

3.6. Conclusions and Policy Implications

The main source of income for women in the study area consist agriculture (mainly crop production) and NTFP activities. With such limited income sources, NTFPs played an important role in contributing to women income (53.76%) and mitigating income inequality. Comparing the Gini index with and without NTFPs production (income), the income disparity lowered from 0.40 to 0.27 in the inclusion of NTFPs. However, the contribution of NTFPs income for women is influenced by several factors. Time spent in NTFPs collection, proximity to the forest, non NTFPs income and distance to market significantly influenced the income women derive

from NTFP activities. Therefore, empowering women through NTFP activities can create significant opportunity for women in terms of income generating and in reducing income inequality if attention is due for the factors that affect NTFPs income in the study area.

Based on the findings of our study, the following policy directions are recommended;

- Income derived from NTFPs extraction contributed significantly to the annual income of sampled women in the study area. Therefore, women economic empowerment policies should give emphasize to the importance of NTFPs by considering it as alternative income generating source for rural women due to its international recognition and usefulness of NTFPs.

- Second, in terms of mitigating income inequality, the result suggested that income inequality can be reduced through the inclusion of NTFPs income. Therefore, inequality should be reduced through policy programs that improve alternative sources of income for women in the study area.

- Distance to market affected the income that women derive from NTFPs. This underscores the importance of improving access to market in order to encourage rural women to diversify into NTFP business activities and increase their income.

- Finally, distance to the forest discouraged the women in the study area from being engaged in NTFP activities. Therefore, the women should be encouraged to domesticate minor forest products in their backyards outside the community forest area.

Acknowledgments

This research is the result of a successful collaborative effort by many individuals and we would like to pay tribute to all individuals who contributed to this work. We are highly indebted to Belachew Adulo, Mulualem and Kasech for their support while collecting the data. We would also like to thank Mizan-Tepi University for providing us the necessary financial support to conduct our work. Our special appreciation goes to Gimbo District Agricultural office for the cooperation to conduct our research in the area. Lastly, we would like to thank Dr. Yared Berhanu.

References

[1] Carr M, (2008). Gender and non-timber forest products: Promoting food security and economic empowerment. International Fund for Agricultural Development (IFAD), Italy.

[2] Hill C, (2011). Enabling Rural Women's Economic Empowerment: Institutions, Opportunities, and Participation.

[3] International Conference on Population and Development (ICPD), 1994. Program of Action

[4] OECD, (2012). Promoting Pro-Poor Growth: The Role of Empowerment. The OECD DAC Network on Gender Equality (GENDERNET)

[5] Esplen E, and Brody A, (2007). Putting Gender Back in the Picture: Rethinking Women's Economic Empowerment. Report prepared at the request of the Swedish International Development Cooperation Agency (Sida); Bibliography No. 19

[6] International Fund for Agricultural Development (IFAD), 2014. The Gender Advantage. Women on the Front of Climate Change.

[7] International Fund for Agricultural Development (IFAD), (2008). Gender and non-timber forest products. Promoting food security and economic empowerment.

[8] Budlender, D, (2010). 'What Do Time Use Studies Tell us about Unpaid Care Work? Evidence from Seven Countries', in Time Use Studies and Unpaid Care Work, ed. Budlender, UNRISD; Hoenig S. A. and Page, A.R.E, 2012, Counting on Care Work in Australia, Report prepared by AECgroup Limited for economic Security4women, Australia; Charmes, J. and J. Unni (2004), "Measurement of work", in G. Standing and M. Chen eds., Reconceptualising Work, International Labour Organization.

[9] Mersha G, (2007). Gender Mainstreaming In Forestry in Africa; Ethiopia. Food and Agriculture Organization of the United Nations Rome, 2007

[10] International Fund for Agricultural Development (IFAD), 2010. Gender and Desertification. Expanding roles for women to restore dry land areas enabling.

[11] Wong J, (2000). The biometrics of non-timber forest product resource assessment: A review of current methodology.

[12] Adepoju, Adebusola A, Salau, Sheu A, (2007). Economic Valuation of Non-Timber Forest Products (NTFPs), Univerisity of Ibbadan, Ibbadan.

[13] Food and Agricultural organization (FAO), (1990). Major Significance of Minor Forest Product: The Local Use and Value of Forest in the West African Humid Forest Zone. Community Forestry Note 6 Rome.

[14] Schreckenberg K, and Marshall E, (2006). Women and NTFPs: Improving income and status? In: Marshall E, Schreckenberg K, and Newton A.C, (eds.) Commercialization of non timberforest products: Factors influencing success. UNEPWCMC, Cambridge. 136p

[15] World Wildlife Fund UK (WWF-UK), (2012). Forest Management and Gender.

[16] Food and Agricultural Organization (FAO), (2014). Women in Forestry: Challenges and Opportunities

[17] FAO, (2013). Forests, food security and gender: linkages, disparities and priorities for action. Background paper for the International Conference on Forests for Food Security and Nutrition, FAO, Rome, 13–15 May, 2013

[18] Mohammed A, Tadesse W, and Abebe Y, (2006). Counting on Forests: Non-Timber Forest Products and Their Role in the Households and National Economy in Ethiopia. Published in Agricultural Economics Society of Ethiopia (AESE): Commercialization of Ethiopian Agriculture, Proceedings of the 8th Annual Conference of Agricultural Economics Society of Ethiopia, February 24-26, 2005, Addis Abeba.

[19] Matheos E, (2001). Inventory of woody species in Bonga Forest. Institute of Biodiversity Conservation and Research. Technical Report No. 1. Addis Abeba.

[20] Tabachnick, B. G., & Fidell, L. S. (1989). Using Multivariate Statistics. 2nd edition. Cambridge, MA: Harper & Row.

[21] Gujarati D, (2004). Basic Econometrics. Fourth edition. New York. The McGraw–Hill Companies.

[22] Wooldridge J, (2005). Introductory Econometrics. A modern Approach. Third Edition. Thomson South Western.

[23] Mulenga B, Richardson R, and Tembo G, (2012). Non-Timber Forest Products and Rural Poverty Alleviation in Zambia, Working Paper No. 62, April 2012

[24] Getachew M, Sjaastad E, and Velded P, (2007). Economic dependence of forest resource: case from Dendi District, Ethiopia. Forest Policy and Economics 9:916-927.

[25] Ermias M, Zeleke E, and Demel T, (2014). Non-timber forest products and household incomes in Bonga forest area, southwestern Ethiopia. Journal of Forestry Research (2014) 25(1): 215–223

[26] Raufu M, Akinniran T, Olawuyi S, and Akinpelu M, (2012). Economic Analysis Of Rural Women Income From Non-Timber Forest Products In Ife South Local Government Area of Osun State, Nigeria. Global Journal of Science Frontier Research Agriculture & Biology; Volume 12 Issue 1 Version 1.0 January 2012

[27] Yemiru T, Roos A, Campbell B, and Bohlin F, (2010). Forest Incomes and Poverty Alleviation under Participatory Forest Management in the Bale Highlands, Southern Ethiopia. International Forestry Review 12(1):66-77.

[28] Bertelsmann Stiftung's Transformation Index (BTI), (2014). Ethiopia Country Report. Gütersloh: Bertelsmann Stiftung, 2014.

[29] Asmamaw A, Pretzsch J, Secco L, and Mohamod T, (2014). Contribution of Small-Scale Gum and Resin Commercialization to Local Livelihood and Rural Economic Development in the Dry lands of Eastern Africa. Forests 2014, 5, 952-977

[30] Ouedraogo B, and Ferrari S, (2012). Incidence of forest income in reducing poverty and inequalities: Evidence from forest dependent households in managed forests' areas in Burkina Faso. Cahiers du GREThA, n°2012-28.

[31] Khanal B, (2013). Determinants of Farmers' Income from Community Forestry in Nawalparasi, Nepal. The Journal of Agriculture and Environment Vol: 14, Jun.2013

[32] Onoja A, and Unaeze H, (2009). Forest Income Determinants among Rural Households of Etche Local Government Area, Rivers State, Nigeria. Journal of Sustainable Development in Africa (Volume 11, No.3, 2009)

[33] Rodríguez F, (2007). Socio- economic Determinants of Non-Timber Forest Products Collection. A case study among indigenous people in Karnataka, India

[34] Quang D, and Anh T, (2006). Commercial collection of NTFPs and Households living in or near the forests; case study in Que, Con Coung and Ma, Toung Doung, Ghe An, Vietnham. Ecological Economics 60 (2006): 65-74

26

Effect of Micro Catchment Techniques on Vegetative Growth of Jatropha (*Jatropha Curcas*)

Azmi Elhag Aydrous[1], Abdel Moneim Elamin Mohamed[2], Abdelbagi Ahmed Abdelbagi[3]

[1]Department of Agricultural Engineering, Faculty of Agriculture, Omdurman Islamic University, Omdurman, Sudan
[2]Department of Agricultural Engineering, Faculty of Agriculture, University of Khartoum, Khartoum, Sudan
[3]Department of Crops sciences, Faculty of Agriculture, Omdurman Islamic University, Omdurman, Sudan

Email address:
azmielhag@yahoo.com (A. E. Aydrous)

Abstract: This study was conducted during two consecutive seasons (2011/2012 and 2012/2013) at the southern Omdurman region near Khartoum New International Airport (KNIA) to evaluate the effect of four rain water harvesting micro catchment techniques on some growth attributes of Jatropha trees. The micro catchments used were semicircular, pits, deep ditch and V-shape in addition to the control. A randomized complete block design with four replications was used to study the effect of these techniques on trees growth. The results of the study showed that plant height, stem thickness (for some cases) and number of leaves/tree for Jatropha trees were significantly increased under the micro catchment techniques as compared to the control, with higher mean obtained by V-shape.

Keywords: Micro-Catchment Techniques, Jatropha

1. Introduction

Water erosion is dominant in semi arid zones. Sudan is suffering from water erosion and water eroded soils in Sudan was estimated to be about 8 million hectares (Akhtar and Menching, 1993). Water harvesting techniques were used in Sudan for increasing agricultural production Omer et al (1997). For trees also Salih and Iinanga (1997) tried this for increasing production of sorghum and gum. The main reducing effect of vegetation on flooding and runoff water erosion is that vegetation roots act as a binding agent to soil particles. In addition, litter and roots which decay in the soil act as a cementing agent that improves soil structure and that vegetation canopy intercepts rain drops and reduces soil erosion caused by rain drop splash (Alaktar and Aga, 1995). Vegetation also reduces runoff speed and thus reduces its capacity to detach and remove soil particles. The alterations it makes to soil through combination of actions such as root binding effect, leaf and litter humus formation, the cover delay and reduce volume of flood and intercept rain drops (Alaktar and Aga, 1995). Forest cover may probably cut at least the peak flow by 80% and run off volume by 40%, besides removal of shrubs and grasses cause potential erosion

risk (Meunive, 1996). Young (1989) reported that 60 -75 % of ground cover maintained through the rainy season was effective against erosion of vertisols in Australia. Also Meunive (1996) indicated that with 40% cover grown mainly in gully reduced sediment yield as much as 92.5 % even during storms. Mechanical means for erosion control can be used such as cut –off drain technique or other methods can be used to reduce erosion. But it should be complemented by plant cover for everlasting and sustainable protection. Plant cover protects soil from direct rain drops and roots of plants bind soil particles and falling litter latter improves physical and chemical soil properties. The objectives of this study were to investigate the effect of the water harvesting techniques on the growth of Jatropha.

2. Materials and Methods

The experimental work was conducted at Jebel Awlia locality 40 kilometers south of Omdurman city and 25 kilometers from the west bank of the White Nile River during 2010-2011 and 2011-2012 rainy seasons. The experiments covered an area of 5 hectares as a part of the area designated for Khartoum New International Airport. The climate of the area is semi desert, which was characterized by high temperature of an average of 45°C during the summer. Wind speed is very high evoking dust. Very sparse herbaceous plants and *Acacia* trees comprise the plant cover which is green during the rainy season. The soil is light to sandy in composition except at lane beds especially Mansourab dam. Soil changes gradually to clay and sand-clay according to level and topography.

Five water harvesting techniques were used as follows:

1. Cultivation on land without water harvesting technique (control) treatment denoted by (T_0).

2. Cultivation with semi- circular water traps, designed with 30 meters diameter and 90 cm height. The distance between one trap and another is 20 meters Jatropha (*Jatropha curcas*) seedlings were sown at the inside the bank of the semi-circular trap with spacing of 4 meters between plants. The water dikes were composed of 3 units, and the dikes were 50 meters from the next unit of dikes (T_1).

3. V-shaped water dikes: Each side was 30 meters long and at the bottom of the V-shape dikes the distance is also 30 meters. The distance between a set of dikes and the other was 20 meters. The water trap was composed of 3shapes at the front and 2shapes at the rear at a distance of 50 meters between the front and rear. The previously mentioned tree species were also sown intent of the dikes at a distance of 4 meters between plants (T_4).

4. Pits: The pits were designed at 5 meters width, 10 meters length and 10 meters between pits. Pits were dug according to the land gradient. Water trap was composed of 3 pits at the front and 2 pits at the rear at a distance of 50 meters between the front and rear pits. The previously mentioned tree species were sown with the same spacing at 4 meters between plants (T_2).

5. Deep ditches: The deep ditches were dug by a motor grader. The length of each ditch was 30 meters and depth of 90cm, at a distance of 20 meters between ditches. The water trap in this design was composed of 3ditches at the front and 2ditches at the rear. The distance between the front and rear ditches was 50 meters, the trees seedlings were sown at the inside of each ditch at spacing of 4 meters between plants, (T_3).

Seedlings:

The seedlings were raised in the Air Port nursery and were transplanted when they were six months old at the onset of the rainy season to give the seedlings the full benefit of the rainy season.

Plant parameters:

The following parameters were taken every month starting after planting.

a) Plant height (cm):

Five plants were chosen at random from each treatment. Plant height was taken from the base of the plant to the top by using a metering device. The mean height of the five plants was recorded.

b) Number of leaves per plant:

Five plants were taken at random from each treatment so as to count the number of leaves per plant, and the mean number of the leaves of the five plants was recorded.

c) Number of branches per plant:

Five plants were taken at random from each treatment so as to count the number of branches per plant, and the mean number of the branches of the five plants was recorded.

d) Stem thickness:

Five plants were chosen at random from each treatment. Stem diameter was measured using a vernia. The mean stem diameter of the five plants was recorded.

3. Results

3.1. Plant Height (cm)

Table 1 (before mortality of control trees) show that V-shape treatment during November and December in both seasons and during January in the first season showed a significantly higher mean of plant height as compared to all other treatments, except with pits during November in both seasons. On the other hand, this treatment during October had a significantly higher mean of plant height only as compared to pits in the first season and the control in the second season, whereas it had no significant differences with all other treatments (Table 1). Meanwhile, both semi-circular and deep ditch treatments in December and January of the first season and during October, November and December of the second season gave a significantly higher mean of plant height than the control treatment, except deep ditch during December of the first season, when there were no significant differences between them (Table 1).

Table 1. *Effect of micro catchment shape on plant height (cm) of Jatropha (before mortality of control trees) during 2011/2012 and 2012/2013 seasons.*

Treatments	Plant height (cm) First season				Plant height (cm) Second season		
	October	November	December	January	October	November	December
T0	28.07 ab	28.77 b	28.78 d	29.47 c	27.72 b	28.05 d	28.16 e
T1	25.33 b	28.73 b	34.50 bc	40.47 b	29.37ab	32.65 b	36.07 c
T2	28.23 ab	31.93 ab	36.90 b	42.33 b	31.24 a	34.99 a	39..95 b
T3	26.07 b	29.33 b	32.63 cd	39.50 b	29.63 a	31.34 c	33.80 d
T4	29.57 a	33.60 a	41.50 a	49.60 a	29.79 a	34.78 a	47.04 a
SE±	0.98	1.13	1.20	1.23	0.58	0.59	0.57

Means with the same latter(s) in the same column are not significantly different at 0.05 level of probability according to Duncan's New Multiple Range Test (DNMRT).

T_0= control; T_1= semi-circular; T_2= pits; T_3= deep ditch; T_4= V-shape.

Table 2 (after mortality of control trees) shows that V-shape treatment during February, March and April in both seasons and during January in the second season reported a significantly higher mean of plant height as compared to all other treatments (semi-circular, pits and deep ditch). Also the Table shows that both semi-circular and pits treatments in this period showed no significant differences between them, except during March and April in the second season, but both of them had a significantly higher mean of plant height as compared to deep ditch during the same period, except with pits during February and March in the first season and during April of the second season.

3.2. Stem Thickness (cm)

In the first season the V-shaped treatment during November, December and January had significantly higher mean of stem diameter as compared to all other treatments, except with the control and pits during November and semi-circular during December (Table 3). Meanwhile the V-shaped treatment during December in the second season significantly increased stem diameter as compared to both control and deep ditch treatments. On the other hand, both control and pits during November in the first season had significantly higher mean of stem diameter as compared to semi-circular,

but during December and January of the same season the reverse was true (Table 3). During December of the second season there were no significant differences between these treatments (Table 3). Meanwhile deep ditch treatment in both seasons for most months reported a significantly lower mean of stem thickness than all the other treatments (Table 3).

Table 4 (after mortality of control trees) reveals that the V-shaped treatment in both seasons for all months had a significantly higher mean of stem thickness as compared to the semi-circular, pits and deep ditch treatments, except with the semi-circular during January and April in the second season. In the first season for all months, the semi-circular reported a significantly higher mean of stem thickness as compared to both pits and deep ditch (Table 4). Whereas in the second season this treatment significantly increased stem diameter as compared to deep ditch during January, February and April, while the reverse was true during March. As for pits and deep ditch treatments (Table 4) shows that there were no significant differences between these treatments in both seasons.

Table 2. *Effect of micro catchment shape on plant height (cm) of Jatropha (after mortality of control trees) during 2011/2012 and 2012/2013 seasons.*

Treatments	Plant height (cm) First season			Plant height (cm) Second season			
	February	March	April	January	February	March	April
T1	46.80 [b]	51.93 [b]	53.77 [b]	43.06 [b]	46.66 [b]	53.51 [b]	55.59 [b]
T2	48.37 [b]	52.23 [b]	53.17 [bc]	44.61 [b]	46.85 [b]	48.82 [c]	50.17 [c]
T3	45.30 [b]	47.70 [b]	48.47 [c]	36.23 [c]	39.83 [c]	41.01 [d]	41.86 [d]
T4	58.00 [a]	64.37 [a]	66.10 [a]	50.14 [a]	54.28 [a]	56.53 [a]	58.79 [a]
SE±	1.28	1.48	1.47	0.61	0.61	0.64	0.63

Means with the same latter(s) in the same column are not significantly different at 0.05 level of probability according to DNMRT.
T_1= semi-circular; T_2= pits; T_3= deep ditch; T_4= V-shape.

Table 3. *Effect of micro catchment shape on stem thickness (cm) of Jatropha (before mortality of control trees) during 2011/2012 and 2012/2013 seasons.*

Treatments	Stem thickness (cm) First season			Stem thickness (cm) Second season			
	October	November	December	January	October	November	December
T0	5.23 [ab]	5.74 [a]	5.79 [b]	5.97 [d]	4.72 [a]	4.78 [a]	4.85 [b]
T1	4.98 [ab]	5.33 [bc]	6.57 [a]	6.93 [b]	5.07 [a]	5.38 [a]	5.94 [a]
T2	5.01 [ab]	5.47 [a]	5.98 [b]	6.23 [c]	4.82 [a]	5.07 [a]	5.47 [ab]
T3	4.74 [b]	4.98 [c]	5.21 [c]	5.89 [cd]	4.62 [a]	4.86 [a]	5.10 [b]
T4	5.27 [a]	5.71 [a]	6.98 [a]	7.43 [a]	4.97 [a]	5.31 [a]	6.10 [a]
SE±	0.16	0.12	0.13	0.13	0.17	0.21	0.23

Means with the same latter(s) in the same column are not significantly different at 0.05 level of probability according to DNMRT.
T_0= control; T_1= semi-circular; T_2= pits; T_3= deep ditch; T_4= V-shape.

Table 4. *Effect of micro catchment shape on stem thickness (cm) of Jatropha (after mortality of control trees) during 2011/2012 and 2012/2013 seasons.*

Treatments	Stem thickness (cm) First season			Stem thickness (cm) Second season			
	February	March	April	January	February	March	April
T1	7.09 [b]	7.18 [b]	7.29 [b]	6.41 [ab]	6.91 [b]	7.40 [c]	7.81 [ab]
T2	6.47 [c]	6.51 [c]	6.57 [c]	5.69 [bc]	5.81 [c]	5.98 [b]	6.39 [bc]
T3	6.23 [c]	6.31 [c]	6.33 [c]	5.44 [c]	5.67 [c]	5.84 [b]	5.99 [c]
T4	7.93 [a]	8.21 [a]	8.87 [a]	6.64 [a]	7.31 [a]	8.02 [a]	8.57 [a]
SE±	0.12	0.13	0.12	0.25	0.25	0.29	0.25

Means with the same latter(s) in the same column are not significantly different at 0.05 level of probability according to DNMRT.
T_1= semi-circular; T_2= pits; T_3= deep ditch; T_4= V-shape.

3.3. Number of Leaves/Tree

Table 5 (before mortality of control trees) shows that in the first season the V-shaped treatment for all months had a significantly higher mean number of leaves/tree as compared

to all other treatments. While, during December in the second season this treatment significantly increased the number of leaves/tree as compared to the control and deep ditch treatment. On the other hand, pits gave a significantly higher mean number of leaves/tree as compared to the control and semi-circular treatments in both seasons for all months, except during December in the second season, when there was no significant difference between it and semi-circular treatment (Table 5). Similarly deep ditch treatment significantly increased number of leaves/tree as compared to control in both seasons for all the months (Table 5).

Table 6 (after mortality of control trees) also shows that the V-shaped treatment significantly increased the number of leaves/tree as compared to all other treatments in both seasons for all months, except with pits during February and March of the first season and January of the second season and with the semi-circular during January and February of the second season.

On the other hand, there were no significant differences between the semi-circular, pits and the deep ditch for the number of leaves/tree in the first season, except during February, in which pits treatment reported a significantly higher mean than both the semi-circular and the deep ditch. Meanwhile in the second season pits had significantly higher mean number of leaves/tree as compared to the deep ditch for all months, except during April, when there was no significant difference between the two treatments.

During February and April of the second season, the semi-circular treatment had a significantly higher mean number of leaves/tree as compared to the pits, whereas there were no significant differences between the two treatments during January and March (Table 6). Also Table 6 shows that there were no significant differences between the semi-circular and the deep ditch for all the months in the first season, but in the second season the former treatment significantly increased the number of leaves/tree as compared to the latter one for all months.

Table 5. Effect of micro catchment shape on number of leaves/tree of Jatropha (before mortality of control trees) during 2011/2012 and 2012/2013 seasons.

Treatments	Number of leaves/tree First season				Number of leaves/tree Second season		
	October	November	December	January	October	November	December
T0	3.33 ^{c}	5.00 ^{c}	8.00 ^{d}	0.67 ^{d}	5.67 ^{a}	7.33 ^{a}	3.33 ^{c}
T1	3.00 ^{c}	5.33 ^{c}	10.67 ^{cd}	18.33 ^{c}	4.33 ^{a}	7.00 ^{a}	12.67 ^{ab}
T2	9.67 ^{b}	12.00 ^{b}	15.33 ^{b}	23.00 ^{b}	7.00 ^{a}	10.67 ^{a}	16.00 ^{a}
T3	8.00 ^{b}	10.66 ^{b}	14.00 ^{bc}	19.33 ^{c}	5.33 ^{a}	9.33 ^{a}	10.00 ^{b}
T4	15.33 ^{a}	19.00 ^{a}	24.00 ^{a}	26.33 ^{a}	7.33 ^{a}	10.33 ^{a}	14.67 ^{a}
SE±	0.95	0.81	1.04	0.99	1.00	0.96	1.41

Means with the same latter(s) in the same column are not significantly different at 0.05 level of probability according to DNMRT.
T_0= control; T_1= semi-circular; T_2= pits; T_3= deep ditch; T_4= V-shape.

Table 6. Effect of micro catchment shape on number of leaves/tree of Jatropha (after mortality of control trees) during 2011/2012 and 2012/2013 seasons.

Treatments	Number of leaves/tree First season			Number of leaves/tree second season			
	February	March	April	January	February	March	April
T1	22.67 ^{b}	13.67 ^{b}	4.00 ^{b}	17.33 ^{a}	18.67 ^{a}	11.00 ^{b}	6.00 ^{b}
T2	27.33 ^{a}	16.33 ^{ab}	2.67 ^{b}	17.00 ^{a}	13.67 ^{b}	8.33 ^{b}	2.67 ^{c}
T3	23.67 ^{b}	14.33 ^{b}	3.00 ^{b}	12.00 ^{b}	8.00 ^{c}	3.33 ^{c}	2.00 ^{c}
T4	29.33 ^{a}	21.00 ^{a}	18.33 ^{a}	16.00 ^{a}	18.67 ^{a}	21.00 ^{a}	22.33 ^{a}
SE±	0.82	1.63	1.04	0.69	0.50	0.84	0.57

Means with the same latter(s) in the same column are not significantly different at 0.05 level of probability according to DNMRT.
T_1= semi-circular; T_2= pits; T_3= deep ditch; T_4= V-shape.

3.4. Number of Branches/Tree

In the second season (before mortality of control trees) the V-shaped treatment had significantly higher mean number of branches/tree as compared to both pits and deep ditch treatments for February, March and April, whereas this treatment had no significant differences with the semi-circular for the same period (Table 7).

Table 7. Effect of micro catchment shape on number of branches/tree of Jatropha (before mortality of control trees) during 2011/2012 and 2012/2013 seasons.

Treatments	number of branches/tree First Season				number of branches/tree Second Season		
	October	November	December	January	October	November	December
T0	1.00 ^{a}	1.00 ^{a}	1.00 ^{a}	1.00 ^{a}	1.00 ^{a}	1.00 ^{a}	1.00 ^{a}
T1	1.00 ^{a}	1.00 ^{a}	1.00 ^{a}	1.00 ^{a}	1.00 ^{a}	1.00 ^{a}	1.00 ^{a}
T2	1.00 ^{a}	1.00 ^{a}	1.00 ^{a}	1.00 ^{a}	1.00 ^{a}	1.00 ^{a}	1.00 ^{a}
T3	1.00 ^{a}	1.00 ^{a}	1.00 ^{a}	1.00 ^{a}	1.00 ^{a}	1.00 ^{a}	1.00 ^{a}
T4	1.00 ^{a}	1.00 ^{a}	1.00 ^{a}	1.00 ^{a}	1.00 ^{a}	1.00 ^{a}	1.00 ^{a}
SE±	0.00	0.00	0.00	0.00	1.00	0.96	1.41

Means with the same latter(s) in the same column are not significantly different at 0.05 level of probability according to DNMRT.
T_0= control; T_1= semi-circular; T_2= pits; T_3= deep ditch; T_4= V-shape.

On the other hand, in February and April of the second season (after mortality of control trees) the semi-circular treatment significantly increased the number of branches/tree as compared to pits and deep ditch treatments which showed no significant differences between them in this period (Table 8).

Table 8. *Effect of micro catchment shape on number of branches/tree of Jatropha (after mortality of control trees) during 2011/2012 and 2012/2013 seasons.*

Treatments	Number of branches/tree First season			Number of branches/tree Second season			
	February	March	April	January	February	March	April
T1	1.00 [a]	1.00 [b]	1.00 [a]	1.33 [a]	2.00 [a]	2.00 [ab]	2.33 [a]
T2	1.00 [a]	1.00 [b]	1.00 [a]	1.00 [a]	1.00 [b]	1.33 [bc]	1.67 [b]
T3	1.00 [a]	1.00 [b]	1.00 [a]	1.00 [a]	1.00 [b]	1.00 [c]	1.00 [b]
T4	1.00 [a]	1.67 [a]	1.00 [a]	1.67 [a]	2.00 [a]	2.67 [a]	2.67 [a]
SE±	0.00	0.16	0.00	0.22	0.02	0.22	0.22

Means with the same latter(s) in the same column are not significantly different at 0.05 level of probability according to DNMRT.
T_1= semi-circular; T_2= pits; T_3= deep ditch; T_4= V-shape.

4. Discussion

The study showed that Jatropha under the control treatment survived only 4 months during the first season and 3 months during the second season, after that they were subjected to permanent wilting. Whereas the Jatropha trees under the micro catchment treatments continued to survive. This may be attributed to the exposure of the trees under the control treatment to a condition of water deficit due to low the moisture content, while under the micro- catchments such condition was avoided through availability of water around the root zone as a result of water collection and storage by these techniques. This is in supported of the statement of Dauda and Baiyeri (2009) saying that micro catchment techniques are appropriate for small scale trees planting in any area which has moisture deficit. Al-seekh and Mohammed (2010) mentioned that water harvesting techniques are effective in increasing soil moisture storage, prolonging the growing season and decreasing the amount of supplemental irrigation required for growing fruit trees.

Also Seidahmed et al., (2012) pointed out that the survival of tree species decreased with time as a result of drought due to short rainy season which affect soil water content. After transplantation, rain water harvesting can be used to speed up tree establishment, deep root development and to reduce the mortality rate of trees.

In general plant height and stem thickness of Jatropha trees during the study periods in both seasons was significantly higher under most of the micro catchment techniques as compared to the control, with higher mean of these characters obtained by V-shape, followed by semi-circular and pits treatments. This may be attributed to the availability of water and nutrients around the root zone of these trees in the micro catchments. Xiao et al., (2005) found that for 4 types of micro catchments, plant height and crown diameter of *Tamarix ramosissima* were significantly higher than the control. The authors attributed this to the effect of these micro catchments on production of more water to be available to the trees. On the other hand, Jianxin et al., (2007) observed that plant height and ear length of Broom – corn millet significantly increased under contour terraces and contour ridge than the control.

The findings of this study also revealed that the monthly increment in plant height under control treatment for the two

Jatropha trees in both seasons was light as compared to the micro catchments, particularly towards January of the first season and December of the second season. Whereas, during October plant height of all treatments including the control was similar, this may infer the importance of availability of water around the root zone and its effect on tree growth.

Number of leaves/tree at most of the study period in both seasons was significantly higher under most of the micro catchment techniques than the control. The significantly higher mean of the number of leaves/tree under the micro-catchment treatments may be due to the effect of these techniques on plant height as well as their effect on prevention of leaf shedding as a result of water deficit. The effect of water deficit on leaf shedding was very clear under the control treatment as shown in the number of leaves during January of the first season and December of the second season, when it was obviously reduced as compared to the previous months. This condition did not occur for the micro catchment techniques at this period.

Almost in both seasons for the study period, the number of branches/tree was not significantly different among treatments including the control except between micro catchment techniques towards the end of the season. The number of branches/tree was limited between 1 – 2. This could be attributed to the fact that the period of the study was not enough for these trees to show this character particularly for the Jatropha in which the emergency of branches under favorable conditions may take a long time.

5. Conclusion

1. The V-shaped water harvesting technique was most effective in enhancing trees growth.

2. There was no significant difference in plant growth parameters (number of branches per plant) in any location due to water harvesting technique.

References

[1]   Aktar, M and Menching, H.G. (1993) Desertification in the Butana Geojournal 31; 41-50.

[2]   Alaktar, M.K and Aga.A.A. (1995) Vegetation and Soil conservation, Halab University Faculty of Agricu-lture Der Elzor Ibn Khaldon Printing Press, Damascus.

[3] Al-Seekh SH. and Mohammad, A.G. (2010). The effect of water harvesting techniques on runoff, sedimentation, and soil properties. College of Agriculture, Hebron University, Hebron, West Bank, Palestine. 14th International Soil Conservation Organization Conference. Water Management and Soil Conservation in Semi-Arid Environments. Marrakech, Morocco, May 14-19, 2010 (ISCO 2010).

[4] Dauda, K.A. and Baiyeri, M.R. (2009). Design and Construction of Negarim Micro Catchment System for Citrus Production. Agricultural Engineering and Water Resources, Institute of Technology, Kwara State Polytechnic, Ilorin, Kwara State, Nigeria.

[5] Jianxin, Z., Dawei Zheng, Yantian Wang, Yu Duan and Yanhua Su. (2007). Two water harvesting type within-field Rainwater harvesting measures and their effects on increasing soil moisture and crop production in north china. College of Resources and Environment, China Agricultural University, Beijing, China, 100094.

[6] Meunive, M. (1996) Forest cover and flood water in small mountain water sheds Unasylva (47)-29.

[7] Omer, A.Mekki and Eltigani, M. Elamin. (1997) Effect of tillage and contour dikking on sorghum establishment and yield on sandy clay soil in Sudan, Soil and tillage research 43-229-240.

[8] Salih, A.A and Inanga.S. (1997) in- situ water harvesting and contour dikking for sorghum production and tree establishment in marginal lands (Abstracts of 1997 meeting of Japanese Association for arid land Studies).

[9] Seidahmed H.A, Salih, A.A and Musnad, H.A., (2012). Rehabilitation of Kerrib Lands in Upper Atbara River using Indigenous Trees and Water Harvesting Techniques, Forestry Research Centre –Soba- ARC, Journal of Science and Technology Vol. 13 Agricultural and Veterinary Sciences (JAVS No.1. 104-109).

[10] Xiao, Yan Li, Lian-You Liu, Shang-Yu Gao, Pei-Jun Shi, Xue-Yong Zou, Chun-Lai Zhang, (2005). Microcatchment water harvesting for growing Tamarix ramosissima in the semiarid loess region of China. The Key Lab of Environment Change and Natural Disaster, Ministry of Education, Beijing 100875, China. Forest Ecology and Management 214 (2005) 111–117.

27

Standardization of Sucrose and 6-Benzyl Aminopurine for *in vitro* Micro Tuberization of Potato

Md. Afzal Hossain[1], Md. Abu Kawochar[2], Abdullah-Al-Mahmud[3*], Ebna Habib Md. Shofiur Rahaman[3], Md. Altaf Hossain[2], Khondoker Md. Nasiruddin[4]

[1]Ministry of Environment & Forests, Bangladesh Secretariat, Bangladesh
[2]Tuber crops research Centre (TCRC), Bangladesh Agricultural Research Institute, Joydebpur, Gazipur, Bangladesh
[3]International Potato Centre (CIP), SWCA, Banani, Dhaka, Bangladesh
[4]Department of Biotechnology, Bangladesh Agricultural University, Bangladesh

Email address:
Mafzalhossain@yahoo.com (Md. A. Hossain), A.Mahmud@cgiar.org (Abdullah-Al-Mahmud)

Abstract: *In vitro* micro tuberization from regenerated plantlets of potato varieties was observed in Biotechnology Laboratory, Department of Biotechnology, Bangladesh Agricultural University. Tubers of potato cultivars Diamant and Cardinal were used as initial experimental materials for meristem culture. The experiment was consisted of three factors; variety (Diamant, Cardinal), sucrose concentration (3%, 6%, 9%, 12%) and 6-benzyl aminopurine (BAP) levels (2.5, 5.0, 7.5 mg L^{-1}). As a whole, 24 treatments were laid out in complete randomized design with three replications. Among the varieties, Diamant required minimum days (6-17) for micro-tuber initiation, produced more number of micro-tubers (4.97) and produced more average weight of micro-tuber (120.39mg) but there had no significant difference. Among the sucrose levels, quickest (6-15 days) micro-tuber initiation, the highest number of micro-tubers vial^{-1} (5.06) and the highest average weight of micro-tuber (137.31 mg) were found in 9% sucrose level. For different BAP levels, quickest (6-15 days) micro-tuber initiation, the highest number of micro-tubers vial^{-1} (5.38) and the highest average weight of micro-tuber (126.31 mg) were found at 5.0 mg L^{-1}. The best combination for minimum duration (6-8 days) of micro-tuber initiation, the highest number of micro-tubers vial^{-1} (6.00) and the highest average weight of micro-tuber (152.01 mg) was in Diamant with 9% sucrose at 5 mg L^{-1}. Concomitantly, the lowest number micro-tubers vial^{-1} (2.00) and the lowest average weight of micro-tuber (89.98 mg) were found in Cardinal cultured with 3% sucrose media where at 7.5 mg L^{-1} BAP and at 2.5 mg L^{-1} BAP, respectively.

Keywords: 6-Benzyl Aminopurine, Microtuberization, Plantlets, Potato

1. Introduction

Potato (*Solanum tuberosum* L.) is an important vegetables crop and is grown in winter only in Bangladesh [15]. Microtubers have become an important mode of rapid multiplication for seed tuber of potato [21]. These micro-tubers are utilized for minitubers production in greenhouse or screen house. Wherever microtuber and minituber production technologies have been implemented, they have halved the field time necessary for conventional method to supply to the commercial growers. Microtubers from meristem grown and/or regenerated microplants are now produced and used in Australia, Brazil, Chile, China, Ecuador, India, Indonesia, Kenya, Korea, Peru, Philippines, Taiwan,

UK, Vietnam and even in Bangladesh as disease free seed [7]. However, the technique is controlled by various physical (light, temperature etc.) and chemical (growth regulators) factors. Among the media components, sucrose played an important role in the induction and development of potato microtubers on in vitro [10] and BAP promoted initiation and growth of micro-tubers [19]. Various sucrose concentrations were tested in the microtuber induction medium where the best results were obtained with 8% [12], with 10% [3], with 12% [20], with 9% and 12% [13] and with 8% and 12% [18]. Tuber induction medium combined with 9% sucrose promoted tuberization and increased microtuber weight more than the tuber induction medium combined with 6% sucrose, BAP and CCC [11]. Better micro-tuber yield was obtained with 6% sucrose and 5 mg L^{-1} BAP under complete dark condition [1]. Microtuber number and fresh weight were greatest with 10 mg L^{-1} BAP in presence of 6% sucrose [16]. Best response to both initiation and production of microtubers were observed in modified MS media containing 5.0 mg L^{-1} BAP with 500 mg L^{-1} CCC [6] and by using 500 mg L^{-1} CCC together with 5.0 mg L^{-1} BAP[9]. Therefore, the present investigation was undertaken to find out the appropriate doses of sucrose and growth regulators and their combination(s) with potato varieties for rapid microtuberization.

2. Materials and Methods

2.1. Design and Treatments of the Experiment

The experiment was carried out in Biotechnology Laboratory, Department of Biotechnology, Bangladesh Agricultural University. The experiment was consisted of three factors; variety (Diamant, Cardinal), sucrose levels (3%, 6%, 9%, 12%) and BAP levels (2.5, 5.0, 7.5 mg L^{-1}). As a whole, 24 were laid out in complete randomized design with three replications.

2.2. Explant, Culture Media and Regeneration of Plantlets

Meristem was used to regenerate for virus free plantlet production. From the regenerated virus free plantlets, plantlets again multiplied through rapid multiplication technique [8]. Required amount of MS medium [14] was prepared and supplemented with mentioned sucrose and BAP levels. Thus, prepared all groups of media were taken to separate test tubes and/or vials according to the number of treatments and replications. Then the test tubes/vials containing the media were sterilized by autoclaving. Stem segments having 1-5 nodes from plantlets of one month old which were collected from previous experiment were cultured for microtuberization. Single stem segments were placed to each test tube containing 10 ml medium and 4-5 stem segments were placed to each vial containing 40 ml medium. When placed in test tube, five test tubes were considered as one replication.

2.3. Incubation

The cultures were incubated at a temperature of $25\pm2°C$ at 16 hrs photoperiod. Days to microtuber initiation were recorded. After two months of explantation, number of microtubers vial^{-1} was calculated on the basis of total number of vials per treatment and average weight of microtuber was calculated on the basis of total number of microtubers produced.

2.4. Data Collection and Statistical Analysis

After two months of explantation, data were recorded on 1) Days to microtuber initiation, ii) Number of microtubers vial^{-1},

iii) Average weight of microtuber. Data were analyzed by statistical software MSTATC and means were adjudged by the Duncan's Multiple Range Test.

3. Results and Discussions

3.1. Main Effect of Variety, Sucrose and BAP on Days to Micro-Tuberization, Number and Weight of Microtubers

Between the varieties, Diamant required comparatively minimum days for microtuber initiation, which started at 6 days and continued up to 17 days, than that of Cardinal where the range was from 8 to 21 days. These findings were supported by Hossain and Sultana (1998) who reported that the response of variety to microtuberization was highly dependent on genetic factors. Microtuber number vial^{-1} did not vary significantly and the result showed that Diamant produced more number of microtubers (4.97) vial^{-1} than that of Cardinal (4.19). The differences of average weight of microtuber were found non-significant where the average weight of microtubers were 120.39 mg and 114.89 mg in case of Diamant and Cardinal, respectively. Considering different concentrations of sucrose, the duration required for microtuber initiation varied with the variation of sucrose levels used in the culture media. Microtuber initiation was quickest (6-15 days) when 9% sucrose was used and this duration gradually increased with both increasing and decreasing the level of sucrose. The present findings were directly in agreement with Jeoung-Lai et al. (1996) who reported that tuber induction medium combined with 9% sucrose promoted tuberization. Microtuber numbers vial^{-1} increased with the increasing of sucrose level up to 9% and then again decreased. The highest number of microtubers vial^{-1} was found in 9% sucrose (5.06) followed by 6% sucrose (4.78), whereas, the lowest was found with 3% sucrose (4.06). The highest average weight of microtuber (137.31 mg) was found in 9% sucrose followed by 12% (119.17 mg) while the lowest (99.01 mg) was in 3% sucrose. The findings were similar to the findings of Jeoung-Lai et al. (1996) who found the heaviest microtubers with 9% sucrose. Among the three concentrations of BAP, duration required for microtuber initiation was 6-21 days at 2.5 mg L^{-1} BAP, 6-15 days at 5.0 mg L^{-1} BAP and 7-20 days at 7.5 mg L^{-1} BAP. The highest number of microtubers vial^{-1} (5.38) was obtained at 5 mg L^{-1} BAP, whereas 2.5 and 7.5 mg L^{-1} BAP produced about similar number of microtubers (4.25 and 4.13 respectively) vial^{-1}. The present finding was supported by Yong et al. (1996) who reported that BAP promoted initiation and growth of microtubers. The highest average weight of microtuber (126.31 mg) was found in 5 mg L^{-1} BAP followed by 7.5 mg L^{-1} BAP (115.29 mg) and the lowest (111.33 mg) was in 2.5 mg L^{-1} BAP. (Table 1)

Table 1. Main effect of variety, sucrose and BAP on days to micro-tuberization, number and weight of microtubers at two months of ex-plantation.

Treatments		Days to micro tuberization	No. microtubers vial^{-1} of 5 plants	Average weight of microtuber (mg)
Variety	Diamant	6-17	4.97 a	120.39 a
	Cardinal	8-21	4.19 a	114.89 a
Sucrose (%)	3	9-17	4.06 b	99.01 d
	6	9-15	4.78 ab	115.10 c
	9	6-15	5.06 a	137.31 a
	12	10-21	4.44 ab	119.17 b
BAP (mg L^{-1})	2.5	6-21	4.25 b	111.33 c
	5.0	6-15	5.38 a	126.31 a
	7.5	7-20	4.13 b	115.29 b
CV (%)		-	16.86	2.13

Figures followed by same letter(s) are statistically similar as per DMRT

3.2. Combined Effect of Variety and Sucrose on Days to Microtuber Initiation and Number and Weight of Microtubers

Considering the variety and sucrose combinations, Diamant with 9% sucrose required comparatively minimum time (6-11 days) for microtuber initiation and maximum (12-17 days) with 12% sucrose (Table 2). Accordingly, Cardinal with 9% sucrose required 8-15 days and with 12% sucrose required 10-21 days for microtuber initiation. The highest number of microtubers vial^{-1} (5.56) was found with Diamant at 9% sucrose followed by Diamant with 6% sucrose (5.33), whereas the lowest number (3.33) was obtained from Cardinal with 3% sucrose (Table 2). Varietal difference on microtubers number with various concentrations of sucrose was observed by Warren and Shirlyn (2000) where number of microtubers increased with the increasing concentration of sucrose up to 16% in the variety Shepody. On the other hand, the highest average weight of microtuber (140.15 mg) was found in Diamant with 9% sucrose, whereas the lowest (95.56 mg) was in Cardinal with 3% sucrose.

3.3. Combined Effect of Variety and BAP on Days to Microtuber Initiation and Number and Weight of Microtubers

In case of variety and BAP combinations (Table 3), Diamant with 5.0 mg L^{-1} BAP required comparatively minimum (6-15 days) and Cardinal with 2.5 mg L^{-1} BAP required maximum (9-21 days) time for microtuber initiation. The maximum number of microtubers vial^{-1} (5.58) was obtained from Diamant at 5 mg L^{-1} BAP, whereas the lowest number (3.67) was obtained from Cardinal at 7.5 mg L^{-1} BAP. Thus, the highest average weight of microtuber (127.96 mg) was found in Diamant with 5 mg L^{-1} BAP followed by Cardinal with 5 mg L^{-1} BAP (124.67 mg) while the lowest weight (107.97 mg) was found in Cardinal with 2.5 mg L^{-1} BAP.

3.4. Combined Effect of Sucrose and BAP on Days to Microtuber Initiation and Number and Weight of Microtubers

In case of sucrose and BAP combination (Table 4), 9%

sucrose performed minimum time (6-10 days) for microtuber initiation with 5.0 mg L^{-1} BAP followed by 2.5 mg L^{-1} BAP (6-12 days), whereas maximum time (15-20 days) was taken for 12% sucrose with 7.5 mg L^{-1} BAP. The highest number of microtubers vial^{-1} (5.67) was obtained with 6% and 9% sucrose at 5 mg L^{-1} BAP followed by 12% sucrose with 5 mg L^{-1} BAP (5.17), whereas, the lowest number (3.33) was found at 3% sucrose with 7.5 mg L^{-1} BAP. The result was directly supported by Texeira and Pinto (1991). Average weight of microtuber (150.38 mg) found to be the best in combination of 9%sucrose and 5 mg L^{-1} BAP, whereas, the lowest (93.81 mg) was in 3% sucrose and 2.5 mg L^{-1} BAP. Variation in microtuber yield were also recorded by Al-Abdallat and Suwwan (2002) where 0, 3, 6, and 9% sucrose and 5 mg L^{-1} BAP were used and better microtuber yield was obtained with 6% sucrose under complete dark condition. [Table 4]

3.5. Combined Effect of Variety, Sucrose and BAP on Days to Microtuber Initiation and Number and Weight of Microtubers

Considering all the three factors (Table 5), it was observed that the best combination was Diamant with 9% sucrose at 2.5 and at 5.0 mg L^{-1} BAP which required minimum duration (6-8 days), whereas, Cardinal with 12% sucrose at 2.5 mg L^{-1} BAP required maximum duration (16-21 days) for microtuber initiation. On the other hand, the highest number of microtubers vial^{-1} (6.00) was obtained from Diamant when cultured on media with 6% and 9% sucrose at 5 mg L^{-1} BAP. At the same time, the lowest number (2.00) was found in Cardinal cultured on media with 3% sucrose and 7.5 mg L^{-1} BAP (Plate 2). The present result may be considered as similar to the previous study of Estrada et al. (1986) where it was observed that MS medium with 5.0 mg L^{-1} BAP, 8% sucrose and 500 mg L^{-1} CCC would able to produce microtubers in a broad range of genotypes. Thus, the highest average weight of microtubers (152.01 mg) was found in Diamant with 9% sucrose and 5 mg L^{-1} BAP followed by Cardinal with 9% sucrose and 5 mg L^{-1} BAP (148.75 mg). Concomitantly, the lowest average weight of microtuber (89.98 mg) was found in Cardinal with 3% sucrose and 2.5 mg L^{-1} BAP. (Table 5)

Table 2. Combined effect of variety and sucrose on days to microtuber initiation and number and weight of microtubers after two months of inoculation.

Combination		Days to microtuber initiation	No. microtubers vial^{-1} of 5 plants	Average weight of microtuber (mg)
Variety	Sucrose (%)			
Diamant	3	9-15	4.78 ab	102.44 e
	6	9-13	5.33 ab	118.55 c
	9	6-11	5.56 a	140.15 a
	12	12-17	4.22 bc	120.41 c
Cardinal	3	10-17	3.33 c	95.56 f
	6	10-15	4.22 bc	111.64 d
	9	8-15	4.56 ab	134.46 b
	12	10-21	4.67 ab	117.93 c
CV (%)	-		16.86	2.13

Table 3. Combined effect of variety and BAP on days to microtuber initiation and number and weight of microtubers after two months of inoculation.

Combination		Days to microtuber initiation	No. microtubers vial^{-1} of 5 plants	Average weight of microtuber (mg)
Variety	BAP (mg L^{-1})			
Diamant	2.5	6-15	4.75 ab	114.69 bc
	5.0	6-15	5.58 a	127.96 a
	7.5	7-17	4.58 ab	118.52 b
Cardinal	2.5	9-21	3.75 b	107.97 d
	5.0	8-15	5.17 a	124.67 a
	7.5	10-20	3.67 b	112.06 cd
CV (%)	-		16.86	2.13

Table 4. Combined effect of sucrose and BAP on days to microtuber initiation and number and weight of microtubers after two months of inoculation.

Combination		Days to microtuber initiation	No. microtubers vial^{-1} of 5 plants	Average weight of microtuber (mg)
Sucrose (%)	BAP (mg L-1)			
3	2.5	12-16	3.83 cd	93.81 h
	5.0	9-12	5.00 ab	105.10 f
	7.5	11-17	3.33 d	98.10 g
6	2.5	9-14	4.50 bc	106.36 f
	5.0	9-13	5.67 a	123.71 d
	7.5	9-15	4.17 bcd	115.22 e
9	2.5	6-12	4.83 abc	128.93 bc
	5.0	6-10	5.67 a	150.38 a
	7.5	7-15	4.67 abc	132.61 b
12	2.5	13-21	3.83 cd	116.23 e
	5.0	10-15	5.17 ab	126.07 cd
	7.5	15-20	4.33 bcd	115.21 e
CV (%)	-		16.86	2.13

Figures followed by same letter(s) are statistically similar as per DMRT

Table 5. Combined effect of variety, sucrose and BAP on days to microtuber initiation and number and weight of microtubers after two months of inoculation.

Treatment combination			Days to micro-tuber initiation	No. microtubers vial^{-1} of 5 plants	Average weight of microtuber (mg)
Variety	Sucrose (%)	BAP (mg L^{-1})			
Diamant	3	2.5	12-15	4.33 b-e	97.64 ij
		5.0	9-12	5.33 abc	108.50 h
		7.5	11-13	4.67 a-e	101.19 i
	6	2.5	9-13	5.00 a-d	111.53 gh
		5.0	10-13	6.00 a	124.86 d
		7.5	9-11	5.00 a-d	119.26 ef
	9	2.5	6-8	5.67 ab	130.36 c
		5.0	6-8	6.00 a	152.01 a
		7.5	7-11	5.00 a-d	138.09 b
	12	2.5	13-14	4.00 cde	119.23 ef
		5.0	12-15	5.00 a-d	126.48 cd
		7.5	15-17	3.67 de	115.51 fg
Cardinal	3	2.5	13-16	3.33 ef	89.98 k
		5.0	10-12	4.67 a-e	101.70 i
		7.5	15-17	2.00 f	95.01 j
	6	2.5	11-14	4.00 cde	101.19 i
		5.0	9-11	5.33 abc	122.56 de
		7.5	13-15	3.33 ef	111.18 gh
	9	2.5	9-12	4.00 cde	127.51 cd

Treatment combination			Days to micro-tuber initiation	No. microtubers vial⁻¹ of 5 plants	Average weight of microtuber (mg)
Variety	Sucrose (%)	BAP (mg L⁻¹)			
		5.0	8-10	5.33 abc	148.75 a
		7.5	10-15	4.33 b-e	127.13 cd
	12	2.5	16-21	3.67 de	113.22 gh
		5.0	10-15	5.33 abc	125.65 cd
		7.5	17-20	5.00 a-d	114.91 fg
CV (%)				16.86	2.13

Figures followed by same letter(s) are statistically similar as per DMRT

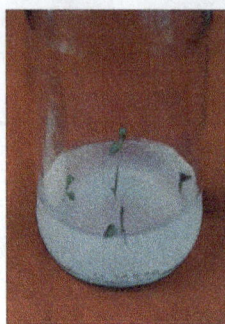
A. During explantation

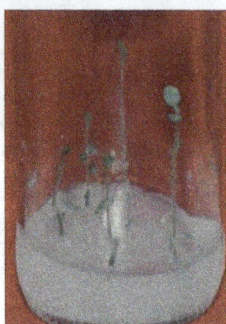
B. Fifteen days after explantation

C. One month after explantation

D. Two months after explantation

Plate 1. Microtuberization stages of potato cv. Diamant.

A. Diamant with 6% Sucrose and 5 mg L⁻¹ BAP

B. Cardinal with 3% Sucrose and 7.5 mg L⁻¹ BAP

A. Cardinal

B. Diamant

Plate 2. Number of microtubers at two months of explantation under different treatments.

Plate 3. Microtubers of two potato varieties after harvest.

4. Conclusion

Microtuber production was affected by different levels of sucrose and 6-benzyl aminopurine (BAP). Genotypic variation was also found among the two genotypes regarding microtuber production. The best combination for rapid microtuber production of potato varieties was Diamant with 9% sucrose at 5 mg L⁻¹ of 6-benzyl aminopurine. Hence the present protocol has the potential for the rapid multiplication of true-to-true type clones without changing the genetic fidelity.

Acknowledgements

The author expresses his deep sense of gratitude to United States Department of Agriculture (USDA) funded project #99/22/USDA and thankful to the authority of Tuber Crop Research Centre (TCRC) of Bangladesh Agricultural Research Institute (BARI), Bangladesh for their help in supplying the seed tubers, several information and sharing the knowledge. It was a part of PhD research works. The author also thanks to all staffs of Biotechnology Laboratory, Department of Biotechnology, Bangladesh Agricultural University, Bangladesh.

References

[1]   Al-Abdallat, A. K., and M. A. Suwwan. 2002. Interactive effects of explant, sucrose and CCC on microtuberization of 'Spunta' potato. Dirasat. Agril. Sci. 29: 19-27.

[2]   Avetisov, V. A., O. S. Melik-Sarkisov and G. I. Sobolkova. 1989. Induction of microtuberization on regenerates from callus of potato. Sel'skokhozyaistvennaya Biologiya. 5: 26-28.

[3]   Cui, C., F. F. F. He, J. C. Wang, Q. Y. Zhou, Y. X. Huang and D. B. Tang. 2001. Effects of photoperiod and carbon sources on the formation of microtubers of potato in vitro. J. Southwest Agril. Varsity. 23: 547-548.

[4] Dodds, J. H., P. Tovar, R. Chandra, D. Estrella and R. Cabello. 1988. Improved methods for in vitro tuber induction and use of in vitro tubers in seed programs. In: *Proc. Symp. on Improved Potato Planting Material,* Asian Potato Assoc., Kunming, China, June, 1988. pp. 157-158.

[5] Estrada, R., P. Tovar and J. H. Dodds. 1986. Induction of in vitro tubers in a broad range of potato genotypes. Plant Cell Tiss. Org. Cult. 7: 3-10.

[6] Haque, M. I., N. B. Mila, M. S. Khan and R. H. Sarker. 1996. Shoot regeneration and in vitro microtuber formation in potato. Bangladesh J. Bot. 25: 87-93.

[7] Hossain, M. A., H. M. Faruquee, N. Islam, M. A. S. Miah and K. M. Nasiruddin. 2005. Agrobacterium-mediated transfer of PsCIPK salt tolerance gene in sugarcane variety Isd 35. Mol. biol. biotechnol. j. 3(1&2): 1-3.

[8] Hossain, M. J. 1995. Effect of population density of cut shoot of potato on growth, tuber yield and multiplication rate. Trop. Sci. 35(2):161-166.

[9] Hossain, M. J. and N. Sultana. 1998. Effects of benzylaminopurine (BAP) and chloro choline chloride (CCC) on *in vitro* tiberization of potato. Bangladesh J. Agril. Res. 23: 685-690.

[10] Islam, M. S. 1995. Indigenous potato varieties of Bangladesh: Characterization by RAPD markers and production of virus free stock. Ph. D. Thesis, Dept. of Horticulture, Bangabandhu Sheikh Mujibur Rahman Agril. University, Gazipur, Bangladesh.

[11] Jeoung-Lai, C., S. M. Kang and Y. W. Choi. 1996. Effects of shoot culture and tuber inducing conditions on in vitro tuberization of potatoes. Proc., 4th Triennial Confc., Asian Potato Association. Philippines. pp. 186-190.

[12] Kabseog, Y. 2004. Relative importance of maltose and sucrose supplied during a 2-step potato microtuberization process. Acta Physiol. Plantarum. 26: 47-52.

[13] Liu, Z. L., Z. L. Wang, S. L. Wang, K. Gao. 2002. Effects of chlorobromoisocyanuric acid on culture of potato (Solanum tuberosum) plantlets in vitro. Chinese Potato J. 16(5): 288-290.

[14] Murashige, T., and F. Skoog. 1962. A revised medium for rapid growth and bioassays with tobacco tissue cultures. Physiologia Plantarum 15:473-497.

[15] Rashid, M. H., S. Akhter, M. Elias, M. G. Rasul and M. H. Kabir. 1993. Seedling tubers for ware potato production: Influence of size and plant spacing. Asian Potato J. 3: 14-17.

[16] Texeira, D. M. C. and J. E. B. P. Pinto. 1991. Minituberization of potatoes at different levels of N, saccharose and BAP. Revista Brasileria de Figiologia Vegetal. 3 (3): 77-83.

[17] Warren, K. C. and E. C. Shirlyn. 2000. Modification of potato microtuber dormancy during induction and growth in vitro or ex vitro. Am. J. Potato Res. 77: 103-110.

[18] Yasmin, L. 2005. Effect of N and K on in vitro microtuberization of potato. M.S. Thesis, Dept. of Biotechnology., Bangladesh Agril. Univ., Mymensingh, Bangladesh. pp. 89.

[19] Yong, L., D. Huiruo, X. Xin, Y. Hongfu, J. Liping, L. Huan and Z. Ying. 1996. Changes in several endogenous phytohormones during in vitro tuberization in potato. In: E. T. Rasco and F. B. Aromin (Eds.). Asian Sweet Potato and Potato Res. and Development, Manila. July 1995-June 1996. 1: 30-37.

[20] Zakaria, M. 2002. Induction and performance of potato microtuber. Ph.D. Thesis, Dept. of Hort., Bangabandhu Sheikh Mujibur Rahman Agricultural University, Salna, Gazipur, Bangladesh.

[21] Zakaria, M., M. M. Hossain, M. A. K. Mian, T. Hossain and M. Z. Uddin. 2008. In vitro tuberization of potato influenced by benzyl adenine and chloro chline chloride. Bangladesh J. Agril. Rer. 33(3): 419-425.

28

Effect of Locally Produced Blood Meal on Growth Performance and Packed Cell Volume of Broiler Chicks

Ufele Angela Nwogor, Ogbu Anthonia Uche, Ebenebe Cordelia Ifeyinwa, Akunne Chidi Emmanuel

Zoology Department, Nnamdi Azikiwe University, Awka, Anambra State, Nigeria

Email address:
ufeleangel@yahoo.com (U. A. Nwogor)

Abstract: This research evaluated the effect of locally produced blood meal on growth performance of broiler chicks. Three experimental diets were formulated; diet one served as control without blood meal, diet 2 contained 100g of blood meal mixed in 500g of chick mash while diet 3 contained 300g of blood meal mixed with 500g of chick mash. Forty-five broiler chicks were randomly assigned to the three treatments having five birds per cage. Each treatment was replicated three times. The experimental diets and portable water were supplied ad libitium throughout the experimental period for 42 days. At the end of the experiment, birds fed diet 2 (100g blood meal and 500g chick mash) gained more weight having mean weight gain of 3.04g than birds fed with diet 3 (300g of blood meal mixed with 500g of chick mash) having mean weight gain of 2.95g and birds fed with diet 1 (control, no blood meal). Also there was significant difference ($P < 0.05$) between the packed cell volume of birds feed with diet 2 and those fed with diet 1 and 3. The result of the research indicated that diet 2 enhances growth and boosts the PCV in broiler chicks. Broilers fed with diet two had the highest mean packed cell volume of 41.40%, followed by those fed with diet 3 which had 40.20% while those fed with diet 1 had 37%.

Keywords: Blood Meal, Weight Gain, PCV and Broiler Chicks

1. Introduction

Poultry has a significant effect on national economy. Report by Agbede and Aletor (2007), showed that about 10% of Nigerian population is engaged in poultry production. Of all poultry business, broiler production is a fast growing agricultural business in developing countries, therefore profitable production of broiler becomes essential and profitable venture. Broiler production involves the keeping of chickens of heavy meat breeds for the purpose of getting good quality meat products usually sold live or processed at ten to twelve weeks of age. The meat is an important source of high quality protein, minerals and vitamins to balance the human diet due to their ability of quick growth and high feed conversion efficiency.

According to the Agriculturist and Nutritionist Ensminger (2002), it has generally agreed that developing the poultry industry in Nigeria is the fastest means of bridging the protein deficiency gap presently prevailing in the country. It is also the promising source of additional income and means of quick returns from investment.

Brooks (2001), highlighted the importance of animal protein to human development, according to him animal protein contains high quality or complete protein in that they supply all the amino acids the body needs to perform body functions. Also protein from this, supply varying amount of other key nutrients including zinc, iron, magnesium, vitamin E and vitamin B. Poultry meat which belongs to a group of white meat has a low fat myoglobin content, which gives it, its color. This white meat produces glycogen, a polysaccharide which plays an important role in the glucose cycle. It is a source of lean protein for people with heart and cholesterol problems. White meat without side effects of extra fats is just the correct food that enhances immunity levels (Toor and Fahimullah 2002).

However, poultry feeds has remained a significant challenge to the poultry industry, this is because the productivity of poultry especially in the tropics has been limited by scarcity and consequent high prices of the conventional protein and energy sources, (Atawodi et al.,

2008). Hence there is need to search for locally available alternative sources of protein for use as feed supplement to poultry. One possible sources of cheap protein to poultry is the blood meal of various livestock species (Odunsi, 2004). According to Donald and Edward (2002), Blood meal is a by-product of the slaughtering industry, used as a protein source in the diets of ruminants and non ruminants. It is a dark chocolate colored powder with characteristic smell. It contains protein and it is one of the richest sources of lysine, a rich source of arginine, methionine, cystine and leucine. When compared with vegetable protein supplements for poultry it is quite high in biological value. Generally, vegetable protein supplements are deficient in two of the essential amino acids which are lysine and methionine whereas blood meal is rich in both of these amino acids. Blood meal improves performance, growth rate and feed intake of various animal productions (Fombad et al., 2004). Blood meal is also available worldwide, but like other animal products its sale and utilization are regulated in some countries for certain species for safety reasons (Fombad et al., 2004). Brookes (2002) reported that good quality feeds creates value by enhancing nutrients utilizing and improve animal conversion. Clearly, there are ways to improve the quality of feed such as the use of blood meal and selection of a reasonable good quality feed is important to the overall profitability of the producers and also gives the greatest opportunity for influencing animal performance beyond nutritional adequacy.

Thus, blood meal fits into the classical strategy for poverty alleviation and sustainable development. The hematological variables and protein levels of the blood of livestock are known to be positively correlated with protein quality and quality of diet (Adeyemi et al., 2000). In addition, different sources of protein may contain different toxic factor and affect differently the packed cell volume and thus the health status of animals (Diarra, 2008).

The various functions of blood are made possible by the individual and collective actions of its constituents- the biochemical and hematological components are influenced by the quality and quantity of feed and also the level of anti-nutritional elements or factors present in the feed (Adeyemi et al, 2000). Components of blood are also valuable in monitoring feed toxicity especially with feed constituent that affect the formation of blood (Oyawoye and Ogunkunle, 2008). Reduction in the concentration of PCV in the blood usually suggests the presence of a toxic factor e.g haemagglutin which has adverse effect on blood formation (Oyawoye and Ogunkunle, 2008). There is evidence in literature that hematological characteristics of livestock suggest their physiological disposition to the plane of nutrition (Madubuike and Ekenyem, 2006). Reduction in packed cell volume is indicative of low protein intake or mild anaemia. Blood chemistry constituents reflect the physiological responsiveness of the animal to its internal and external environments which include feed and feeding (Iheukwumere and Okoli, 2002). Broilers fed with good quality feed reach maturity with weight 2.2 kg in 8 weeks; but when feed is poor, it extends to 12 weeks with lower weight (Randy, 2002).

Blood parameters have been shown to be major indices of physiological, pathological and nutritional status of an organism. The level of the blood parameters along with the nutrients retention could be an effective method of evaluating the nutritive value of an ingredient. Therefore evaluating the PCV of the broiler chicks will indicate the effectiveness of the blood meal administered to them.

2. Materials and Methods

2.1. Procurement of Experimental Animal

A total of 45 broiler chicks at four weeks old, already vaccinated for New castle and Gumboro were used for the experiment. The birds were randomly selected into three (3) treatment groups of five birds each. Each group was replicated three times. The birds were allowed to acclimatize in their new environment for one week.

2.2. Experimental Treatments

Pelletized broiler finisher feed was used to mix blood meal. Blood meal diet was prepared using fresh blood from the slaughter of cattle livestock species. Diet 1 is the control which contained only the chick mash and this was used to feed birds in treatment 1 (T1). Diet 2 contained 100g of blood meal mixed with 500g of chick mash this is used to feed birds in treatment 2 (T2). Diet 3 contained 300g of blood meal mixed with 500g of chick mash, this was used to feed birds in treatment 3 (T3).

2.3. Data Analysis

The weight of the birds was taken weekly using a sensitive weighing balance. The PCV was determined using microhaematocrit centrifuge. The result of the experiment was analyzed using Analysis of variance (ANOVA). The comparison of mean was separated using a post Hoc test (Least Significant Difference), (William and George, 2008).

3. Result

Figure 1 shows that the broiler chicks fed with treatment 2 (100g blood meal and 500g of chick mash) had the highest mean weight gain (3.04g) than other treatments; treatment 1(control) had 2.78g while treatment 3(300g blood meal and 500g of chick mash) had 2.95g. Statistically there was a significant difference ($P < 0.05$) between the weight gain of the broiler chicks fed with treatment 2 and the ones fed with treatments 1 and 3 using LSD multiple comparison.

Figure 2 shows the mean values of Packed Cell Volume (PCV) of the broiler chicks, from the figure, it was observed that the chicks fed with treatment 2 had the highest PCV (41.40%) followed by broiler chicks fed with treatment 3 (40.20%), while the ones fed with treatment 1 (control) had the lowest PCV (37%).

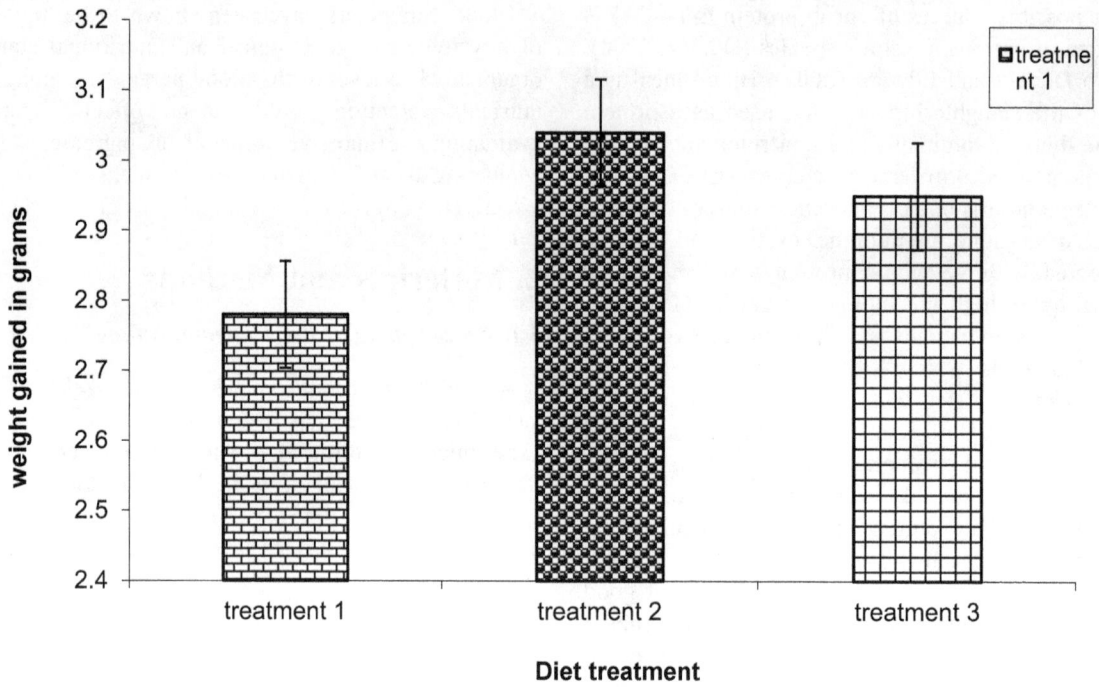

Figure 1. *Mean weight gain of Broiler chicks.*

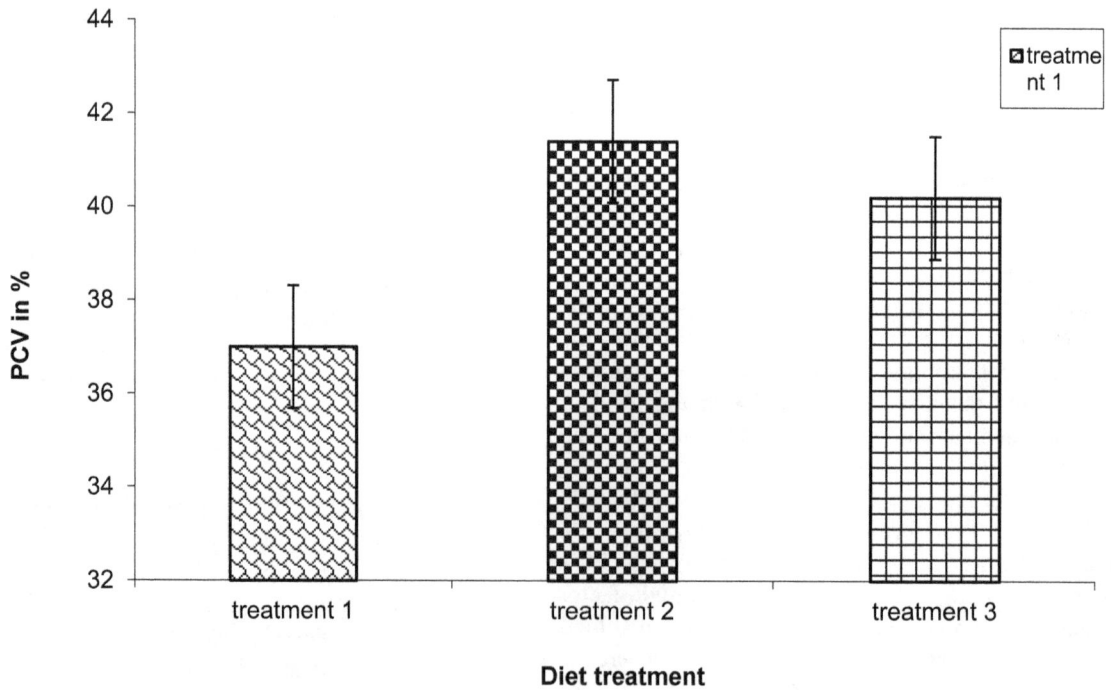

Figure 2. *Mean values of PCV of Broiler chicks.*

4. Discussion

From the result above, it was observed that there was a significant difference ($P < 0.05$) in weight gain in the broilers fed with blood meal and broilers fed with the control diet.

These results are in agreement with those of Toor and Fahimullah (2002), who reported that addition of blood meal, resulted in significant improvement in growth rate and showed an increased weight gain. It was also observed that the packed cell volume of birds fed with blood meal had a higher percentage (41.40% and 40.20%) than other birds fed with control diet (37%). These results are in agreement with those Gous and Morris, (2005), who reported that the higher values of the packed cell volume may be attributed to nutritional content of feed. Although the blood meal enhanced growth performance of the broiler chicks, it should be noted that the broilers fed with treatment 2 (100g of blood meal) had the best performance both in the growth, that is weight gain (3.04g) and Packed Cell Volume (41.40%). This shows that if the blood meal is too much, the feed will not be palatable for the broilers to feed on. Therefore the inclusion of blood meal should be in moderate amout so as to make the feed palatable to the birds and at the same time enhance their growth rate and weight gain as well.

5. Conclusion

The result obtained from the research shows that blood meal when given in moderate amount enhanced growth performance of the broilers and it also enhanced the PCV of the broilers. It is therefore recommended that farmers should include blood meal in their feed production for the broiler birds, at most 100g of blood meal in 500g of chick mash.

References

[1] Adeyemi, O.A., Fasina, O.E. and Balogun, M.O. (2000). Utilizating full fat of jatropha seed in broiler diets: Effects on Haematological Parameters and Blood Chemistry. Proceedings of the 25th Annual Conference Nigerian Society of Animal Production, Michael Okpara University of Agriculture, Umudike, Nigeria. 3:163-166.

[2] Agbede, J.O., and Aletor, V. (2007) The Performance, Nutrient Utilization and Cost Implications of Feeding Broiler Finisher Conventional or Underutilized Resources. Applied Tropical Agriculture. 2:57-62.

[3] Atawodi, S.E., Mari, D., Atawodi, J.C. and Yahaya, Y. (2008). Assessment of Leucaena and leucocephala leaves as feed supplement in laying hens. African Journal of Biotechnology. 7 (3): 317-321.

[4] Brooks, M.C. (2002). Nutrition research techniques for domestic and wild animals. Animal Science Department. Utah State University, Logaus, U.S.A. 2:44-50.

[5] Brooks, M.C. (2001). Effect of Protein on Human Growth and Development. International Journal of Nutrition. 25:46-55.

[6] Diarra, S.S. (2008). Utilization of Saseme (Sasemum indicum).Seed meal as a source of methionine by broilers and layers. Ph.D. Thesis. Abubakar Tafawa Balewa University, Bauchi State. Pp. 1-59.

[7] Donald, P. and Edward, F. (2002). Animal Nutrition. 4th Edn. Published in the United States with John Wiley and Sons. Inc. New York, 455-483.

[8] Ensminger, W.I. and Akubilo, C.O. (2002). Thermal Analysis and Evaluation of Protein Requirement of a Passive Solar Energy Poultry Chick Brooder in Nigeria. Journal of Renewal Energy. 9:1-7.

[9] Fombad, G.O., Michel, J., and Changneu, A.M. (2004). The effects of dietary protein independent of essential amino acids on growth and body composition in genetically lean and fat chickens. British Poultry Science. 41: 214-218.

[10] Gous, R. M. and Morris, T. R. (2005). Nutritional Intervention in Alleviating the Effects of high temperature in broiler production. World Poultry Science. 61: 463-475.

[11] Iheukwumere, F.C. and Okoli, I.C. (2002). Preliminary Studies on Raw Napoleana imperialis as feed ingredient. Performance and Blood Chemistry of Weaner Rabbits. Tropical Animal Production. 5:100-110.

[12] Madubuike, F.N and Ekenyem, B.U. (2006). Haematology and Serum Biochemistry Characteristics of Broiler Chicks fed varying dietary level of Ipomoea asarifolia Leaf Meal. International Journals of Poultry Science. 5:9-12.

[13] Odunsi, A. A. (2004). Blend of Bovine Blood and Rumen Digesta as a Replacement for Fishmeal and Groundnutcake in Layer Diets. International Journal of Poultry Science. 2 (1): 58-61.

[14] Oyawoye, E.O. and Ogunkunle, M. (2008). Physiological and biochemical effects of raw jack beans on broilers. Proceedings of Annual Conference of Nigerian Society of Animal Production, 23:141-142.

[15] Randy, N. (2002). Nutrient Requirements of Domestic Animals. Nutrient Requirements of Poultry, 8th ed., National Academy Press, Washington DC, USA. Pp. 231-338.

[16] Toor, A. and Fahimullah, A. (2002). Effects of Different Levels of Blood Meal on the Performance of Broiler Chick. M.Sc. Thesis. University of Agriculture Faisalabad, Pakistan. Pp. 1-52.

[17] William, A.C., and George, W.S. (2008). Statistical Methods, 6th Ed., The Iowa State University Press. Ames, Iowa, USA. Pp. 167-263.

29

Optimization of Minituber Size and Planting Distance for the Breeder Seed Production of Potato

Md. Altaf Hossain[1], Abdullah-Al-Mahmud[2], Md. Abdullah-Al-Mamun[3], Md. Shamimuzzaman[4], Md. Mizanur Rahman[5]

[1]Tuber Crops Research Centre (TCRC), Bangladesh Agricultural Research Institute (BARI), Joydebpur, Gazipur, Bangladesh
[2]International Potato Center (CIP), USAID Horticulture Project, Bangladesh
[3]Department of Agriculture Extension (DAE), Rangpur, Bangladesh
[4]Department of Crop Sciences, University of Illinois at Urbana-Champaign, Urbana, Illinois 61801, USA
[5]Department of Horticulture, Bangabandhu Sheikh Mujibur Rahman Agricultural University, Joydebpur, Gazipur, Bangladesh

Email address:

Altafmy@yahoo.com (Md. A. Hossain), A.Mahmud@cgiar.org (Abdullah-Al-Mahmud), Mamun.dae@gmail.com (Md. Abdullah-Al-Mamun)

Abstract: Six grades of potato minitubers (<5 mm, 5-10 mm, 10-15 mm, 15-20 mm, 20-25 mm and > 25 mm) and four planting distance (25 cm, 20 cm, 15 cm and 10 cm) with a potato variety Diamant were taken in an study during 2013-14 at the Tuber Crops Research Centre of Bangladesh Agricultural Research Institute, Gazipur, Bangladesh. The objective was to observe the effect of minituber grades and planting distance on growth, seed yield, increase ratio and seed potential of potato. The largest minitubers (>25 mm) planted at widest distance (25 cm) produced maximum number of tubers per plant (18.7). The highest number of tubers per m^2 (306.7) was obtained with largest minitubers (>25 mm) planted at the closest plant spacing 10 cm, while it was lowest (64.0) in smallest size minituber (<5 mm) with the widest distance 25 cm. A significant increase ratio was found ranged from 14 (largest minituber with the closest planting distance) to 297 (smallest minituber with widest planting distance). The maximum percentage (53%) of 'A grade' seed (28-55mm size) was obtained from the pea size (5-10 mm) minituber size planted at 15 cm distance. The highest seed potential (39.8) was in >25 mm size minituber planted at 10 cm distance. The lowest (4.4) was in <5 mm size minituber when planted at 25 cm distance. Seed sizes increasing from <5 mm to >25 mm had significant increase ratio ranged from 12 to 269. The highest economic return (9.4) would occur for the pea size (5-10 mm) minituber when planted at 15 cm spacing.

Keywords: Minituber, Breeder Seed and Potato

1. Introduction

Unavailability of certified seed tubers is a major constraint to potato production in Bangladesh. This compels most farmers to use planting materials from informal sources such as previous harvests from own field, local markets and neighbours. Bangladesh Agricultural Research Institute (BARI) has the national mandate to produce basic seed tubers (Breeder's seed) but can only supply less than 1% of the national requirements. BADC and other private seed potato producer can supply maximum 5% quality seed (Hossain et al., 2008). It is recorded that in 2012-2013, BADC supplied 19,322 M tons of seed potato which is only 4.12% of total quality seed [1].

Minitubers are usually defined as the progeny tubers produced on in vitro derived plantlets [17]. The size of minitubers may range from 5-25 mm and a range in weight between 0.1-10 gm and sometimes higher. Larger mini-tubers also have become common ([8],[17]). Minituber production has significantly reduced the number of generations required to produce commercial seed potatoes. This has reduced exposure to pathogens during field multiplication, resulting in healthier tubers in seed crops. Experience in production of single-hill, first generation, seedling screening material in the variety development program has shown that minitubers as small as 1-2 g can produce viable productive plants [13]. The optimizing of plant density is one of the most important subjects of potato production, because, it affects to seed cost, plant development, yield, and quality of the crop [4]. In practice, plant density in potato crop is manipulated through the number and size of the seed tubers planted [3]. Widely spaced plants allow better separation at harvest to isolate tubers from individual hills. The low population provides individual plants an advantage in access to moisture, nutrients, and sunlight. To optimize production from pre-nucleus minitubers, plants populations should closely mirror populations typically used for seed production [16]. Therefore, many studies have been conducted to establish the optimal combination of seed size and planting distance for a certain environment ([18], [5], and [4].

A lot of research work has been done to evaluate the performance of potato seed tubers but little information exists on the field performance of minituber for the breeder's seed production of potato. Therefore, the present study was conducted to evaluate the field performance of different size potato minituber and planting distance for the production of breeder's seeds of potato.

2. Materials and Methods

2.1. Site, Soil and Season of the Experiment

The experiment was conducted at the net house of Tuber Crops Research Centre, BARI, Gazipur during November, 2013 to March 2014. The location of the experimental site was to the 34 km north from Dhaka city (24.38^0 N latitude and 90.13^0 longitudes) at 8.4m above the sea level. The soil of the experimental field was grey terrace contained pH 6.4. This area is moderately drought prone, and face drought both winter and late winter season. The experimental site is situated in a sub- tropical climate zone and characterized by no rainfall during December to March.

2.2. Planting Materials and Date of Planting

Diseases free well sprouted seed potato minitubers of Diamant variety were used as planting material for the experiment. Mini-tubers were planted on 6th November, 2013.

2.3. Crop Management

The field was ploughed 3-4 times to a depth of 25 cm. Full doses of well rotten cow dung (10 t ha^{-1}), TSP (220 kg ha^{-1}), MP (270 kg ha^{-1}), Gypsum (120 kg ha^{-1}), Boric acid (6 kg ha^{-1}) and half doses of Urea (175 kg ha^{-1}) were applied at the time of final land preparation. The rest half dose of Urea (175

kg ha^{-1}) was applied as side dressings at 30 DAP followed by earthing-up and light irrigation. First earthing up was done at 30 DAP when the plant attained a height of about 15-20 cm from the base, second earthing-up was done after 20 days of first earthing up. Before first earthing up, Urea was applied. Irrigation was applied 3 times. First one was applied just after planting, second one was just after earthing up at 30 DAP, and last one was on 55 DAP. During land preparation, Furadan 5G was applied @10 kg ha-1 as basal during land preparation and Admire (0.2%) was sprayed in two installments at 45 and 60 DAP to control insects. The crops were also sprayed alternatively with Dithane-M 45 (0.2%) and Secure (0.1%) at 15 days interval to prevent the late blight infection of potato. The field was netted during the entire growing period to protect the plants from the insect infestation specially aphids which is the vector of different viruses. Seeds were planted at row distance of 60 cm row and planting distance of 25, 20, 15 and 10 cm. Haulm pulling was done at 75 DAP by hand. Hardening and setting up of skins of tubers were allowed for 10 days under the soil there after crop was harvested at 85 DAP. Tubers were collected carefully with the help of spade without any injury.

2.4. Design and Treatments of the Experiment

The experiment was laid out in two factors Randomized Complete Block Design (RCBD) with three replications. First of all the entire experimental field was divided into three blocks, representing three replications. Each block again divided into twenty four unit plots. The treatment was assigned randomly to unit plots of each block. The size of a unit plot was 3.0 m × 2.4 m. There were six grades of minitubers based on minituber diameter (S_1/Under size = <5 mm, S_2/pea size=5-10 mm, S_3/small size=10-15 mm, S_4/medium size = 15-20 mm, S_5/large size = 20-25 mm and S_6/extra-large= > 25 mm) based on minituber diameter and four planting distance (D_1=25 cm, D_2=20 cm, D_3=15 cm and D_4=15 cm) which formed twenty four treatment combinations. Treatment combinations were as follows-

S_1D_1, S_1D_2, S_1D_3, S_1D_4, S_2D_1, S_2D_2, S_2D_3, S_2D_4, S_3D_1, S_3D_2, S_3D_3, S_3D_4, S_4D_1, S_4D_2, S_4D_3, S_4D_4, S_5D_1, S_5D_2, S_5D_3, S_5D_4, S_6D_1, S_6D_2, S_6D_3 and S_6D_4.

2.5. Climatological Data

Air temperature and humidity, precipitation, evaporation, soil temperature and ground water table were recorded throughout the crop period (Table 1).

Table 1. Climatological data of 2013-14 crop season.

Month	Air Temperature ($^{\circ}$ C)			Humidity (%)	Rain Fall (mm)
	Max.	Min.	Av.		
2013-14					
November	26.60	22.43	24.52	80.47	8.44
December	19.90	15.45	17.68	89.05	0.00
January	15.20	11.58	13.39	90.80	0.00
February	23.85	19.08	21.47	89.89	8.43
March	31.16	26.09	28.62	76.70	29.84

2.6. Data Collection

Data on different growth and yield contributing characters were recorded from the sample plants of each plot during the course of experiment. The sampling was done randomly. The plants in the outer row were excluded during random selection. Five plants were randomly selected from each plot to record the data on the following parameters: Plant emergence, plant height, leaf area, number and weight of tubers per plant, yield (kg m^{-2}), and percentage of different grades of tuber by number, seed potentials and seed increase ratio.

2.7. Statistical Analysis

To find out the significance of experimental results, the collected data on different parameters were analyzed statistically by using MSTAT-C program. The mean for all the treatments were calculated and analysis of variance for each parameter was performed by F-test. The mean separation was done by DMRT at 5% level of probability.

3. Results and Discussion

3.1. Plant emergence (%)

Analysis of variance indicated that the interaction effect of minituber size and planting distance had significant influence on emergence rate at 30 DAP (Table 2). The highest emergence rate was found in large minituber with 25 cm planting distance (94.1%), which was statistically identical to other planting distance and extra-large. The lowest emergence rate (61.8%) was obtained from under-size minituber at 10 cm planting distance, which was statistically at par with 15 cm planting distance (62.7%). In an average emergence performance of the under-size minituber at any planting distance is very low compared to other sizes. These results support the findings of Rykbost and Charlton (2004); Karafyllidis et al. (1997) and El Amin et al. (1996).

Table 2. *Interaction effect between minituber size and planting distance on percent plant emergence*

Minituber size	Planting distance (cm)				Mean
	25	20	15	10	
Under size	69.9	68.9	62.7	61.8	65.8
Pea	79.4	75.8	77.8	78.2	77.8
Small	82.5	80.4	83.8	83.2	82.5
Medium	89.4	82.9	87.6	84.9	86.2
Large	94.1	89.2	90.5	91.4	91.3
Extra-large	91.1	89.4	88.1	88.8	89.4
Mean	84.4	81.1	81.8	81.4	
LSD (0.05)					
Planting distance (P)	8.72				
Minituber size (M)	7.99				
P x M	21.35				

3.2. Plant Height (cm)

The tallest plant (85.9 cm) was produced by extra-large size

minituber when planted at 25 cm distance; shortest (50.2 cm) plant was produced by under-size mintuber with 25 cm (Table 3). Jagroop et al. (1993) found taller plants with large size normal seed tubers planted at closer spacing. Probably the plant height was the highest in the larger size minituber due to the presence of more reserve food which caused rapid growth of plants earlier. Similar findings have also been reported by Zakaria (2003) but his research was on the effect different size of microtuber on plant height.

Table 3. *Interaction effect between minituber size and planting distance on plant height (cm)*

Minituber size	Planting distance (cm)				Mean
	25	20	15	10	
Under size	50.2	59.8	58.4	57.8	56.6
Pea	71.7	66.5	64.3	70.7	68.3
Small	65.7	68.7	62.0	67.7	66.0
Medium	70.4	69.5	65.5	70.7	69.0
Large	71.1	71.0	71.3	65.3	69.7
Extra-large	85.9	72.2	79.8	68.8	76.7
Mean	69.2	68.0	66.9	66.8	
LSD (0.05)					
Planting distance (P)	2.97				
Minituber size (M)	4.85				
P x M	9.69				

3.3. Leaf Area (cm^2)

Generally, yield was positively correlated to leaf area and was increased linearly as leaf area increased. Planting distance had significant difference on the leaf area production of plants derived from minituber (Table 4). Extra-large minituber contributed the highest leaf area values across all population density levels in comparison to other seed minitubers. The trend was such that the bigger the seed piece, the greater the leaf area. There was increase in leaf area with increase in minituber size and planting distance. The leaf area was highest in largest minituber with planting distance 25 cm but lowest in smallest mini-tuber with closest planting distance 10 cm. These results are in conformity with the findings of Akhtar et al. (2010). Lower leaf area index and radiation interception in small seed size undoubtedly reduced production of assimilates.

Table 4. *Interaction effect between minituber size and planting distance on leaf area*

Minituber size	Planting distance (cm)				Mean
	25	20	15	10	
Under size	837.7	814.3	797.0	785.7	808.7
Pea	924.0	891.3	849.0	831.0	873.8
Small	966.0	948.3	933.7	912.0	940.0
Medium	1042	1031	1008	991.0	1018.0
Large	1233	1210	1173	1150	1191.5
Extra-large	1414	1385	1353	1280	1358.0
Mean	1069.5	1046.7	1019.0	991.6	
LSD (0.05)					
Planting distance (P)	120.6				
Minituber size (M)	130,2				
P x M	124.8				

3.4. Tuber Number per Plant and per m²

Minituber size and planting distance interacted significantly and affected the number of tuber per plant (Table 5 & 6). The tuber number per plant increased with increase in minituber size in each planting spacing. The extra-large minituber planted at widest distance (25 cm) produced maximum number of tuber per plant (18.7) which was statistically similar to 20 cm (18.0)) and 10 cm (18.4) of same groups respectively. Number of tubers per plant was the lowest in smallest minituber with closest planting distance 10 cm (8.10). Larger minituber have higher amount of reserve food and interplant competition for space, light, water and nutrient is less in the wider spacing that can contribute the increase number of tuber per plant. The number of tubers per m2 was increased with increase in minituber size with closer planting distance (Table 16). The highest number of tuber per m2 was obtained with extra-large minitubers planted at the 10 cm distance (306.7), while it was lowest in under-size minituber with the widest distance 25 cm (64.0). The results were in conformity with the findings of Haverkort et al. (1991) who found increasing number of tubers per plant with increase in size of microtuber. The same information was reported by Rykbost and Charlton (2004) and Karafyllidis et al. (1997). These results are in conformity with the findings of Tuku (2000) who reported that higher yield was associated with proper nutrients and water availability to the plant and more tuber weight. Gopal et al. (2007) after conducting similar study also proposed that selection for tuber yield can be practiced at the minituber level in potato breeding processes. Zkaynak & Samanci (2006) worked on field performance of three weight classes of small minitubers ranging from 6.0 - 18.0 g was studied in two years at different planting dates. The heavy minitubers gave higher values than light minitubers for tuber yield, tuber weight, tuber number and stem number.

Table 5. Interaction effect between minituber size and planting distance on tuber number per plant

Minituber size	Planting distance (cm)				
	25	20	15	10	Mean
Under size	9.6	9.6	10.8	8.1	9.5
Pea	11.5	12.9	10.3	10.0	11.2
Small	12.5	10.7	10.4	9.6	10.8
Medium	11.5	10.4	11.5	13.9	11.8
Large	13.1	14.8	13.7	15.7	14.3
Extra-large	18.7	18.0	14.4	18.4	17.4
Mean	12.8	12.7	11.9	12.6	
LSD (0.05)					
Planting distance (P)	0.85				
Minituber size (M)	1.40				
P x M	2.09				

Table 6. Interaction effect between minituber size and planting distance on tuber number per m²

Minituber size	Planting distance (cm)				
	25	20	15	10	Mean
Under size	64.0	84.8	120.8	151.0	105.2
Pea	76.4	114.1	114.0	166.7	117.8
Small	83.5	94.1	90.3	173.4	110.3
Medium	76.4	91.8	127.3	231.0	131.6
Large	87.1	130.7	152.4	262.1	158.1
Extra-large	124.5	158.9	160.0	306.7	187.5
Mean	85.3	112.4	127.5	215.2	
LSD (0.05)					
Planting distance (P)	14.42				
Minituber size (M)	18.28				
P x M	7.57				

3.5. Tuber Weight per Plant and per m²

Minituber size and planting distance interacted significantly and there was a significant influence on tuber weight per plant (Table 7 & 8)). Tuber yield per plant increased significantly with increase in minituber size and planting distance. The maximum tuber weight per (985.0 g) plant was obtained from the extra-large minituber with planting distance 25 cm and minimum tuber weight per plant (119.0 g) was found from the under-size minituber with closest planting distance 10 cm which was statistically similar to other spacing of the same size. Different trend was observed in weight of tuber per m². The highest weight of tubers per m² (8.50 kg) was obtained from the largest minitubers with closest planting distance 10 cm which was at par with large size minituber with closest planting distance 10 cm (8.29 kg). The lowest weight of tuber per m2 (0. 99 Kg) was found in under-size with widest spacing 25 cm.

Table 7. Interaction effect between minituber size and planting distance on tuber weight per plant (g)

Minituber size	Planting distance (cm)				
	25	20	15	10	Mean
Under size	148.4	147.2	126.0	119.4	135.3
Pea	719.6	641.0	615.7	469.7	611.5
Small	670.2	584.4	552.7	476.9	571.1
Medium	688.2	600.2	495.9	390.5	543.7
Large	811.1	757.6	651.7	497.3	679.4
Extra-large	985.0	831.3	694.9	509.9	755.3
Mean	670.4	593.6	522.8	410.6	
LSD (0.05)					
Planting distance (P)	3.09				
Minituber size (M)	3.88				
P x M	7.57				

Table 8. Interaction effect between minituber size and planting distance on tuber yield (kg m⁻²)

Minituber size	Planting distance(cm)				
	25	20	15	10	Mean
Under size	0.99	1.30	1.40	1.99	1.42
Pea	4.80	5.66	6.84	7.83	6.28
Small	4.47	5.16	6.14	7.95	5.93
Medium	4.59	5.30	5.51	6.51	5.48
Large	5.41	6.69	7.24	8.29	6.91
Extra-large	6.57	7.34	7.72	8.50	7.53
Mean	4.47	5.24	5.81	6.85	
LSD $_{(0.05)}$					
Planting distance (P)	0.22				
Minituber size (M)	0.27				
P x M	0.53				

3.6. Seed Potential

From the calculation, the seed potentials of minituber of the different sizes planted at different spacing ranged from 4.4 to 39.8 (Table 9). The highest seed potential (39.8) was found in extra-large size minituber planted at 10 cm distance. The lowest seed potential (4.4) was in under-size minituber when planted at 25 cm distance. However, the yield (both number and weight per plant) of under-size and extra-large size minituber was negligible.

3.7. Increase Ratio

Results observed in this trial demonstrated that extremely high increases ratios as minituber size decreased and a large reduction in this ratio as minituber size increased (Table 9). Effects of increasing minituber size on yield are attributed to a combination of increases in both number and size of daughter tubers. Seed sizes increasing from under-size to extra-large size had significant increase ratio ranged from 12 to 269. Rykbost and Charlton (2004) reported 65 to 317 increase ratios from the minituber size ranged from 1.2 g to 13.6 g. They also reported that a typical seed increase expectation is 15 or 20 to 1 but in the irrigated production, the increase ratio is likely to be 20 to 1. The similar results were found by Masarirambi *et al.* (2012) and Islam *et al.* (2012).

Table 9. Interaction effect between minituber size and planting distance on seed potential and increase ratio

Minituber size	Seed potential				Increase ratio			
	Planting distance (cm)				Planting distance (cm)			
	25	20	15	10	25	20	15	10
Under size	0.99	1.30	1.40	1.99	130	119	93	72
Pea	4.80	5.66	6.84	7.83	269	223	210	155
Small	4.47	5.16	6.14	7.95	90	75	72	60
Medium	4.59	5.30	5.51	6.51	47	41	34	26
Large	5.41	6.69	7.24	8.29	26	25	21	16
Extra-large	6.57	7.34	7.72	8.50	22	19	16	12
Mean	4.47	5.24	5.81	6.85	97.33	83.67	74.33	56.83

3.8. Economic Analysis

Significant variation in partial budget analysis was observed in different treatment combinations (Table 10). In seed production of potato from mini tuber total variable cost (TVC) was highest in extra-large minituber when planted at 10 cm distance (Tk.161.88).The highest gross net return Tk. 224.09 was found in the same treatment but it BCR was 1.4 which was lower than other treatment combinations. The lowest net return Tk. 18.57 was found in under-size planted at 25 cm distance and its BCR was also lowest (1.5). Closer planting required more labour involvement and higher seed rate/ha which resulting high TVC and lower the BCR in breeder seed production by using mini tuber as has been reported by Mamun (2012). The result suggests that the highest economic return (9.4) would occur for the 1-4 g small minituber when planted at 15 cm spacing if the price per kilogram of the minitubers is equal for all sizes. Pricing compensation for larger seed sizes would need to be large for much lower production potential. Production of basic seed from minituber is very costly.

Table 10. Partial budget analysis of potato for different treatment combinations of minituber size

Treatment		Total material cost (Tk.)	Total non-material cost (Tk.)	Total variable cost (Tk.)	Gross return (Kg/m²)		Net return (Tk/m²)	BCR
Minituber size	Planting distance				Seed	Non-seed		
Under size	60 × 25	5.33	6.75	12.08	13.01	5.56	18.57	1.5
	60 × 20	5.60	6.95	12.55	15.76	7.62	23.38	1.9
	60 × 15	5.89	7.15	13.04	15.57	8.81	24.38	1.9
	60 × 10	6.58	7.35	13.93	18.01	13.90	31.90	2.3
Pea	60 × 25	7.84	6.75	14.59	107.48	12.17	119.65	8.2
	60 × 20	8.92	6.95	15.87	118.21	17.20	135.41	8.5
	60 × 15	10.06	7.15	17.21	139.78	21.81	161.59	9.4
	60 × 10	12.84	7.35	20.19	155.22	26.56	181.78	9.0
Small	60 × 25	14.51	6.75	21.26	107.56	8.85	116.41	5.5
	60 × 20	17.75	6.95	24.70	119.92	11.63	131.55	5.3
	60 × 15	21.17	7.15	28.32	144.08	13.37	157.45	5.6
	60 × 10	29.51	7.35	36.86	181.45	19.02	200.47	5.4

Treatment		Total material cost (Tk.)	Total non-material cost (Tk.)	Total variable cost (Tk.)	Gross return (Kg/m²)		Net return (Tk/m²)	BCR
Minituber size	Planting distance				Seed	Non-seed		
Medium	60 × 25	24.51	6.75	31.26	112.79	8.30	121.09	3.9
	60 × 20	30.99	6.95	37.94	130.62	9.46	140.08	3.7
	60 × 15	37.83	7.15	44.98	136.75	9.52	146.27	3.3
	60 × 10	54.51	7.35	61.86	156.94	12.79	169.73	2.7
Large	60 × 25	44.52	6.75	51.27	127.21	11.70	138.91	2.7
	60 × 20	57.48	6.95	64.43	161.00	13.23	174.23	2.7
	60 × 15	71.16	7.15	78.31	170.31	15.63	185.94	2.4
	60 × 10	104.52	7.35	111.87	190.13	19.52	209.65	1.9
Extra-large	60 × 25	64.53	6.75	71.28	161.96	14.79	176.75	2.5
	60 × 20	83.97	6.95	90.92	179.18	15.66	194.84	2.1
	60 × 15	104.49	7.15	111.64	187.80	15.80	203.61	1.8
	60 × 10	154.53	7.35	161.88	207.88	16.21	224.09	1.4

Breeder's Seed price = 30 Tk/kg and non-seed price = 10 Tk./kg

4. Conclusion

Breeder seed production of potato was affected by minituber size and planting distance. Larger size of minituber produced more number of tubers with increased yield when it was planted in greater distance. But, the higher seed yield potential was found in larger sized with closer planting of minituber. So, it can be concluded that pea size minituber with 15 cm planting distance might have the highest economic return. However, based on the yield and net return, gross return, small size minituber at 10 cm planting distance may be used for cost effective production of breeders' seeds of potato.

Appendixes

Appendix 1. Meteorological conditions of the experimental site during January 2012 to December 2013

Month	Air Temperature (°C)		Humidity (%)	Rain Fall (mm)
	Max.	Min.		
January	21.42	16.77	89.20	1.68
February	24.33	19.68	85.25	0.00
March	28.81	24.60	81.61	13.63
April	31.95	27.12	82.83	38.68
May	31.47	25.89	83.84	162.43
June	34.00	25.80	84.53	247.34
July	32.19	25.94	85.07	363.60
August	31.16	25.94	86.29	590.30
September	31.70	27.47	86.57	206.46
October	29.74	26.70	85.29	182.43
November	26.60	22.43	80.47	8.44
December	19.90	15.45	89.05	0.00
Ave./total	28.61	23.65	85.00	1814.94

Source: Weather station, BARI, Gazipur

Appendix 2. Meteorological conditions of the experimental site during January 2013 to December 2014

Month	Air Temperature (°C)		Humidity (%)	Rain Fall (mm)
	Max.	Min.		
January	15.20	11.58	90.80	00.0
February	23.85	19.08	89.89	8.43
March	31.16	26.09	76.70	29.84
April	32.13	28.2	76.53	57.11
May	31.40	27.90	82.0	252.23
June	32.06	29.26	85.96	369.53
July	32.32	27.67	83.45	269.13
August	32.0	25.90	85.58	138.65
September	30.38	26.53	89.46	212.65
October	30.67	27.06	87.41	187.01
November	27.76	23.76	85.66	00.0
December	24.80	16.58	90.70	55.19
Ave./total	29.46	24.81	85.34	1579.77

Source: Weather station, BARI, Gazipur

References

[1]   AIS. 2014. Seed supplied by BADC in 2012-13. *Krishi* Diary. Agriculture Information Service, Khamarbari, Farmgate, Dhaka, Bangladesh. P.7.

[2]   Akhtar, Parveen., S. J. Abbas, M. Aziz, A. H. Shah and N. Ali. 2010. Effect of Growth Behavior of Potato Mini Tubers on Quality of Seed Potatoes as Influenced by Different Cultivars. Pak. J. Pl. Sci. 16 (1): 1-9.

[3]   Allen, E. J. and D. C. E. Wurr. 1992. Plant density. In: P. M. Harris (Ed.), The Potato Crop. The scientific basis for improvement. Second edition. Chapman and Hall, London, UK, pp. 292-333.

[4]   Bussan, A.J., P.D. Mitchell, M.E. Copas and M.J. Drilias, 2007. Evaluation of the effect of density on potato yield and tuber size distribution. Crop Sci. 47: 2462–2472.

[5]   Creamer, N.G., C.R. Crozier and M.A. Cubeta, 1999. Influence of seed piece spacing and population on yield, internal quality and economic performance of Atlantic, Superior and Snowden potato varieties in Eastern North Carolina. *American J. Potato Res.*, 76: 257–261.

[6]   El-Amin, S. M., B. Adam, E. Varis and E. Pehu. 1996. Production of seedling tubers. Experimental Agric. 32 (4): 419-426.

[7]   Gopal, J., R. Kumar and G. S. Kang. 2007. The effectiveness of using a minituber crop for selection of agronomic characters in potato breeding programmes *Potato Journal*, 34 (1 & 2): 145-151.

[8]   Hassanpanah D., A. A. Hosienzadeh and N. Allahyari. 2009. Evaluation of planting date effects on yield and yield components of Savalan and Agria cultivars in Ardabil region. *Journal of Food, Agriculture & Environment,* 7 (3&4): 525-528.

[9]   Haverkort, A. J., M. Van de Waart & J. Marinus.1991. Field Performance of Potato Microtubers as Propagation Materials. Potato Research. 34: 353-364.

[10] Hossain, M. A., A. U. Hauque, M. S. Alam, M. Hossain, M. M. Khatun, M. M. Hasan and S. N. Begum. 2008. Disease free

Minituber Production of Potato Using Tissue Culture Methods (In Bangla), TCRC, BARI, Joydebpur, Gazipur-1701, Bangladesh. P.1.

[11] Islam, M. S., S. Moonmoon, M. Z. Islam, H. Waliullah and M. S. Hossain. 2012. Studies on Seed Size and Spacing for Optimum Yield of Potato in Northern Region of Bangladesh. Bangladesh J. Prog. Sci. & Tech. 10(1): 113-116.

[12] Karafyllidis, D.I., D.N. Georgakis, N.I. Stavropoulos, E.X. Nianiou, and I.A. Vezyroglou. 1997. Effect of planting density and size of potato seed-minitubers on their yielding capacity. Acta Hort. 462:943–949.

[13] Kenneth, A. R. and B. A. Charlton. 2005. Effects of prenuclear minituber seed size on production of Wallowa Russet Seed. Annual Report. Klamath Experiment Station. USA. 31-38 pp.

[14] Mamun, A. A. 2012. Effect of Planting time and spacing of top shoot cutting for breeder's seed production of potato. MS thesis. BSMRAU. Gazipur-1706. 1-66 pp.

[15] Masarirambi, M. T., F. C. Mandisodza, A. B. Mashingaidze and E. Bhebhe, 2012. Influence of plant population and seed tuber size on growth and yield components of potato (*Solanum tuberosum*). Int. J. Agric. Biol. 14: 545–549.

[16] Rykbost, K. A. and B. A. Charlton.2004. Effects of Prenuclear Minituber Seed Size on Production of Wallowa Russet Seed. Annual Report. Klamath Experiment Station (KES), Klamath Falls, Oregon, USA. pp 38-43.

[17] Struik, P. C. 2007. The canon of potato science: Minitubers. Potato Res. 50(3-4):305-308.

[18] Sultana N and Siddique A. 1991. Effect of cut seed piece and plant spacing on the yield and profitability of potato. Bangladesh Horticulture. 19(1): 37-43.

[19] Tuku, B. T. 2000. The utilization of true potato seed (TPS) as an alternative method of potato production. Indonesian J. Agric. Sci. 1 (2): 29-38.

[20] Zakaria, M. 2003. Induction and Performance of Potato Microtuber. Ph D Dissertation. Department of Horticulture. BSMRAU, Gazipur-1701. Bangladesh. 144-159 pp.

[21] Zkaynak E. and B. Samanci. 2006. Field performance of potato minituber weights at different planting dates. Archives of Agronomy & Soil Scienc. 52 (3): 333-338.

30

Establishment of Africa Red Mahogany (*Khaya anthoteca*) Pre-Inoculated with Arbuscular Mycorrhizae Fungi (AMF) and Compost Application on an Ex-Coal Mined Site

Philip Worlanyo Dugbley[1, *], Irdika Mansur[1, 2], Basuki Wasis[1]

[1]Department of Silviculture, Faculty of Forestry, Bogor Agricultural University, Bogor-Indonesia
[2]SEAMEO Biotrop, Jalan Raya Tajur 6K, Bogor, Indonesia

Email address:
philidug87@yahoo.com (P. W. Dugbley), irdikam@gmail.com (I. Mansur)

Abstract: Coal mining provides a means for creating wealth and significantly contributes to export earnings, economic activity and employment whilst supporting regional development. However, coal mining is one of the most severe disturbances in terrestrial ecosystems. Thus, the removal of the natural vegetation and upper soil horizons for mining exploration hinders the establishment and survival of plant and soil microbial communities. Revegetation of ex-coal mined lands is therefore required to enable the recover, as close as possible, to its previous integrity. The establishment of tree species capable of protecting the underlying soil and its micro-fauna and flora is one way of achieving this aim. This study therefore aims to investigate the effect of arbuscular mycorrhizae fungi (AMF) pre-inoculation and compost application on the growth performance of the Africa red mahogany, *Khaya anthoteca* on an ex-coal mined site. The field design for this study was the completely randomized design (CRD) in factorial experiment. Four (4) levels of each factor namely compost and AMF were used with sixteen treatment combinations and each treatment replicated four times giving sixty four (64) experimental units. The results indicated that compost has significant effect (P<0.001) on height, diameter and leaf increment with steady increment during this study. There was no significant effect of mycorrhizae treatment as well as interaction between both factors (AMF and compost) on the growth of *K. anthoteca*. However, compost composition from a mixture of *Salvinia natans* and that prepared from the paddy husk (C3) recorded the highest increment in height of 9.31 cm while compost from *S. natans* only (C1), rice hull compost; herein known as paddy husk compost (C2) and control (C0) recorded increments of 9.00 cm, 5.78 cm and 4.47 cm respectively. The arbuscular mycorrhizae fungi played a role in the survival of the species on the field. There was percentage difference of between 18.5-37.5% over the control treatment. AMF from *Glomus manihotis* had the highest percentage survival of 81.25% whiles the control treatment of mycorrhizae had the lowest percentage of 43.75%. Plants are also able to withstand harsh environmental conditions through fungi-plant symbiosis enhancing the chances of survival on the field and thus, aiding the plant establishment. The study concludes that AMF and compost applications are feasible and sound technologies for the establishment of *K. anthoteca* on ex-coal mined sites.

Keywords: Coal Mining, *Khaya anthoteca*, Plant Growth, Arbuscular Mycorrhizae Fungi (AMF), Compost

1. Introduction

Coal mining provides means for creating wealth and enhances export earnings, economic activities as well as employment whilst supporting regional development. For example, the mining sector of Indonesia contributes to the nation's economy currently, for about 11.54% of total GDP [23].However, coal mining is one of the most severe disturbances in terrestrial ecosystems. It causes large-scale deforestation and land degradation with complete loss of topsoil. Thus, the removal of the natural vegetation and upper soil horizons for exploration and mining that hinders the establishment and survival of plants and soil microbial communities [7]. Mining also results in the formation of artificial habitats that are microbiologically poor, requiring human intervention for their proper restoration [27].

Revegetation of coal mined lands is therefore required to enable the re-use of such lands for other purposes. The establishment of tree species capable of protecting the underlying soil and its micro-fauna and flora is one way of achieving this aim.

Khaya anthoteca commonly referred to as the Africa red mahogany is a species belonging to the family Meliaceae. This species is heavily exploited, particularly in East and West of Africa. The species is used in high class cabinetwork and for the production of veneers and any application where good quality, medium weight hardwood is needed [13]. It also weathers well and is resistant to borers and termites. The dense crown makes it suitable as a shade tree and also popular for windbreaks as well as for aesthetic purposes. It has successfully been introduced in South Africa, Cuba and Puerto Rico and on a limited scale in Indonesia and Peninsular Malaysia where it has been used in plantations and also in Taungya systems.

Mycorrhizal fungi have over the years played critical roles in nutrient cycling and the functioning of ecosystems. According to [29], mycorrhiza fungi are the main pathway through which most plants obtain mineral nutrients and as such, are critical in terrestrial ecosystem functioning. In this mutualistic symbiosis, plants exchange photosynthates, not only for mineral nutrients, but also for increased resistance to disease, drought and extreme temperatures. Thus, plants are able to withstand harsh environmental conditions through fungi-plant symbiosis. It also been reported that mycorrhizal fungi are removed entirely in newly graded lands and always requires inoculation if the objective is a functional terrestrial ecosystem. Eroded land is also in nearly the same condition. The arbuscular mycorrhizal fungi (AMF) can therefore be integrated in soil management. These are structures resulting from the symbiosis between these fungi and plant roots, occurring in most soils and colonize roots of many plant species and directly involved in plant mineral nutrition. The symbiotic root-fungal association increases the uptake of less mobile nutrients [21], essentially phosphorus (P) but also of micronutrients like zinc (Zn) and copper (Cu), the symbiosis has also been reported as influencing water uptake. AMF can also benefit plants by stimulating the production of growth regulating substances, increasing photosynthesis, improving osmotic adjustment under drought and salinity stresses as well as increasing resistance to pests and soil borne diseases [1] and benefits have mainly been attributed to improved phosphorous nutrition.

Furthermore, the addition of composted residue is an effective treatment for increasing rhizosphere aggregate stability. Mycorrhiza is increasingly important for improving the growth of seedlings following the addition of composted residue to soil under severe climatological conditions as reported by [6]. They also reported that high proportion of stable aggregates of soil is mainly attributed to a higher microbial activity of root biomass and particularly to the presence of vesicular arbuscular mycorrhiza in the rhizosphere aggregates. At the same time, reforestation techniques based on the addition of composted residue and

mycorrhizal inoculation in the nursery could be used as a tool for improving soil structure, with subsequent improvement in plant growth. This study therefore aims to investigate the effect of arbuscular mycorrhizae fungi (AMF) pre-inoculation and compost applications on the growth performance on vegetatively propagated *K. anthoteca* (Africa red mahogany), on an ex-coal mined site located in South Sumatra of Indonesia.

2. Materials and Methods

2.1. Study Area

This research was conducted in a coal mining site namely; PT. Bukit Asam located in South Sumatera of Indonesia. South Sumatra falls within coordinates 2°45′S 103°50′E of Indonesia. The area is characterized by a bimodal rainfall pattern with the major wet season between May and July. This area experiences a short dry season in August and a long one between December and March. The annual rainfall ranges between 2000 mm and 3000 mm. The study was conducted from August to December, 2014.

2.2. Research Materials and Tools

Materials and tools used for the study included; record log sheets, clip board, pencil, tape measure, electronic caliper, meter rule, analog weighing scale, non-poisonous maker, Microsoft excel and Minitab Version 16. Three-month old *K. anthoteca* seedlings were obtained from Carita, Banten province of West Java, Indonesia.

2.3. Experimental Design and Layout Procedures

Thedesign for this field research was the Factorial experiment in Completely Randomized Design (CRD). Two factors; mycorrhizae and compost with four levels of each factor were used giving a total of sixteen (16) treatment combinations and each treatment combination replicated four (4) times. In total, sixty-four (64) experimental units were used for this study. The specified treatments (M) for mycorrhizae were as follows;M_0 (control; without mycorrhiza inoculation); M_1(±50 single spores of *Gigaspora margarita* mycorrhizae); M_2 (±50 spores of *Glomus manihotis* mycorrhizae); M_3 (±25 spores each of *Gigaspora margarita* and *Glomus manihotis* mycorrhizae). The treatments (C) for compost were; C_0 (control; no compost); C_1(±5 Kg of *Salvinia natans* compost); C_2(±5 Kg of paddy husk compost); C_3(±2.5 Kg each of paddy husk and *Salvinia natans* composts). Planting space of 4 x 4 m (625plants/ha) was used and the various treatment combinations were randomly allocated to the plots with each plot having one of the treatment combinations. Reference points of 5 cm above the soil surface of the seedlings were marked with non-poisonous indelible ink marker to provide consistency at the point of height and diameter measurements.

2.4. Data Collection and Analysis

The plant parameters measured were the height (cm), stem diameter (mm), leaf counts and percentage survival. According to [17], height and stem diameter are some of the frequently used methods of measuring the growth of multi-cellular living systems and also often advantageous to use several characteristics for the same system. An indelible ink was used to mark each plantlet 5cm above the soil, where the diameter and height readings were taken since irregularity of the soil around the seedlings can affect the recording. Initial measurements of plant parameters were recorded two days after transplanting and fortnightly thereafter over sixteen weeks (4 months) of the field research. Data on percentage survival was collected after the field experiment. Data on plant parameters were subjected to a two-way analysis of variance (ANOVA) at a significance level of 5% (alpha ≤0.05). Microsoft excel was used in data entry and computation for the increment of the various plant parameters measured (data management). All statistical analyses were performed using the Minitab statistical analysis package V.16. (Minitab, Inc.). Tukey's Honest Significant Difference test was used for multiple comparison tests where treatment means differed significantly. The mathematical model for the research design is as follows;

$$Yijk = \mu + \tau i + \beta j + (\tau\beta)ij + \varepsilon ijk$$

Where:

Yijk= the (ijk)th observation for i=1, 2, 3 and 4;j = 1, 2, 3 and 4; and k =1, 2, 3 and 4

μ= parameter common to all treatments (the mean)

τi= the ith treatment effect of factor A (mycorrhizae)

βj= the jth treatment effect of factor B (compost)

$(\tau\beta)ij$ = the (ij)th interaction effect of factor A and B

εijk= random error component with normal distribution

3. Results and Discussion

3.1. Plant Growth

Growth is the irreversible increase in the size of a plant and has also been considered as the product of its physiological processes. Plants have indeterminate growth, thus, they have the capacity to grow from the apical meristem indefinitely. Growth generally occurs in cycles such as seasonally or daily. Factors affecting the growth of plants are broadly categorized into genetic and environmental factors as well as the interaction between these factors.

3.2. Growth of K. anthoteca on an Ex-Coal Mined Site

The growth of plants depends on the availability of nutrients from the soil. Thus, it is important that the soil should continuously provide nutrients for the growth and development of plants [25].

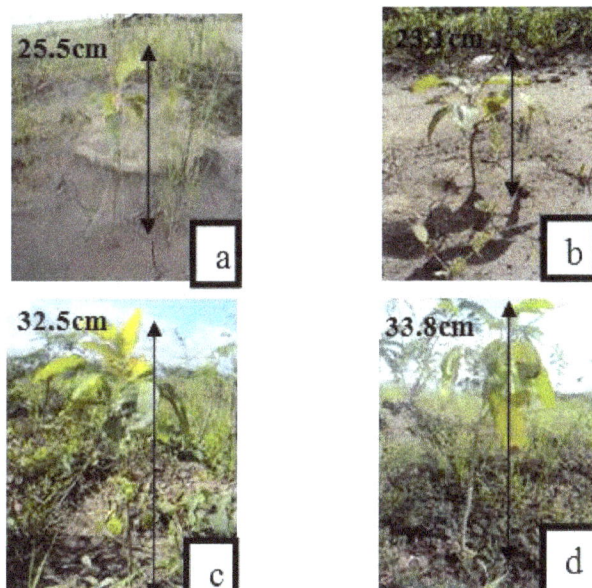

Figure 1. Growth of Khaya anthoteca on the field: a) Paddy husk compost b) Without Compost c) Salvinia natans compost d) Mixed compost.

Compost application to *K. anthoteca* had a significant effect on the height growth of the species. Analysis of variance (ANOVA) tested at α ≤ 5% showed a significant effect (P<0.001) of compost on the mean height increment. There was no significant effect (P>0.05) of mycorrhizae treatment as well as the interaction between both factors (Table 1). Compost composition from a mixture of *S. natans* and that prepared from rice hull (herein known as paddy husk compost) recorded the highest increment in height of 9.31 cm while compost from *S. natans* only (C1), paddy husk compost (C2) and control (C0) had increments of 9.00cm, 5.78 cm and 4.47 cm respectively. However, there was no difference between increments of C3 and C1 (Table 4). An increment of between 29.31-100.22% of the compost treatment over the control was recorded. The use of compost to enhance the growth of plants has been reported and figure 1 shows the final heights of the plants at the time of termination of this study with the various compost treatments. Compost application significantly enhances mitigate the negative effect of a delay in sowing. The use of compost, therefore, is a sound technology for combating soil degradation and often leads to humus accumulation in the soil and thereby supplying nutrients in several binding forms to the soil matrix. Compost also has dynamic impact on exchange processes between root system and sorption complex. This has mainly been attributed to the humified OM and the colloid properties of the humic substances [22].

However, the nutrient effect of organic composts depends largely on the transformation processes that take place in the soil after their application. This applies especially to nitrogen (N), for which its plant availability is tightly related to microbial transformations. The results of this study are in conformity to earlier reports that, compost application significantly enhances the growth of plants with respect to height. Effect of compost prepared from organic material, thus, *S. natans* at a rate of 3.13 tons/ ha on the increment of *K. anthoteca* height was significantly different from that observed both in the control and compost from a less organic material such as paddy husks (rice hull) compost treatments. Laboratory analysis of paddy husk compost (C2) and composite soil of the research site showed low levels of organic carbon and nitrogen of 5.23%; 1.05% and 0.39%;

0.07% respectively. On the other hand, compost prepared from *S. natans* had 15.94% of organic carbon and 1.15% of total nitrogen. According to the European Commission, (2001), the degradability and C/N ratio of the organic matter determines whether it comes to a net mineralization or immobilization of N after the application of an organic fertilizer.

Table 1. Analysis of composts

No	Parameter	*Salvinia natans* compost	Rice hull (paddy husk) compost[b]
1	Moisture Content (%)	78.27	40.28
2	pH	6.96	6.55
3	Ash content (%)	49.61	72.35
4	Organic material (%)	19.32	5.34
5	Organic carbon (%)[a]	15.94	5.23
6	C/N	13.86	4.98
7	N (%)[a]	1.15	1.05
8	P (mg/100g)	19.34	11.84
9	K (mg/100g)	835.84	158.67
10	Mn (ppm)[a]	914	50.32
11	Fe (ppm)[a]	12349	

[a]Sample analyzed at SEAMEO Biotrop. [b]Sample analyzed by PT. Bukit Asam (South Sumatra)

Again, phosphorus (P), another important and macro nutrient for plant growth responsible for photosynthesis and energy transfer was relatively high (19.34 mg/100 g) in compost from *Salvinia natans* as compared to that of community compost from paddy husk of 11.84 mg/100 g (Table 1). Also, according to the [30], compost products contain a considerable variety of macro and micronutrients. Although often seen as a good source of nitrogen, phosphorous, and potassium, compost also contains micronutrients essential for plant growth. Since compost contains relatively stable sources of organic matter, these nutrients are supplied in a slow-release form. It also improves the cation exchange capacity of soils, enabling them to retain nutrients longer. It allows tree crops to more effectively utilize nutrients while reducing nutrient loss by leaching.

Table 2. Two-way analysis of variance for height increment (cm)

Source	Df	SS	MS	F	P-Value
Mycorrhiza	3	3.66	1.22	0.45 [ns]	0.72
Compost	3	272.71	90.90	33.50*	0.00
Interaction	9	22.09	2.45	0.90 [ns]	0.52
Error	48	130.26	2.71		
Total	63	428.72			

* Significant at alpha (α) $\leq$ 5%[ns] not significant

Microorganisms can also promote root activities as specific fungi work symbiotically with plant roots, assisting them in the extraction of nutrients from soils. In a symbiotic association, fungus colonizes plant root hairs through the cortex cells and acts as an extension of the root system. This type of association is characterized by the formation of arbuscles (finely branched hyphal structures) in the region of the root cortex that may function as nutrient organs or nutrient exchange sites between the symbionts as well as fungal multiplication. Effect of AMF inoculation to *K. anthoteca* for height increment was not significantly different from the control treatment although the means of the inoculated propagules were relatively higher compared to the control. *Glomus manihotis* inoculation recorded a height increment of 7.51±0.62 cm whiles *Gigaspora margarita*, mixed culture and the control treatment recorded increments of 7.18±0.69 cm, 7.06±0.646 cm and 6.84±0.70 cm respectively. According to [8] the AMF genera *Gigaspora* and *Scutellospora* produce only arbuscles with extensive intraradical and extraradical hyphal networks whereas *Glomus, Entrophospora, Acaulospora*, and *Sclerocystis* in addition, also produce vesicles.Arbuscules are usually short-lived and begin to collapse after a few days while vesicles are storage structures and can remain in roots for months or even years [4].

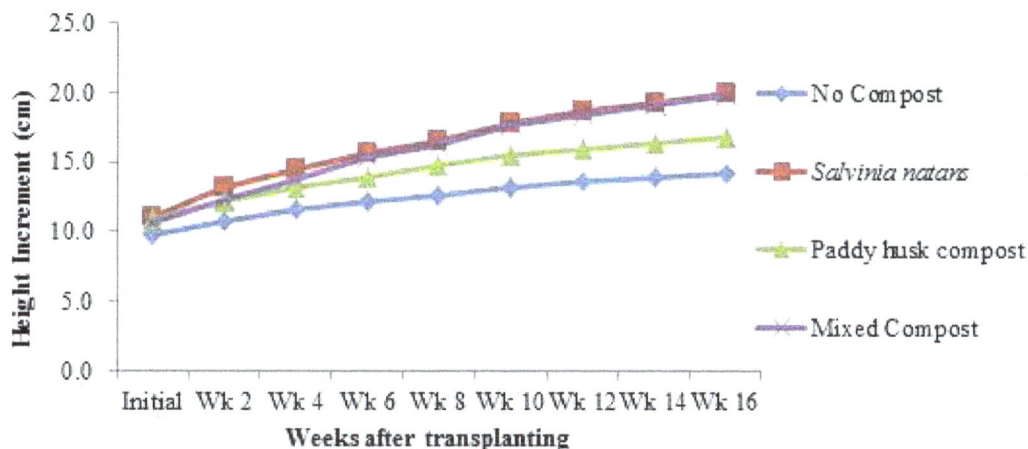

Figure 2. *Effect of compost treatment on height increment (cm).*

3.3. Diameter Increment

Increase in plant diameter or girth has often been considered as a secondary growth. However, with the addition of organic materials this process can be hastened. Analysis of Variance (ANOVA) tested at $\alpha \leq 5\%$ showed a significant effect ($P \leq 0.000$) of compost on the mean plant diameter increment with steady increment for the study period. There was no significant effect ($P = 0.148$; 0.188) of mycorrhizae treatment as well as the interaction between both factors on diameter increment (Table 3). In terms of diameter increment, compost from *S. natans* recorded the highest mean increment of 3.61±0.238 mm although there was no significant difference between the means of *S. natans* (C1) and mixture compost (C3) according to Tukey's Honest Significant Difference (HSD) method at 95.0% confidence (Table 5). There was however percentage increment of between 9 and 48% of compost treatment over the control experiment.

Although soil may vary considerably in structure and in physical, chemical and biotic properties, the rate of growth of a seedling is influenced by those properties of the soil. From the soil, the plant derives its nutrients and it is a storehouse for water and oxygen, all of which are necessary for the physiological processes associated with growth. Hence the relative abundance of these factors in a particular soil, determine the rate at which the seedling will grow [3]. Plants need a number of essential elements to enable them to grow and reproduce. Plant macronutrients (primary nutrients) including Nitrogen (N), Phosphorus (P) and Potassium (K). A laboratory test of the composts used in this study had varied levels of these nutrients (Table 1) and they are needed in relatively large quantities for plants metabolic processes.

The formation of mycorrhizae induces great changes in the physiology of the roots, in the internal morphology of the plant and in the mycorrhizosphere, thus, the soil surrounding the roots [16]. The symbiotic association of AMF and plant roots has been considered to be the oldest symbiosis of plants and is suspected to ecologically be the most important symbiotic relationship between microorganisms and higher plants [24]. The fungi of vesicular-arbuscular mycorrhizae

colonize considerable portions of the root system and in spite of the carbon drain they impose on the host plant, their presence within the root tissues can positively influence several aspects of the host plant's physiology. In the majority of cases, improved phosphate uptake is the primary cause of growth and yield enhancements in the mycorrhizal plants. Although according to this study mycorrhizae treatment did not significantly affect the diameter growth of *K. anthoteca*, *Glomus* Spp recorded the highest increment and this is due to the fact that mycorrhizae produced by *Glomus* Spp often result in simultaneous growth in 2 directions. Also, the mycorrhizal roots have different phosphate absorption kinetics and lower threshold values than non-mycorrhizal roots. The external hyphae developing around mycorrhizae explore a large volume of soil and absorb available phosphate beyond the depletion zone at the root surface. Phosphate accumulating in the external fungal hyphae is translocated to the internal mycelium by a well-developed transport system and transferred to the host tissues mainly across the intracellular arbuscules [11]. Certain specialized enzyme activities are specifically associated with this alternative pathway of phosphate nutrition in mycorrhizal plants. Improved phosphate nutrition is not always sufficient to explain the observed effects of vesicular-arbuscular mycorrhizae on the host plant's physiology [11].

Table 3. *Two-way analysis of variance for diameter increment (mm)*

Source	Df	SS	MS	F	P-Value
Mycorrhizae	3	3.11	1.04	1.86[ns]	0.15
Compost	3	17.44	5.81	10.44 *	0.00
Interaction	9	7.34	0.82	1.46[ns]	0.19
Error	48	26.72	0.56		
Total	63	54.61			

* Significant at alpha (α) $\leq 5\%$[ns] not significant

According to [4], the primary aim of mycorrhizal inoculation is to increase the yield of plants grown for plantation forestry. These growth responses however, depend on the mycorrhizal dependency of the host plant species, soil properties; especially the availability of nutrients such as P and the capacity of the fungi to provide benefits to the host plant. Since there is relatively no difference between the amount of phosphorus in the soil and applied composts the effect of mycorrhizae on the diameter increment of *K. anthoteca* may not be remarkable. Although there was no significant different among the means of the mycorrhizae treatment on diameter increment of *K. anthoteca*, *Glomus* spp had the highest mean increment of 3.25± 0.17 mm compared to *Gigaspora* spp, mixed culture and non-mycorrhizal plants recording increment of 2.65±0.12 mm. Aside the factors mentioned above, soil temperature and moisture contents are other factor limiting the full potential of AMF [4]. The study site is quite critical with some extreme environmental conditions such as high temperatures leading to low soil moisture due to high evaporation rates even after watering. According to [20] of all the factors controlling seedling growth, water is the most critical. Water is the vehicle for all physiological and biochemical processes through which life is maintained. In the plant, opposing effect of transpiration and water absorption controls water. Whenever transpiration is greater than absorption, the plant becomes dehydrated. A decrease in hydration of protoplasm of cells in the meristematic tissues usually results in cessation or checking of cell division or cell enlargement or both. There is considerable variation between and within species of fungi in their response to those factors. The capacity of fungi to produce soil hyphae is thought to be a major determinant of mycorrhizae fungus benefits to their host and would result from inherent properties of the fungi and interactions with soil conditions [5].

3.4. Leaf Count

Unlike animals, plants only have 3 organs namely; the roots, the stems and the leaves with the stems and leaves together forming the shoot of the plant. Leaves differ from stems in that, they do not have an apical meristem, thus, leaves are determinate (limited in their growth), while stems are indeterminate (theoretically capable of growing forever). In this study, analysis of variance of compost application showed a significance effect (Table 4). The same was observed for both height and diameter increments with significant differences among the various compost types

(Tukey HSD, $p<0.05$). The growth of plants depends on the availability of nutrients from the soil. Prolonged uptake of nutrients by either growing plants or excavation of the land for any purposes depletes soil of vital nutrients, adversely affecting the growth of plants [25]. In this study, plant biomass (leaf count) was substantially increased when green waste compost (*S. natans*) was used as the organic matter component instead of paddy husk. It has been shown previously that increasing the organic fraction of a substrate increases plant growth [19]. Compost application in the research showed a significant increment in the number of leaves (Fig. 3). However, compost prepared from green waste such as *S. natans* was significantly different from the other compost (Table 5).

Table 4. *Two-way analysis of variance for leaf increment*

Source	Df	SS	MS	F	P-Value
Mycorrhiza	3	19.81	6.60	2.60^{ns}	0.06
Compost	3	135.06	45.02	17.71^*	0.00
Interaction	9	13.06	1.45	0.57^{ns}	0.81
Error	48	122.00	2.54		
Total	63	289.94			

* Significant at alpha (α) $\leq 5\%$ ^{ns}Not significant

As indicated in other plant parameters earlier, mycorrhizae effect in this study did not significantly affect the growth of *K. anthoteca* on the field. In recent times, there have been a number of cases which indicate that AM colonization does not result in any increases in growth or in total plant P, and sometimes the AM plants are smaller than the non mycorrhizal controls [15, 28]. There is therefore a continuum of responses from strongly positive to negative, indicating considerable 'functional diversity' in AM symbioses [14]. The AM-responsiveness in terms of plant growth is determined by properties of the plant genome such as development of extensive root systems and long root-hairs that enhance P uptake by the plant when it is non-mycorrhizal, the AM fungal genome, thus, inherent extensiveness of external hyphae and other features and plant-fungus genomic interactions [28, 29].

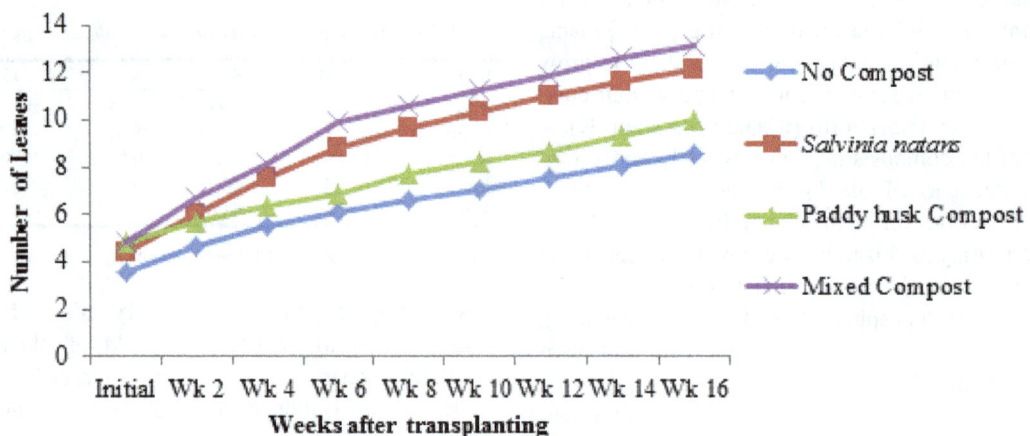

Figure 3. *Effect of compost treatment on leaf counts*

Table 5. Grouping information for compost treatment means for measured plant parameters using Tukey's HSD method at 95% confidence.

Compost Treatments	N	Mean ± SE Mean		
		Height (cm)	Diameter (mm)	Leaf Count
C0	16	4.47±0.22[b]	2.44±0.15[b]	5.00±0.29[b]
C1	16	9.00 ±0.45[a]	3.61±0.24[a]	7.69±0.45[a]
C2	16	5.78±0.42[b]	2.53±0.19[b]	5.19±0.32[b]
C3	16	9.31±0.47[a]	3.42±0.20[a]	8.25±0.51[a]

Means that do not share a letter in the same column are significantly different at a confident of 95% ($\alpha \leq 5\%$)

3.5. Survival of Khaya anthoteca under Field Conditions

After four months of transplanting vegetatively propagated *K. anthoteca* to the field, non-mycorrhizal (NM) treatment recorded the highest mortality of 56.25% with only 43.75% plants surviving while *K. anthoteca* propagules pre-inoculated with *Glomus manihotis* had the highest percentage survival of 81.25%. There were significant differences between NM plants and AMF inoculated plants as well as between the various AM treated plants on the percentage survival of *K. anthoteca* according to a t-test analysis for independent samples. However, there were no significant differences between compost and mycorrhizal applications on the percentage survival of *K. anthoteca* P (T ≤ t, 0.90) although the mean percentage survival of mycorrhizal treatment was higher (64.06%) than the application of compost treatment (62.5%). Association of fungi with plant roots has long been documented and they function as plants root hair and acting as an extension of the root system [18]. The beneficial effects of AM fungi result from one or several mechanisms. With mycorrhizal colonization in the roots, there is increased absorption surface area, greater soil area exposed for longevity of absorbing roots, better utilization of low-availability nutrients and better retention of soluble nutrients, thus reducing reaction with soil colloids or leaching losses [18, 26]. The arbuscular mycorrhizae fungi increase establishment, nodulation and atmospheric nitrogen fixation capacity in [30]. Mycorrhizae influence the colonization of roots by other microorganisms, and reduce the susceptibility of roots to soil-borne pathogens such as nematodes or phyto-pathogenic fungi [26]. The study area for this research is characterized by high temperatures with low soil moistures hence propagules with AMF inoculation were more tolerant as compared to NM. According to [18] AM also modify soil-plant-water relations, thus promoting better adaptation of plants to adverse conditions, such as drought, salinity or heat stress. At elevated heavy metal concentrations in soils, mycorrhizal fungi have been shown to detoxify the environment for plant growth. Many studies have again, indicated that AM symbioses can significantly alter plant water relations, but the reported effects have not been consistent between different investigations and mechanisms are not clear. Nonetheless, an extensive review by [2] covering hundreds of studies, highlights a number of trends in AM compared to non-mycorrhizal (NM) plants growing under water restrictions in pot experiments. These studies include increased drought tolerance, greater depletion of soil water, higher stomatal conductance and transpiration, better supply of diffusion-limited nutrients in dry soil and lower drought stress (assessed as reduced concentrations of xylem abscisic acid in AM plants). Such differences suggest that AM plants are under less stress in dry conditions than NM controls [9]. Some studies have also indicated that, improved nutrition of AM compared with NM plants may be the main basis for improved drought tolerance. About 80% of the studies assessed by [2] also revealed AM plants grow better under drought stress than NM plants.

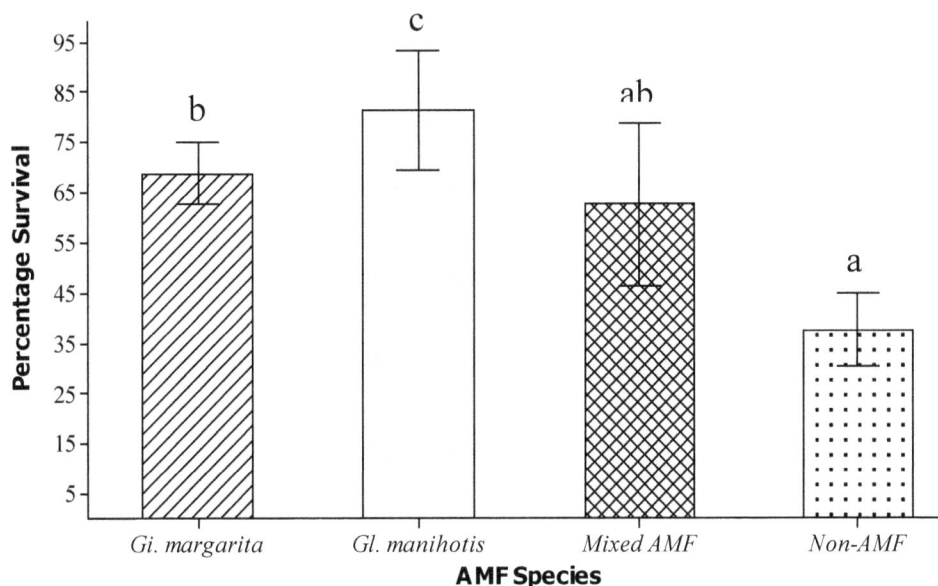

Figure 4. Effect of mycorrhizae treatment on percentage survival of Khaya anthoteca

In this study, *Glomus manihotis* inoculation recorded a higher percentage survival compared to *Gigaspora margarita* and mixed inoculums (*Glomus manihotis* and *Gigaspora margarita*) inoculations. A study conducted by [12] on the effect of inoculation of two arbuscular mycorrhizal fungi, alone and in combination on the establishment and growth of *Acacia auriculiformis* in a wasteland soil, studied under nursery and field conditions showed that *Acacia auriculiformis* exhibited a maximal mycorrhizal dependency of 79.6% on dual inoculation. Mycorrhizal dependency differed with AM fungal isolates and age of the plant. Under field conditions, AM colonization of *A. auriculiformis* enhanced tree survival rates (85%) after transplanting. Arbuscular mycorrhiza-colonized plants showed significant increase in height, biomass production and girth as compared to non-mycorrhizal plants. The study concluded that, mycorrhizal symbiosis play vital role in helping *A. auriculiformis* to establish and thrive in alkaline wasteland soils. Hence, the nursery inoculation programs with selected AM fungal species or combination of synergistically interacting species may be helpful to produce vigorous seedlings to survive under wasteland stress soil [12]. In this study however, the species used for the pre-inoculation were from two different genera, thus, *Glomus* and *Gigaspora* and a mixture of these two microorganisms therefore may therefore not be synergic and competing for the same resources under critical field conditions.

4. Conclusions

Mycorrhizal fungi play critical roles in nutrient cycling and terrestrial ecosystem functioning. Mycorrhizae improve plant growth through increase supply of nutrients to plant, enhance water uptake efficiency. Again, some of the benefits of inoculating with mycorrhizal fungi include; improved plant establishment rates, increased drought resistance, decreased transplant shock and improved survival. Furthermore, the addition of composted residue is an effective treatment for increasing rhizosphere aggregate stability. *K. anthoteca* is an important tropical tree species with high economic value and this study has shown that the species can be established on ex-coal mined sites. Base on the outcome of this study therefore, stem cutting of *K. anthoteca* pre-inoculated with AMF and compost applications can be established on ex-coal mined sites. Compost made from organic materials such as *S. natans* is preferable for the growth and development of the plants with application rate of 5 Kg/plant and the propagules at heights of at least 30 cm tall. Nonetheless, other aquatic weeds for composting and subsequent application should also be experimented. Again, the results of this study should be ascertained through the establishment of *K. anthoteca* in other mining sites to widen its application for revegetation. Finally, it must however be emphasized that, *K. anthoteca* can also be established on the field (ex-coal mined sites)

without mycorrhizae inoculations. On the other hand, inoculation will help facilitate the reintroduction of these soil microorganisms into the ecosystem.

Acknowledgment

We are grateful to God for His divine protection. Special thanks to all staff of PT. Bukit Asam (Tanjung Enim, South Sumatra) especially Mr. Muhamad Bagir and Dedy Rosa for their immense assistance during the field work

References

[1] Al-Karaki GN. 2006. Nursery inoculation of tomato with arbuscular mycorrhizal fungi and subsequent performance under irrigation with saline water. *Scientia Horticulture* 17:109

[2] Augé RM. 2001. Water relations, drought and vesicular arbuscular mycorrhizal symbiosis. *Mycorrhiza* 11:3–42. doi:10.1007/s005720100097

[3] Brady NC, Weil RR. 2008. The Nature and Properties of Soils. 14th Ed. Prentice Hall. NJ Pp 462.

[4] Brundrett M, Bougher N, Dell B, Grove T. and Malajczuk N. 1996. *Working with Mycorrhizae in Forestry and Agriculture*. ACIAR Monograph. Pirie Printers, Canberra, Australia Pp 179-185

[5] Brundrett MC. 1991. Mycorrhizae in natural ecosystems. In: Macfayden A, Begon M, Fitter AH (ed.), Advances in Ecological Research. Vol. 21 Academic Press, Londen, 171-313.

[6] Caravaca F, Herna´ndez T, Garcı´a C, Rolda´n A. 2002. Improvement of rhizosphere aggregate stability of afforested semiarid plant species subjected to mycorrhizal inoculation and compost addition. *Geoderma (Elsevier)* 108 133– 144

[7] Cunha LO, Fontes MAL, Oliveira AD, Oliveira-Filho AT. 2003. Análise multivariada da vegetac, ão como ferramenta para avaliar a reabilitac, ão de dunas litorâneas mineradas em Mataraca, Paraíba, Brasil. Revista Árvore 27, 503–515.

[8] Douds DD, Millner PD. 1999. Biodiversity of arbuscular mycorrhizal fungi in agroecosystems. *Agric. Ecosyst. Environ.* 74, 77–93

[9] Duan XG, Neuman DS, Reiber JM *et al,*. 1996. Mycorrhizal influence on hydraulic and hormonal factors implicated in the control of stomatal conductance during drought. *Journal of Experimental Botany* 47:1541–1550. doi:10.1093/jxb/47.10.1541

[10] European Commission. 2001. Applying Compost; benefits and needs. Seminar proceedings Brussels, 22-23 November 2001. Brussels, Belgium. P 128.

[11] Gianinazzi-Pearson V, Gianinazzi S. 1983. The physiology of vesicular-arbuscular mycorrhizal roots. *Plant and Soil, Springer* 71: 1-3, Pp 197-209

[12] Giri B, Kapoor R, Agarwal L, Mukerji KG. 2004. Preinoculation with Arbuscular Mycorrhizae Helps Acacia auriculiformis Grow in Degraded Indian Wasteland Soil. *Communications in Soil Science and Plant Analysis*. Marcel Dekker, Inc. 35:1, 2 Pp. 193–204

[13] Hawthorne, W. 1998. *Khaya anthotheca*. The IUCN Red List of Threatened Species. Version 2014.3. www.iucnredlist.org.Downloaded on 24 January 2015.

[14] Jakobsen I, Smith SE, Smith FA. 2002. Function and diversityof arbuscular mycorrhizae in carbon and mineral nutrition. In: van der Heijden MGA, Sanders IR (eds) Mycorrhizal ecology. Springer-Verlag, Berlin, Heidelberg, pp 75–92.

[15] Johnson NC, Graham JH, Smith FA. 1997. Functioning of mycorrhizal associations along the mutualism-parasitism continuum. *New Phytologist.* 135:575–586. doi:10.1046/j.1469- 8137.1997.00729.x

[16] Martin F, Perotto S, Bonfante P. 2007, Mycorrhizal fungi: A fungal community at the interface between soil and roots, pp. 201–236. *In* R. Pinton, Z. Varanini, and P. Nannipieri (Eds.), *The rhizosphere: Biochemistry and organic substances at the soil-plant interface.* Marcel Dekker, New York.

[17] Mohr H. and Schopfer P. 1995. Plant physiology, Springer-Verlag, Berlin Heidelberg, Hong Kong. Pp 294.

[18] Muchovej RM. 2004. Importance of mycorrhizae for agricultural crops. SS-AGR-170, Agronomy Department, Florida Cooperative Extension Service, Institute of Food and Agricultural Sciences, University of Florida.

[19] Nagase A, Dunnett N. 2011. The relationship between percentage of organic matter in substrate and plant growth in extensive green roofs. Landsc. Urban Plan. 103, 230–236.

[20] Nwoboshie LC. 1982. Tropical silviculture*: principles and techniques.* Ibadan University Press. 333pp.

[21] Ortas I. 2006. *Mycorrhizae inoculated seedling production systems inorganic farming under greenhouse and field conditions.* 5th International Conference on Mycorrhizae ICOM5 Mycorrhiza for Science and Society, 23-27 July, Granada, Spain.

[22] Ouédraogo EA. Mando, Zombré NP. 2001. Use of compost to improve soil properties and crop productivity under low input agricultural system in West Africa. *Agriculture, Ecosystems and Environment (Elsevier)* 84 259–266.

[23] Pamerindo Indonesia. 2015. A Pamerindo Indonesia Trade Event. *The 17th International Mining and Minerals Recovery Exhibition and Conference,* Jakarta International Expo, Kemayoran Jakarta-Indonesia. 9-12 September, 2015.

[24] Paszkowski U. 2006. A journey through signaling in arbuscular mycorrhizal symbioses. *New Phytol.* 172, 35–46.

[25] Russell EW 1998. *Soil condition and plant growth.* 11th edition, Longman Publication, UK. Pp 34.

[26] Selvaraj T, Chelleppan P. 2006. Arbuscular mycorrhizae: a diverse personality, *Central Eur. J. Agr.* 7, 349–358.

[27] Singh V, Singh TN. 2004. Environmental impact due to surface mining in India. *Minetech,* 25, 3-7

[28] Smith FA, Grace EJ, Smith SE. 2009. More than a carbon economy: nutrient trade and ecological sustainability in facultative arbuscular mycorrhizal symbioses. *New Phytol.*doi:10.1111/j.1469 8137.2008.02753.x

[29] Smith SE, Read DJ. 2008. Mycorrhizal Symbiosis, 3rd ed. Elsevier and Academic, New York, London, Burlington, San Diego.

[30] The United States Composting Council. 2008. USCC Factsheet: Compost and Its Benefits. Ronkonkoma, NY. Pp 1-2

[31] Turk MA, Assaf TA, Hammed KM, Al-Tawaha AM. 2008. Significance of mycorrhizae, *World J. Agric. Sci.* 2, 16–20.

Salicylic Acid, Phosphorous Acid and Fungicide Sumi 8 Effects on Polyphenol Oxidases Activities and Cassava Resistance to Anthracnose

Seu Jonathan Gogbeu[1, *], Koffi Mathurin Okoma[2], Koua Serge Beranger N'Goran[3], Dénézon Odette Dogbo[4]

[1]Laboratory of Plant Physiology and Pathology, Department of Agroforestry, University Jean Lorougnon Guede, Daloa, Cote d'Ivoire
[2]Central Laboratory of Biotechnology, National Agronomic Research Centre (CNRA), Abidjan, Cote d'Ivoire
[3]National Centre of Floristic, Department of Biosciences, University Felix Houphouet-Boigny, Abidjan, Cote d'Ivoire
[4]Laboratory of Biology and Plant Production Improvement, Department of Sciences and Nature, University Nangui-Abrogoua, Abidjan, Cote d'Ivoire

Email address:

jgogbeu@yahoo.fr (S. J. Gogbeu), okomakoffi@yahoo.fr (K. M. Okoma), kouaberanger@yahoo.fr (K. S. B. N'Goran), denezon@yahoo.fr (D. O. Dogbo)

Abstract: In Côte d'Ivoire, cassava contributes enormously to improve food security of population by increasing national production and financial resources of vulnerable households. But, plant is attacked by several diseases including anthracnose. This study was done to improve its resistance to anthracnose by stimulating its natural defense following treatment plants with salicylic acid, phosphorous acid and fungicide Sumi 8 as elicitors. Polyphenol oxidases were chosen as resistance marker. Results showed that in the three cultivars (*yacé*, *TMS30572* and *I88/00158*), *yacé* was more susceptible to anthracnose ($p<0.05$; F = 6.83). After treatments, cassava resistance against anthracnose has been improved. Polyphenol oxidases activities were more stimulated in presence of elicitor's phosphorous acid and salicylic acid. Native-PAGE of polyphenol oxidases revealed 11 isoenzymes including 7 new isoenzymes detected in elicited plants, treated plants contaminated or uncontaminated by *Colletotrichum gloeosporioïdes*, pathogen of anthracnose. Recent isoenzymes were specific for each cultivar. Their appearance was correlated with plant resistance to *C. gloeosporioïdes*. In these plants, in particular those germinated directly in elicitation medium, anthracnose symptoms were lessened. These elicitors were thus induced and/or stimulated cassava defense especially polyphenol oxidases activities.

Keywords: Anthracnose, Cassava, Elicitors, Polyphenol Oxidases

1. Introduction

Polyphenol oxidases [PPO (EC. 1.14.18.1 and EC. 1.10.3.2)] are plant enzymes using molecular oxygen to oxidize phenolic compounds to quinones [1]. These reactions commonly contribute to tissue browning of fruits and vegetables, and consequently to the deterioration of consumer products quality. However, according to Mayer [2], Constabel and Barbehenn [3], role of PPO plants is associated with the defense against pathogenic microorganisms. In fact, role attributed to PPO is related to modification of endogenous phenolic compounds by these enzymes, in particular quinones that are toxic to pathogenic microorganisms [4] [5]. In tomato (*Solanum lycopersicum* L) plants infected by *Fusarium oxysporium*, PPO activities increased both in susceptible and resistant plants [6]. Other studies have indicated that induction of PPO activities could be caused by elicitors [7] [8]. Thus, in cassava processing plants by salicylic acid and phosphorous acid, similar responses were evoked [9] [10]. Therefore, Flurkey and Inlow [11] show that induction of PPO genes in response to plant hormones clearly suggests that these enzymes are released during plant resistance. Based on these results, PPO can be mentioned as marker for plant defenses. In order to establish a link between production of PPO under effect of elicitor's salicylic acid, phosphorous acid or fungicide Sumi 8 and resistance of cassava to anthracnose, this study was conducted. Indeed, cassava in Côte d'Ivoire contributes enormously to improve food security of populations by increasing national production and financial resources of vulnerable households. With production estimated at over 2 million tonne, cassava is the second food after yams [12]. But despite these production efforts, cassava is prone to many diseases including anthracnose [13]. However, in developing countries such as Côte d'Ivoire, application of fungicides is only means of protection of plants against diseases. But, the misuse of these products leads to the induction of long-term pathogenic microorganisms resistance and environmental damage through the toxic residues accumulation [14] [15], which could act negatively on the health of consumers. Natural defense induction by elicitor application then appears as a means to fight against pathogenic microorganisms and especially environment preservation. Or in cassava, work on PPO activities induction and plant resistance to pathogens has not been discussed in the literature.

2. Materials and Methods

2.1. Plant and Experimental

Yacé, TMS30572 and I88/00158 plants aged of six week were used for experimentation. Yacé is a traditional cultivar commonly grown in Côte d'Ivoire. By cons, TMS30572 cultivar was introduced by International Institute of Tropical Agriculture (IITA, Nigeria). As for I88/00158 cultivar, it has been improved by National Centre of Agricultural Research (CNRA, Côte d'Ivoire) and now popularized as Bocou2.

Cassava plants were obtained from hydroponic according to Gogbeu et al [10] method. Cuttings sterilized with alcohol 70% (v/v) were placed in two germination medium: nutrient medium containing phosphorus (P_2O_3) and dolomite ($CaMg(CO_3)_2$) at dose of 80 mg L^{-1} each, namely M_0 medium, and M_0 medium supplemented with 1 mM of salicylic acid (SA, SIGMA), 1 mM of phosphorous acid (PA, SIGMA) or 0.5 mM of fungicide Sumi 8 (Syngeta Society) qualified M_{SA}, M_{PA} and M_S medium respectively. For each cultivar, plants were divided into two blocks according to contamination mode: block uncontaminated plants [plants from M_0 medium (3 plants) and plants from M_{SA}, M_{PA} and M_S medium (3 plants / medium) that have not been contaminated with C. gloeosporioïdes] and block contaminated plants [plants from M_0 medium (3 plants / contamination time), plants from M_{SA}, M_{PA} and M_S medium (3 plants / medium /contamination time) and plants from M_0 medium then transferred to M_{SA}, M_{PA} or M_S medium (3 plants /medium/contamination time) infected with C. gloeosporioïdes].

2.2. Estimation of Pathogen Propagation Speed

Pathogen was isolated from stems of cassava diseased plants. These stems were disinfected with alcohol 70% (v/v) and quickly flamed under a laminar flow hood. Samples of 0.5 cm collected around necrotic area were placed in Petri dishes containing PDA medium previously prepared. After 3 days of incubation at 28 °C in dark, fruiting bodies were collected using a sterile needle and transplanted into new PDA medium. After 5 to 6 subculture, pure cultures of fungi were obtained and stored at 4 °C in refrigerator. Plant contamination was performed according to Makambila [16] method. Stems of selected plants were pricked in part not yet lignified [2/3 upper stem] using a thin heated to red needle. On the 3rd day after injection, fungi (mycelial and conidia) were given in capsule form (1 mm²) collected by scraping on culture medium. After various treatments, room humidity was maintained by daily watering. Distance traveled by fungi within stem was determined at 12th Day after plant contamination. To do this, contaminated stems were cracked in length and distance traveled by pathogen within stem was measured using a ruler [17].

2.3. Extraction and Assay of Polyphenol Oxidases Activities

Polyphenol oxidases were extracted using Gogbeu et al [18] method with some modification. Extraction buffer of PPO varied according to cultivar. There are 0.2 M sodium phosphate buffer pH 5 [PPO extracted from I88/00158 (iPPO)] and pH 6 [PPO extracted from TMS30572 (t_{30}PPO)], and 0.2 M Tris-HCl pH 7.5 [PPO extracted from yacé (yPPO)]. To do this, one g of limbs was ground in 10 mL of extraction buffer supplemented with Triton X-100 (100/1, v/v) and the whole was centrifuged at 15000 x g for 30 min at 4 °C. Supernatant was recovered and pellet was taken up in 5 mL of extraction buffer and then ground and centrifuged as before. Combined supernatants formed extracted PPO.

Enzyme activity was assayed in 3 mL of reaction mixture, consisting of 50 µL of enzymatic extract and 100 mM of dopamine [yPPO and iPPO] or pyrocatechol [t_{30}PPO]. After 5 min incubation at 30 °C (t_{30}PPO), 35 °C (yPPO) and 40 °C (iPPO), reaction mixtures were cooled in a controlled bath regulated at 4 °C. PPO activities were determined by measuring absorbance (spectrophotometer, Milton Roy) at 420 nm (yPPO, t_{30}PPO) or 470 nm (iPPO) against a control containing no substrate. PPO activities were expressed in absorbance per minute per milligram of protein (ΔDO/min/mg prot.). Maximum stimulation of PPO activities was expressed as difference between high enzyme activity and PPO activities extracted from plants from M_0 medium.

2.4. Native-PAGE of Polyphenol Oxidases

Polyacrylamide gel electrophoresis was performed according to Laemmli [19] for separating PPO isoenzymes of cassava leaves in nondenaturing conditions. It was performed with a discontinuous buffer system using 4% stacking gel and 10% resolving gel. Resolving gel was prepared by mixing 33.3 mL of acrylamide / bis-acrylamide (30% T, 2.6% C) to 40.2 mL of distilled water and 25 mL of 1.5 M Tris-HCl buffer pH 8.8. After 15 min, 500 µL of 10% (w/v) ammonium persulphate and 50 µL of TEMED were added. Whole was mixed and poured between two glass plates, separated on both sides by spacers which assembly is placed on a support. Gel is covered with 1 mL of n-butanol. After 90 min of polymerization, n-butanol is rinsed thoroughly and 4% stacking gel is poured above resolving gel. This gel was prepared by mixing 1.3 mL of acrylamide / bis-acrylamide (30% T, 2.67% C) to 6.1 mL of distilled water, 2.5 mL of 0.5 M Tris-HCl buffer pH 6.8, 50 μL of 10% (w/v) ammonium persulphate and 10 μL of TEMED. Enzyme samples (32 μL enzyme extract and 8 μL of 0.1% bromophenol blue) were applied to well spaces in stacking gel. Migration was performed at 18 °C. It was carried out first at constant current of 9.8 mA and then increased to 12 mA and it was stopped when bromophenol blue reached the bottom of resolving gel. Revelation of PPO was to put essentially highlight functional proteins by Wu and Duan [20] method. Thus, after migration, gel removed from plates was immersed for 120 min into solution containing 0.2 M sodium phosphate buffer pH 5 and 100 mM dopamine for iPPO; 0.2 M sodium phosphate buffer pH 6 and 100 mM pyrocatechol for t_{30}PPO or 0.2 M Tris-HCl pH 7.5 and 100 M dopamine for yPPO. Isoenzyme bands were photographed. Calculated frontal reports were classified isoenzymes.

2.5. Amount of Protein

Amount of protein was determined using dye-binding method of Bradford [21], with bovine albumin as the standard, measuring optical density at 595 nm.

2.6. Statistical Analysis of Data

SPSS version 11.5 software was used to compare data. Analysis of variance (ANOVA) with one and two classification criteria was made. Difference between means at 95% confidence level calculated using Duncan test.

3. Results and Discussion

3.1. Colletotrichum Gloeosporioides Propagation Speed Within Cassava Stems

Analysis of table 1 shows that average length of fungi propagation in control plants (PC) varied among cultivars (p < 0.05; F = 7.13). It was 3.63; 3.03 and 1.40 cm respectively for *yacé*, *I88/00158* and *TMS30572*. When these plants were elicited and contaminated (PE), pathogen growth within stem was delayed in *yacé*. By cons in *TMS30572* and *I88/00158*, speed was slightly important [PA (1.17 cm) and Sumi 8 (1.50 cm) for *TMS30572*; SA (3 cm) and PA (2.53 cm) for *I88/00158*]. In contrast, plants (PT) obtained from cuttings directly into elicitation medium prevented pathogen growth. Indeed, apart from *I88/00158* where *C. gloeosporioides* progression speed was important for plants from elicitation PA (F = 14.29, P <0.001) medium, first symptoms observed at 5th Day could not develop (Table 1). These results suggest that all cultivars are susceptible to anthracnose. In terms of progression speed of pathogen within stem, *yacé* would be more susceptible than *I88/00158* and *TMS30572*. But after elicitation of *yacé* plants, pathogen growth was slower. So, we can say SA, PA and Sumi 8 helped improve cassava

defense system against anthracnose. These phenomena were more pronounced with results recorded in PT. Indeed, in the latter, pathogen propagation speed was significantly reduced; this would correspond to fungi arrest mechanism in migration process. Latter was confined at inoculation site. This result could be explained by the fact that these plants had already set up their defense system after different treatments. Similar results were obtained by several authors [22] [23]. They advocated pretreatment method with elicitors in plants as a means to fight against plant diseases. In bean (*Vigna mungo*), pretreatment of plant with SA 24 hours before virus (urdbean leaf crinke virus) inoculation, has helped reduce disease symptoms while contaminated plants at same time to SA treatment developed significant symptoms [23]. Our results indicate that in cassava, a long period of pretreatment may be recommended to reduce anthracnose impact.

3.2. Native-PAGE of Polyphenol Oxidases

In order to identify all functional PPO isoenzymes in cassava leaves, native-PAGE was performed. This showed existence of 4 isoenzymes constituent rated PPO_1, PPO_2, PPO_3 and PPO_4 in studied cultivars. When plants were stressed (elicitation and / or inoculation with pathogen), 7 new forms were induced: PPO_5, PPO_6, PPO_7, PPO_8, PPO_9, PPO_{10} and PPO_{11}. Indeed, in *yacé*, 4 distinct bands representing different forms of PPO were revealed (Figure 1). Among these isoenzymes, PPO_1 and PPO_3 were constantly present whatever treatment undergone by plants. In contrast, in PC contaminated, at 12th Day (line 3) and plants elicited with salicylic acid (SA) (lines 4 and 5) and phosphorous acid (PA) (lines 6 and 7), PPO_7 has been detected. In *yacé* plant treated, PPO_7 was found in all plants. After contamination, PPO_6 was demonstrated in all plants except those having germinated in fungicide Sumi 8 (lines 16 to 18). In *TMS30572*, 5 isoenzymes rated PPO_1, PPO_3, PPO_5, PPO_9 and PPO_{11} were found (Figure 2). Isoenzymes PPO_1 and PPO_3 are present in all plants. By against, PPO_9 and PPO_{11} were identified only in plants treated with SA (lines 10 to 12) and PA (lines 13 to 15) (Figure 2). PPO_5 form is specifically appeared at 12th Day in contaminated PC (line 3), plants elicited with SA and contaminated (lines 4 and 5), plants treated with SA and contaminated (lines 11 and 12) and plants treated with PA and contaminated at 5th Day (Figure 2). Four isoenzymes namely PPO_2, PPO_4, PPO_8 and PPO_{10} were observed in *I88/00158* (Figure 3). PPO_2 and PPO_4 are present in all plants regardless of treatment received. Isoenzymes PPO_8 and PPO_{10} are found in most elicited and treated plants. However, some differences were observed. PPO_8 was absent in plants elicited with Sumi 8 and contaminated (lines 8 and 9) as well as in plants treated with Sumi 8 and contaminated at 12th Day (line 18). This form was also absent in PC (line1) and contaminated witnesses during 5 days (line 2). As for isoenzyme PPO_{10}, it was detected in all treated plants except plants treated with Sumi 8 and contaminated (lines 17 and 18) and those treated with PA and contaminated at 5th Day (line 14). In elicited plants, PPO_{10} was identified by the presence of AP (lines 6 and 7) and SA after 5 days of infection (line 4)

(Figure 3). Examples of PPO isoenzymes induction were mentioned by Niranjan-Raj *et al.* [24], Karthiekeyan *et al.* [23] and, Wu and Duan [20] in bean (*V. mungo*) and millet (*Glycine max*). These authors noted the timeliness and amplification of PPO activities in resistant varieties compared to susceptible varieties. Thus, in millet, five isoenzymes were identified in resistant varieties whereas in susceptible varieties, they are four [24]. Fifth PPO has been induced during *Sclerospora graminicola* inoculation. In *V. mungo*, new forms were induced after treatment plants with SA and benzothiadiazole [23].

3.3. Polyphenol Oxidases Activities

Polyphenol oxidases (PPO) activities measured in cassava plant leaves subjected to different treatments was expressed in percentage, taking as 100%, enzyme activity obtained with plants from uncontaminated medium (M_0; 0 Day) (Table 1). In response to *C. gloeosporioides*, plants from M_0, M_{AS}, M_{AP} and M_S medium have reacted differently. Indeed, in PC (M_0 uncontaminated), PPO activities remained constant for all cultivars during experiment. When these PC were contaminated, PPO activities increased rapidly at 5th Day before declining towards the end of experiment. Percentage of PPO stimulation was higher among *TMS30572* (39%), average for *yacé* (21%) and lower in *I88/00158* (15%) (Table 1). When plants were elicited and contaminated by fungi (M_{0SA}, M_{0PA} and M_{0S}) in the presence of PA, maximum stimulation of enzyme was 60 and 68% respectively in *yacé* and *TMS30572*. At *I88/00158*, it was 39%. In the presence of SA, value has hovered around 50% for *yacé* (52%) and *TMS30572* (49%). For *yacé* and *TMS30572*, PPO activities increased rapidly before falling at 12th Day for SA and PA (Table 1). In uncontaminated treated plants group (M_{SA}, M_{PA} and M_S uncontaminated), PPO activities were well above that of non-infected PC (M_0 uncontaminated) exception of plants resulting from Sumi 8 medium (Ms) (Table 1). In this environment, Sumi 8 negatively influenced enzyme activity evolution in *yacé*. When treated plants were infected (M_{SA}, M_{PA} and M_S contaminated), PPO activities varied among cultivars and elicitors (Table 1). Maximum stimulation of PPO exceeded 50% in all cultivars treated with SA. Greatest stimulation of enzyme was among *I88/00158*. It was 112%. In the presence of PA, significant stimulation was achieved in *TMS30572* (83%), followed *yacé* (64%). At *I88/00158*, value was 10%. For treated plants with Sumi 8, maximum stimulation of PPO did not exceed 15% (Table 1). These results show that increase of PPO isoenzymes was responsible for amplitude of their activity. In many plant tissues, increasing PPO genes in response to hormones related to defense such as salicylic acid and jasmonic acid is implicitly linked to the resistance of these pathogens [2] [11]. In cassava, induced isoenzymes identified were specific to cultivars. In contrast, intensity of reaction was more enhanced with treatment and / or contamination with *C. gloeosporioides*. Signal produced by pathogen or elicitor SA, PA or Sumi 8 was essential to trigger the synthesis and accumulation of defense gene products in treatment plants

[25] [26] [27]. Beneficial role of Sumi 8 in improving cassava defense can therefore lead to the synthesis and accumulation of phenolic compounds although PPO activities were low in the presence of the latter [10]. According to these authors, metabolic pathways leading to phenol synthesis would be induced in cassava as suggested by Rodriguez *et al.*[28] work. On cassava tubers, they indicated accumulation of three hydroxycoumarins which scopoletin. It had an antimicrobial function.

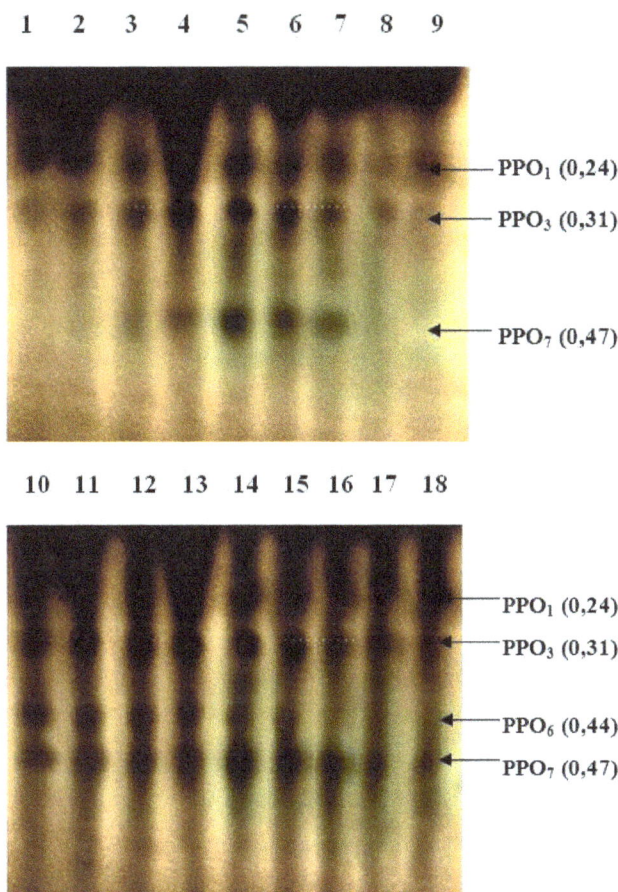

Figure 1. *Native-PAGE of polyphenol oxidases extracted of cassava yacé leaves treated with salicylic acid (SA), phosphorous acid (PA) and fungicide Sumi 8 and contaminated by Colletotrichum gloeosporioides (C.gl.).*

A : 1- control plant; 2- contaminated plant by C.gl at 5th Day; 3-contamined plant by C.gl at 12th Day; 4-elicited plant with SA and contaminated by C.gl, at 5th Day; 5- elicited plant with SA and contaminated by C.gl, at 12th Day; 6- elicited plant with PA and contaminated by C.gl, at 5th Day; 7- elicited plant with PA and contaminated by C.gl, at 12th Day; 8- elicited plant with Sumi 8 and contaminated by C.gl, at 5th Day; 9- elicited plant with Sumi 8 and contaminated by C.gl, at 12th Day.

B : 10- treated plant with SA; 11- treated plant with SA and contaminated by C.gl at 5th Day; 12-treated plant with SA and contaminated by C.gl at 12th Day; 13-treated plant with PA; 14-treated plant with PA and contaminated by C.gl, at 5th Day; 15-treated plant with PA and contaminated by C.gl, at 12th Day; 16-treated plant with Sumi 8; 17-treated plant with Sumi 8 and contaminated by C.gl, at 5th Day; 18-treated plant with Sumi 8 and contaminated by C.gl at 12th Day.

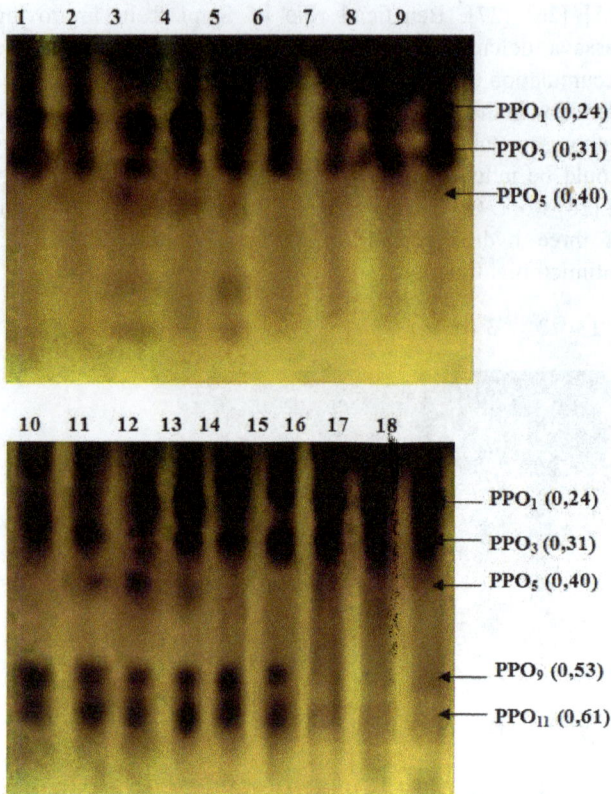

Figure 2. Native-PAGE of polyphenol oxidases extracted of cassava TMS30572 leaves treated with salicylic acid (SA), phosphorous acid (PA) and fungicide Sumi 8 and contaminated by Colletotrichum gloeosporioides (C.gl.).

A : 1- control plant; 2- contaminated plant by C.gl at 5th Day; 3-contamined plant by C.gl at 12th Day; 4-elicited plant with SA and contaminated by C.gl, at 5th Day; 5- elicited plant with SA and contaminated by C.gl, at 12th Day; 6-elicited plant with PA and contaminated by C.gl, at 5th Day; 7- elicited plant with PA and contaminated by C.gl, at 12th Day; 8- elicited plant with Sumi 8 and contaminated by C.gl, at 5th Day; 9- elicited plant with Sumi 8 and contaminated by C.gl, at 12th Day.
B : 10- treated plant with SA; 11- treated plant with SA and contaminated by C.gl at 5th Day; 12-treated plant with SA and contaminated by C.gl at 12th Day; 13-treated plant with PA; 14-treated plant with PA and contaminated by C.gl, at 5th Day; 15-treated plant with PA and contaminated by C.gl, at 12th Day; 16-treated plant with Sumi 8; 17-treated plant with Sumi 8 and contaminated by C.gl, at 5th Day; 18-treated plant with Sumi 8 and contaminated by C.gl at 12th Day.

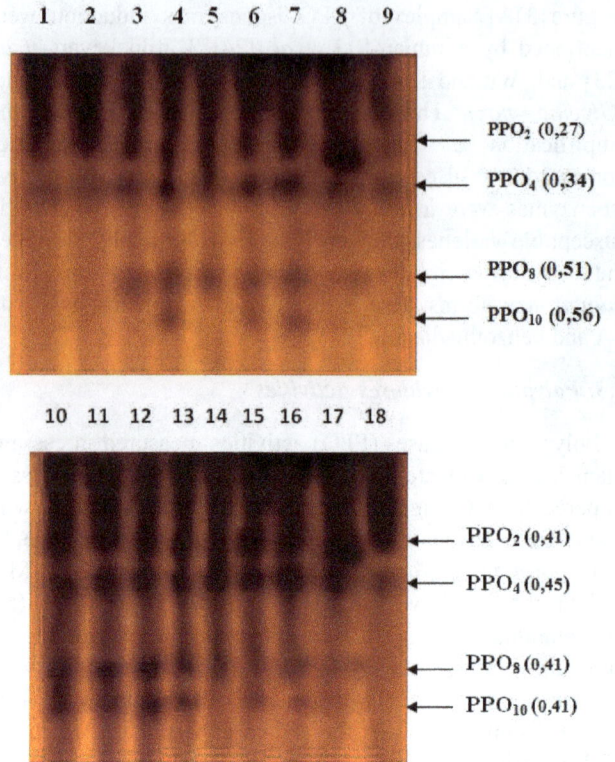

Figure 3. Native-PAGE of polyphenol oxidases extracted of cassava I88/00158 leaves treated with salicylic acid (SA), phosphorous acid (PA) and fungicide Sumi 8 and contaminated by Colletotrichum gloeosporioides (C.gl.).

A : 1- control plant; 2- contaminated plant by C.gl at 5th Day; 3-contamined plant by C.gl at 12th Day; 4-elicited plant with SA and contaminated by C.gl, at 5th Day; 5- elicited plant with SA and contaminated by C.gl, at 12th Day; 6-elicited plant with PA and contaminated by C.gl, at 5th Day; 7- elicited plant with PA and contaminated by C.gl, at 12th Day; 8- elicited plant with Sumi 8 and contaminated by C.gl, at 5th Day; 9- elicited plant with Sumi 8 and contaminated by C.gl, at 12th Day.
B : 10- treated plant with SA; 11- treated plant with SA and contaminated by C.gl at 5th Day; 12-treated plant with SA and contaminated by C.gl at 12th Day; 13-treated plant with PA; 14-treated plant with PA and contaminated by C.gl, at 5th Day; 15-treated plant with PA and contaminated by C.gl, at 12th Day; 16-treated plant with Sumi 8; 17-treated plant with Sumi 8 and contaminated by C.gl, at 5th Day; 18-treated plant with Sumi 8 and contaminated by C.gl at 12th Day.

Table 1. Effects of salicylic acid, phosphorous acid and fungicide Sumi 8 on Colletotrichum gloeosporioïdes (cm) propagation inside cassava stem at 12th Day after plant contamination.

Cassava cultivars		Elicitors					
		Salicylic acid			Phosphorous acid		Sumi 8
	PC	PE	PT	PE	PT	PE	PT
yacé	3,63a1	1,17a2	0,10a3	1,40a2	0,10a3	1,90a2	0,10a3
	±0,85	±0,35	±0	±0,75	±0	±0,87	±0
bonoua2	3,30a1	1,03a2	0,13a3	1,10a2	0,10a3	1,70a2	0,10a3
	±0,75	±0,11	±0,05	±0,30	±0	±0,78	±0
TMS30572	1,40b1	0,77a2	0,17a3	1,17a12	0,10a3	1,50a1	0,10a3
	±0,62	±0,15	±0,05	±0,05	±0	±0,10	±0
I88/00158	3,03a1	3,00b1	0,20a2	2,53b1	0,87b3	0,93a3	0,10a2
	±0,20	±0,45	±0,10	±0,51	±0,80	±0,15	±0

PC: control plant; PE: elicited plant; PT: treated plant; each value is the average of 3 replicates ± standard deviation.
For each column, means followed a single alphabetical letter are not statistically different for a threshold of 5% according to the test Dancun.
For each line, means followed by the same figure not statistically different for a threshold of 5% according to the test Dancun.

Table 2. Activities of polyphenol oxidases (%.) extracted of cassava leaves after treatment with salicylic acid (SA), phosphorous acid (PA) or fungicide Sumi 8 and/or inoculated by Colletotrichum gloeosporioides at 0, 5th and 12th Days after plant contamination.

Cassava cultivars		Culture medium										
		Uncontaminated				Contaminated						
		PC	PT			PC	PE			PT		
		M0	MSA	MPA	MS	M0	M0SA	M0PA	M0S	MSA	MPA	MS
yacé	0j	100	108	121	85	100	100	100	100	108	121	85
	5j	100	108	121	85	121	152	161	84	161	164	75
	12j	100	108	121	85	113	129	150	105	139	134	91
	0j	100	115	104	101	100	100	100	100	115	104	101
I88/	5j	100	115	104	101	115	138	128	132	212	107	107
	12j	100	115	104	101	107	113	139	127	141	110	105
	0j	100	106	115	101	100	100	100	100	106	115	101
TMS30	5j	100	106	115	101	139	149	168	109	174	183	115
	12j	100	106	115	101	130	140	153	105	116	164	109

PC: control plant; PE: elicited plant; PT: treated plant; SA: salicylic acid, PA: phosphorous acid, *TMS30*: TMS30572, *I88/*: I88/00158. M_0: nutrient medium; M_{SA}, M_{PA} and M_S: nutrient medium respectively containing SA, PA and Sumi 8; M_{0SA}, M_{0PA} and M_{0S}: germination in the nutrient medium and then transfer in the nutrient medium supplemented with SA, PA or Sumi 8. 100% = *1247,61 (yacé); 1130,83(TMS30572) and 930,83 (I88/00158) ΔDO min^{-1}mg^{-1}prot.*

4. Conclusion

This work aims to investigate alternatives to use of pesticides as the only means of struggle against cassava enemies. Results of this study have demonstrated that several PPO isoenzymes were induced after different treatments with elicitors SA, PA and Sumi 8. SA and PA involved more amplification of PPO activities. By cons, fungicide Sumi 8 inhibits the activity of the enzyme. But it also improves the strength of cassava anthracnose.

References

[1] Thipyapong P, Stout MJ and Attajarusit J (2007). Functional analysis of polyphenol oxidases by antisense/sense technology. *Molecules* 12: 1569-1594.

[2] Mayer AM (2006). Polyphenol oxidases in plants and fungi: going places ? A review. *Phytochemistry* 67(21): 2318-2331.

[3] Constabel C.P. and R. Barbehenn. 2008. Defensive roles of polyphenol oxidase in plants. In A. Schaller (ed.), Induced Plant Resistance to Herbivory. Springer, New York.

[4] Shahidi F and Naczk M (1995). Phenolic compounds in fruits and Vegetables. In Food phenolics: Sources, Chemistry, Effects and Applications; Technomic Publishing: Lancaster, PA 75-107.

[5] Weir TL, Park SW and Vivanco JM (2004). Biochemical and physiological mechanisms mediated by allelochemicals. *Current Opinion in Plant Biology* 7(4):472-479.

[6] Gyanendra kumar Rai, Rajesh kumar, Singh J, Rai P.K. and Rai S.K (2011). Peroxidase, polyphenol oxidase activity, protein profile and phenolic content in tomato cultivars tolerant and susceptible to *Fusarium oxsyporum* f.sp.*lycopersici*. *Pakistan Journal of Botany* 43(6): 2987-2990

[7] Ogawa D, Nakajima N, Sano T, Tomaoki M, Aono M, Kubo A, Kanna M, Ioki M, Kamada H and Saji H (2005). Salicylic acid accumulation under O_3 exposure is regulated by ethylene in tobacco plants. *Plant Cell Physiol*ogy 46: 1062-1072.

[8] Ojha S and Chatterjee NC (2012). Induction of resistance in tomato plants against Fusarium oxysporium f. sp. Lycopersici mediated through salicylic acid and trichoderma harzianum. *Journal of Plant Protection Research* 52(2): 220-225.

[9] Dogbo DO, Békro-Mamyrbekova JA, Békro Y-A, Sié RS, Gogbeu SJ and Traoré A (2008). Influence de l'acide salicylique sur la synthèse de la phénylalanine ammonia-lyase, des polyphénoloxydases et l'accumulation des composes phénoliques chez le manioc (*Manihot esculenta* Crantz). *Science & Nature* 5(1):1-13.

[10] Gogbeu S.J, Dogbo D.O, Zohouri GP, N'zue B, Bekro Y-A and Békro-Mamyrbekova J.A (2012). Induction of polyphenol oxidases activities and phenolic compounds accumulation in cells and plants elicited of cassava (*Manihot esculenta* Crantz). *Journal of Scientific Research and Reviews* 1(1): 7 – 14.

[11] Flurkey W.H, Inlow J.K. (2008). Proteolytic processing of polyphenol oxidase from plants and fungi. *Journal of Inorganic Biochemistry* 102 : 2160–2170

[12] FAO (2010). Statistical databases. Rome (Italy).http://www.fao.org

[13] Prusky D, Freeman S and Dickman MB (2000). *Colletotrichum*: Host specificity, pathology and host-pathogen interaction. St Paul USA: APS Press. 400 p.

[14] Buhot N (2003). Rôle des élicitines et des protéines de transfert de lipides dans l'induction de la résistance des plantes à leurs agents pathogène. Thèse de Doctorat unique. *Université de Bourgogne* 278 p

[15] Thakore Y (2006). The biopesticide market for global agricultural use. *Industrial Biotechnology* 2(3): 294-208.

[16] Makambila (1983). Epidémiologie de l'anthracnose du manioc. In : Plantes-racines tropicales : culture et emplois en Afrique, actes du second symposium triennal de la société internationale pour les plantes-racines tropicales, Douala, Cameroun, 75-80.

[17] Gogbeu S.J, Sekou D, Kouakou K.J, Dogbo D.O and Bekro Y-A (2015). Improvement of Cassava resistance to *Colletotrichum gloeosporioïdes* by Salicylic acid, Phosphorous acid and Fungicide Sumi 8. *International Journal of Current Microbiology and Applied Sciences* 4(3): 854-865

[18] Gogbeu S.J, Dogbo D.O, Gonety T.J, N'zue B, Zohouri G.P and Boka A (2011). Study of some characteristics of soluble polyphenol oxidases from six cultivars callus of cassava (Manihot esculenta Crantz). *Journal of Animal and Plant Sciences*, 9(3): 1169- 1179

[19] Laemmli UK (1970). Cleavage of structural protein during the assembly of the head of bacteriophage T4. *Nature* 227:680-685.

[20] Wu HY and Duan YX (2011). Defense response of soybean (*Glycine max*) to soybean cyst nematode (*Heterodera glycines*) race3 infection. *The Journal of Animal and Plant Sciences* 21(2): 165-170.

[21] Bradford MM (1976). A rapid and sensitive method for the quantification of microgram quantities of protein utilizing the principle of protein binding. *Analytical Biochemistry* 72: 248-254

[22] Reuveni R and Reuveni M (1998). Foliar-fertilizer therapy – a concept in integrated pest management. *Crop Protection* 17(2): 111-118.

[23] Karthikeyan G, Doraisamy S and Rabindran R (2009). Induction of systemic resistance in blackgram (*Vigna mungo*) against urdbean leaf crinkle virus by chemicals. *Archives of Phytopathology and Plant Protection* 42(1):1-15

[24] Niranjan-Raj S, Sarosh BR and Shetty HS (2006). Induction and accumulation of polyphenol oxidases activities as implicated in development of resistance against pearl millet downy mildew disease. *Functional Plant Biology* 33:563-571

[25] Conrath U, Beckers GJM, Flors V, Garcia-Augustin P, Jakab G, Mauch F, Newman MA, Pieterse CMJ, Poinssot B, Pozo MJ, Pugin A, Schaffrath U, Ton J, Wendehenne D, Zimmerle L, Mauch-Mauch B and Grp PAP (2006). Priming: Getting ready for battle. *Molecular Plant- Microbe Interctactions* 19(10):1062-1071

[26] Kishimoto K, Matsui K, Ozawa R and Takabayashi J (2006). ETR1-, JAR1-and PAD2-dependent signaling pathways are involved in C6-aldehyde-induced defense responses of *Arabidopsis*. *Plant Science* 171(3): 415-423.

[27] Ton J, D'Alessandro M, Jourdie V, Jakab G, Karlen D, Held M, Mauch-Mani B and Turlings TCJ (2007). Priming by airborne signals boosts direct and indirect resistance in maize. *Plant Journal* 49(1):16-26.

[28] Rodriguez MX, Buschmann H, Iglesias C and Beeching JR (2000). Production of antimicrobial compounds in cassava (*Manihot esculenta* Crantz) roots during post-harvest physiological deterioration. In: Carvalho LJCB, Thro AM, Vilarinhos AD, eds. *Cassava Biotechnology*. IVth International Scientific Meeting *Cassava Biotechnology* Network. Embrapa, Brasilia 320-328

32

Influence of Groundnut and Machine Characteristics on Motorised Sheller Performance

Wangette Isaac S., Nyaanga Daudi M., Njue Musa R.

Department of Agricultural Engineering, Egerton University, Njoro, Kenya

Email address:
wangettesi@yahoo.com (Wangette, I. S.), dmnyaanga@gmail.com (Nyaanga, D. M), musanjue@yahoo.com (Njue, R. M)

Abstract: Groundnut shelling is a fundamental process in post-harvest management. Manual shelling is inefficient and laborious with low throughput. Motorised shellers experience less than 100% shelling efficiency and varying levels of kernel damage. From the research, throughput per unit power consumption and shelling efficiency increased with reduction in % moisture content (mc), with maximum outputs realized at 6%. Kernel mechanical damage decreased with increase in % mc up to a minimum at between 15% and 18% mc then increased marginally with further rise in mc. Meanwhile, throughput per unit power consumption increased with bulk density of the groundnut variety being shelled. In addition, kernel to pod diameter ratio had a significant influence on the outputs under study. All the three output parameters under review rose exponentially with increase in feed rate. Throughput per unit power consumption and shelling efficiency rose steadily with increase in shelling speed with the highest values obtained at a shelling speed of about 12 m/s. Kernel mechanical damage remained low (less than 4%) for speeds below 8 m/s, and then rose sharply with further increment in speed. All the output parameters increased with reduction in concave clearance with maximum values obtained at 10 mm clearance. Steel and rubber paddles yielded the highest throughput per unit power consumption. At low shelling speeds (less than 8 m/s), rolling rubber and steel pipes resulted in lowest shelling efficiency and kernel mechanical damage but at higher speeds they resulted into both highest shelling efficiency and kernel mechanical damage.

Keywords: Groundnut Sheller, Input Characteristics, Taguchi DOE, S/N Ratio, Optimization, Sheller Performance

1. Introduction

1.1. Background Information

Groundnut (Arachis hypogaea) is the sixth most important oil crop in the world (Ikechukwu et al., 2014). The major groundnut producing countries include India, China, Nigeria, Senegal, Sudan, Burma and the United States of America, with a world average yield of 1.4 metric tonnes per hectare (t/ha) (Madhusudhana, 2013). Shelling of groundnuts is a fundamental process as it allows the kernel and hull to be available for use. It constitutes about 38% of postharvest costs (Butts et al., 2009). Traditional shelling methods have been found to be inefficient, laborious, time consuming and result in low output (Gitau et al., 2013). Hence, there is need for motorised shellers. Abubakar and Abdulkadir (2012) categorized factors that affect groundnut shellers into three types. First are machine based that include cylinder speed, concave clearance and fan speed. Next are crop factors such as moisture content, size and orientation. Last are operational based factors like feed rate and operator's skill and experience. Performance of groundnut shellers is evaluated by determining the effect of these characteristics on some measurable dependent variables. The most often used parameters include throughput, shelling efficiency, winnowing or cleaning efficiency and mechanical damage.

Studies to determine optimum operating conditions for shellers have been done using different designs of and varied results have been obtained. Gamal et al (2009) investigated the effect of moisture content on groundnut maximum stress, deformation and toughness. Helmy et al (2007) modified a rotary sheller into a reciprocating one and determined optimum shelling speed and feed rate as 1.4 m/s and 160 kg/h respectively. Adedeji and Ajuebor (2002) determined the best shelling speed, concave clearance and feed rate for a motorised groundnut sheller and Oluwole et al (2007) evaluated the influence of moisture content, impeller angulation and impeller slots on performance of a centrifugal Bambara groundnut sheller. There has been limited research work on comprehensive groundnut sheller performance that involves the combined influence of four or more machine, nut and operational factors. Research involving many factors and levels lead to large numbers of experimental runs that result into high costs and is time consuming. In addition, using a one factor at a- time method when dealing with several variables fails to consider any possible factor interactions, hence it is less efficient than other methods based on statistical approach to design (Ballal et al., 2012).

1.2. Statement of the Problem

Challenges in groundnut shelling include tedious and time consuming methods among manual shellers; to kernel damage and incomplete shelling in motorised types. Current performance evaluation of groundnut shellers is based on machine throughput; shelling and winnowing efficiencies and kernel mechanical damage. At present a key factor to consider in machine operations is its rate of energy use, determined by considering the machine power consumption against its throughput. Research data on power consumption in groundnut shellers is limited. Hence, a gap exists in the evaluation of groundnut sheller performance. Research on nut moisture content required for optimum shelling performance has yielded varying results, with values of 5% , 13% and 15% being suggested (Adedeji and Ajuebor, 2002; Akcali et al., 2006; and Nyaanga et al., 2007). Results from concave clearance tests recommend differing optimum values of 12 mm, 18 mm and 30 mm (Adedeji and Ajuebor, 2002; Helmy et al., 2007 and Rostami et al., 2009). This indicates inconsistence in recommendations for both moisture content levels and concave clearance for optimum shelling. In addition, information on influence of shelling blades on performance in groundnut shellers is limited.

1.3. Objectives

1. To determine the influence of groundnut moisture content, variety and feed rate on throughput per unit power consumption, shelling efficiency and kernel mechanical damage in a motorised sheller.

2. To determine the influence of concave clearance, shelling speed and shelling blade type on throughput unit power consumption, shelling efficiency and kernel mechanical damage in a motorised sheller.

1.4. Research Questions

1. What is the influence of groundnut moisture content, variety and feed rate on throughput per unit power consumption, shelling efficiency and kernel mechanical damage in a motorised sheller?

2. What is the influence of concave clearance, shelling speed and shelling blade type on throughput per unit power consumption, shelling efficiency and kernel mechanical damage in a motorised sheller?

2. Literature Review

2.1. Groundnut Characteristics

The major groundnut characteristics are their morphology, physical and mechanical properties like cracking stress, moisture content, variety, size dimensions, coefficient of friction and angle of repose. Physical properties of seeds in machine design are recognized as important parameters to be determined along with the machine parameters. They are useful in solving many of the problems associated with machine design and also in analysis of the behavior of products during agricultural processing. Dimensions such as geometric mean diameter, arithmetic mean diameter, aspect ratio and sphericity describe the size and shape of the seed which influence its behavior such as flowability (Amoah, 2012).

2.1.1. Variety

Studies on groundnut physical properties of various varieties have been carried out and their findings published. Examination of some physical properties of Turkish groundnut varieties yielded results in Table 2.1.

Table 2.1. Physical properties of Turkish groundnuts.

Kernel solid density (g/cm³)	Shell solid density (g/cm³)	Kernel bulk density (g/cm³)	Shell bulk density (g/cm³)	Angle of repose	Coefficient of friction
0.88-0.93	0.27-0.30	0.54-0.59	0.066-0.077	29⁰	0.23-0.76

Source: Akcali *et al.* (2006)

Akcali et al. (2006) determined the size of groundnuts by measuring their principal axial dimensions. The average major, intermediate and minor diameters of kernels were found to be 8.54 mm, 6.93 mm and 3.55 mm respectively. Angle of repose

of kernels on wooden surface was found to be 17⁰. Characteristics of two groundnut varieties used in an experiment had average measurements as given in Table 2.2.

Table 2.2. Physical properties of two groundnut varieties.

Variety	Pod diameter (mm)	Pod length (mm)	Kernel diameter (mm)	Kernel weight (g)	Moisture content of pods at shelling (%)
Manipinta (Red)	12.6	31.9	7.8	1.4	12-22
Chinese (White)	11.6	26.3	7.5	1.0	12-28

Source: Bobobee (2002)

Table 2.3. Physical properties of ICRISAT groundnut varieties in Kenya.

	Axial Dimensional length (mm)	Major diameter (mm)	Minor diameter (mm)	Bulk density (kg/m³)	1000 pod weight (g)	Angle of repose (⁰)
Maximum range	25.10-34.55	12.85-15.90	11.65-13.50	760.40-680.70	572.10-591.40	36-32
Minimum range	14.95-16.05	8.65-10.65	8.55-8.85	508.50-410.60	560.90-569.30	26-25

Source: Gitau *et al.* (2013)

A study by Gitau *et al.* (2013) on the physical characteristics of groundnut varieties developed by the International Crops Research Institute for the Semi-Arid Tropics (ICRISAT) in Kenya yielded results as shown in Table 2.3.

Results from the same research indicated that large sized varieties resulted in higher shelling efficiencies than their small sized counterparts for the same concave clearance and feed rate.

2.1.2. Moisture Content

Moisture content of seed refers to the amount of water contained in the seed (Amoah, 2012). According to Armitage and Wontner (2008), too moist cereals and oilseeds can be subject to mould growth and mycotoxin production, mite infestations and sprouting. Conversely, over-dried grain before or during storage can result in splitting and cracking, low quality and wastage in energy utilization. As such, moisture content of grain is one of the most important parameters considered when deciding the quality and prize of grain at the stage of harvesting, storage, processing and marketing (Rai *et al.*, 2005).

Gitau *et al.* (2013) showed that shelling efficiency increased with decrease in moisture content for all groundnut varieties studied. Gamal *et al.* (2009) found out that increase in moisture content leads to an increase in the major, minor and intermediate diameters of groundnut kernel. Results from experiments for Bambara nuts indicated that moisture content had higher effect on performance than feed rate. Thus, percentage seed damage increases with increase in moisture content while shelling efficiency decreases with increase in moisture content (Atiku *et al.*, 2004). Nyaanga *et al.* (2007) gave a probable explanation of the effect; that as the moisture content increases, the efficiency decreases since the pods become friable, tending to flex instead of cracking and breaking hence leading to a higher percentage of unshelled groundnuts.

The value of optimum moisture content for shelling varies across researchers. Gitau *et al.* (2013); and Akcali *et al.* (2006) gave a figure of 5% while Nyaanga *et al.* (2007); and Adedeji and Ajuebor (2002) proposed 13% and 10-15% respectively (unless specified, all moisture contents are expressed on a wet basis). Moisture conditioning can be carried out to obtain different desired levels of moisture contents. The formula employed in the process according to Gamal *et al.* (2009) is as follows:

$$Q = \frac{W_i(M_f - M_i)}{(100 - M_f)} \qquad (2.1)$$

Where;

Q = Mass of water to be added (kg)
W_i = Initial mass of the sample (kg)
M_i = Initial moisture content of the sample (%)
M_f = Final or desired moisture content of the sample (%)
Experiments on choice of moisture content for optimum shelling have been carried out in this research in a bid to breach the gap between varying results.

2.1.3. Feed Rate

Nyaanga *et al.* (2007) determined that feed rate increased with concave clearance. This was explained by the fact that the bigger the opening in the chamber the more pods that can be shelled per revolution. Trials on a manual sheller showed that in both rubber tyre and wood paddle shellers, feed rate of between 50-100 kg/hr at an average of 75 rpm does not significantly affect shelling performance (Chinsuwan, 1983). According to Amodu (2012), energy consumption in soy bean and cowpeas threshing is directly proportional to feed rate and tip cylinder speed irrespective of concave clearance. In the case of castor oil, shelling capacity, here referred to as mass flow rate, was found to increase with increase in cylinder speed (Balami *et al.*, 2012).

The determination of influence of groundnut feed rate on power consumption, shelling efficiency and kernel damage was part of this research work.

2.2. Machine Characteristics

2.2.1. Concave Clearance

A decrease in concave clearance leads to an increase in shelling efficiency and kernel damage. While this general trend was observed by all researchers on the subject, their values of concave clearance for optimum shelling were different. Nyaanga *et al.* (2007) observed that the efficiency of the sheller increases from 73.6% at concave clearance of 20 mm to a peak of 79.8% at 30 mm then decreases to 73.2% at 40 mm clearance. Experiments in Thailand showed that shelling efficiencies and kernel damage decrease with increase in clearance. Clearances of between 7 mm to 15 mm were used in a TPI sheller for Taina and other groundnut varieties local to Thailand (Chinsuwan, 1983). Less damage could be obtained with a larger clearance but shelling efficiency would be substantially decreased. The same conclusions were reached by Rostami *et al.* (2009), who observed that shelling efficiency decreased as clearance increased and damage rapidly decreased as clearance increased from 8 to 12 mm and gradually decreased as clearance increased from 12 to 20mm. Helmy *et al.* (2007) concluded that the optimum shelling efficiency of 95.44% at a feed rate of 80 kg/h could be obtained with a clearance of 18 mm, while Adedeji and Ajuebor (2002), gave a range of 30 mm- 40 mm. Bobobee

(2002) while working on a variable speed motor arrived at a concave clearance of 16-18 mm at 180-220 rpm.

In this work further research has been conducted to determine concave clearances that result in optimal sheller power consumption, shelling efficiency and kernel mechanical damage.

2.2.2. Type of Sieve

There are two types of sieves in common use, namely; the wire mesh sieve and the slotted grate sieve. The sieve size is chosen depending on the size of groundnuts to be shelled. The wire mesh size used in the experiments for manual shellers in Thailand was 11 mm by 11 mm (Chinsuwan, 1983). In the development of a groundnut sheller with a capacity of 35 kg/h, a concave made of round steel bars of 5 mm diameter was employed (Park *et al.*, 1990). Helmy *et al.* (2007) concluded that the performance of the wire mesh sieve was better than the slotted grate as shown in Table 2.4.

Table 2.4. Comparison of effects of sieve type on shelling.

Type of concave sieve	Shelling capacity (kg/h)	Shelling efficiency (%)	Percentage breakage (%)
Wire mesh	86	83-89	3.7-6.7
Slotted grate	60	82-84	8.4-12.6

Source: Helmy *et al.* (2007)

2.2.3. Type of Shelling Blades

Groundnut pods are shelled when they get embedded in the space between shelling blades and the concave sieve where they are acted upon by shearing, impact or, frictional forces either singly or in combination with each other. The characteristics of the blades that most likely affect shelling performance include; their material type, shape and the number mounted on the cylinder. Helmy *et al.* (2007) found out that the shelling efficiency, using rubber covered drum was less than that of both steel and wooden drum. For shellers utilizing paddles, kernel damage due to wood paddles was found to be substantially less than that due to rubber covered paddles, while the difference in shelling efficiency was relatively small (Chinsuwan, 1983). Gitau *et al.* (2013) determined that shelling efficiency was higher in steel rod beater shellers than wooden beater shellers. As far as the number of beaters is concerned, Helmy *et al.* (2007) determined that an increase in number of drum beaters from 4 to 8 increased shelling efficiency at low drum speeds of 1.83 m/s and 4.58 m/s. Studies by Kamboj *et al.* (2012) on pea shelling, employing L-shaped blades to provide maximum rubbing action, resulted in minimum kernel damage compared to that by both impact and shearing actions. Centrifugal impellers or rollers rotating in counter directions are utilized in shelling kernels with hard pods or coats such as Bambara nuts (Siebenmorgan *et al.*, 2006). Research on the influence of various shapes of blades on shelling performance of groundnuts has been included in this work.

2.2.4. Shelling Speed

Nyaanga *et al.* (2007) observed that shelling efficiency increased to a maximum with increase in speed but decreased with further increase in speed. Rostami *et al.* (2009) concluded that shelling efficiency increased with speed but had no significant effect on kernel damage. This agrees with results obtained from performance evaluation on a *Prosopis africana* pod thresher which showed that threshing efficiency and seed loss increased with increase in cylinder speed and fan speed (Ishola, 2011). Amodu (2012) found out that though visible grain damage to groundnut kernels at high speeds may be below 5 %, internal damage to the grains could be very high as determined by germination tests.

Definite speed values for optimum shelling performance are machine based. For instance, studies involving the use of a variable speed motor indicated that speeds of 180-200 rpm produce an output range of 240-250 kg/h with a breakage rate of 10-14% in a pneumatic drum sheller (Bobobee, 2002). Analysis of shelling speeds in castor oil fruits showed that the machine performed best at 240 rpm (Balami *et al.*, 2012). The best performance for experiments conducted by Adedeji and Ajuebor (2002) on groundnut shelling was achieved at 260 rpm and 150 kg (pods)/h feed rate. In carrying out in-field groundnut shelling tests, Butts *et al.* (2009) utilized a cylinder rotating within a range of 160rpm-300 rpm.

In this research experiments have been done to determine speed levels that yield optimum shelling efficiency, kernel mechanical damage and throughput per unit power consumption in a motorised groundnut sheller.

2.3. Sheller Design Formulae

In the design and development of a sheller, several quantitative models can be employed in sizing various parts. Following is a description of formulae for a groundnut sheller as provided by Akcali (1996) and Khurmi and Gupta (2009).

2.3.1. Motor Power Rating

Power requirement is determined using the following expressions;

$$P_r = a_0 + a_1 \frac{R_1}{R_2}(1 - K_Y)\rho_{bg} \qquad (2.2)$$

$$F = P_r A \qquad (2.3)$$

$$T = FR \ (2.4) \quad P_s = \frac{2\pi NT}{60} \qquad (2.5)$$

$$P_m = \frac{P_s}{system\ efficiency * motor\ efficiency} \qquad (2.6)$$

Where; P_r = pressure exerted, a_0 = pressure strain coefficient at the surface of the beater

a_1 = pressure strain coefficient, R_1 = radius of the beater

R_2 = the radius of the concave from the centre of the beater, and approximated as;

$$R_2 = R_1 + \beta_s \qquad (2.7)$$

Where;

β_s = average pod size,

$$\rho_{bg} = bulk\ density\ of\ groundnuts$$

$$K_Y = \frac{shelled\ groundnut\ weight}{input\ groundnut\ pods\ weight} \qquad (2.8)$$

A= surface area in contact with shelling blades in one revolution

F= Cracking force

T=Torque required

R = radius of beater

P_s= Power transmitted by the shelling shaft and

P_m= Motor Power requirement.

For groundnuts, a_0= 2.4 and a_1= 10.1 (Nyaanga et al., 2007)

2.3.2. Pulley and Belt Formulae

$$\frac{N_1}{N_2} = \frac{d_2}{d_1} \qquad (2.9)$$

$$L = \pi(r_1 + r_2) + 2x + \frac{(r_1 - r_2)^2}{x} \qquad (2.10)$$

Where;

N_1 = Speed in rpm of pulley 1

N_2 = Speed in rpm of pulley 2

d_1 = Diameter of pulley 1

d_2 = Diameter of pulley 2

r_1 = Radius of pulley 1

r_2 = Radius of pulley 2

x = Centre distance between pulley 1 and 2

L = Length of belt connecting pulley 1 and 2.

2.3.3. Shaft Diameters

Shaft diameters are determined by the following formulae;
According to Guest's theory,

$$T_e = \sqrt{(M_{max})^2 + (T_{torq})^2} = \frac{\pi \tau d^3}{16} \qquad (2.11)$$

while the expression for Rankin's theory is:

$$= \frac{1}{2}\left\{ M_{max} + \sqrt{(M_{max}^2 + T_{torq}^2)} \right\} = \frac{\pi \sigma d^3}{32} \qquad (2.12)$$

Where;

T_e = equivalent twisting moment

Me = equivalent bending moment

M_{max} = maximum bending moment on shaft

T_{torq}= torque acting on the shaft,

And

$$T_{torq} = (T_1 - T_2)R \qquad (2.13)$$

Where;

T_1 = tension in the tight side of belt

T_2 = tension in the slack side of the belt

R = radius of the pulley

τ = allowable shear stress of the shaft material

σ = allowable normal stress of the shaft material

d = diameter of the shaft

3. Research Methodology

Design of the groundnut sheller was accomplished using equations 2.2 to 2.13 in the literature review section.

3.1. Shelling Shaft Speed

The design began with selection of a desired output of shelled kernels per unit time. A kernel throughput of 200 kg/h was deemed adequate for experimental purposes and translates into shelling of 500 kg kernels in 2.5 hours, an amount equivalent to the average Kenyan groundnut yield per hectare as indicated in the introduction section. Determination of shelling shaft speed, N_{ss}, in revolutions per minute (rpm) was done by considering groundnut characteristics of volume and both bulk and solid densities of pods, kernels and shells. In addition, the width of the sieve in the shelling chamber was set to enable computation of the volume of pods shelled per revolution. The formula used was as follows;

$$N_{ss} = \frac{Volume\ of\ shelled\ pods\ per\ minute}{Volume\ of\ shelled\ pods\ per\ rev.*cracking\ efficiency} \qquad (3.1)$$

N_{ss} was determined as 350 rpm.

3.1.1. Power Requirement

First, pressure (P) exerted in shelling was calculated using formula 2.2. Now, for groundnuts, $a_0 = 2.4$, $a_1 = 10.1$, $R_1 = 0.2$ m, $\beta_S = 0.0126$ m, $R_2 = 0.2126$ m, $K_Y = 2/3$ and ρ_{bg} =172.44 N/m³; resulting in $P = 548.54$ N/m². Cracking force per revolution, F, was determined by applying formula 2.3, while contact area (A) was computed as shown below;

$$A = C_{1/2}wn_c \qquad (3.2)$$

Where:

$C_{1/2}$= half circumference of shelling cylinder

w = width of sieve

n = number of shelling blades

wn_c= cracking efficiency

Contact area was determined as 0.424 m². Determined values of P and A were inserted in formulae 2.4 to 2.6, and system and motor efficiencies of 75% and 80% respectively applied. This results yielded motor power capacity of 2.84 Kw (equivalent to 3.8 hp). A motor with a power rating of 4.4 hp was deemed adequate.

3.1.2. Pulley Sizes

Pulley sizes were determined by using the velocity ratio formula 2.9. Diameter of pulley on shelling shaft was set as 500 mm while that of pulley on motor shaft as 125 mm. The dimensions of the intermediate pulleys were varied according to the shelling speeds desired.

3.1.3. Shaft Diameters

Shaft diameters were determined by applying the equivalent twisting moment formulae 2.11 and 2.12. The diameter of fan shaft was computed as 25 mm while that of the sheller shaft as 35 mm. The resultant dimensions of the major machine components are shown in Table 3.1.

Table 3.1. *Dimensions of sheller components.*

Shelling blades radius (mm)	Concave radius (mm)	Motor power rating (hp)	Shelling shaft pulley diameter (mm)	Motor shaft pulley diameter (mm)	Fan shaft diameter (mm)	Shelling shaft diameter (mm)
200	213	4.4	500	125	25	35

3.2. Sheller Performance

Groundnuts were made ready for experiments by sorting and cleaning by hand-removal of defective pods and unwanted materials like soil and stone particles. The nuts were then dried in sunlight to a moisture content of 6%. For each experiment unit, the following quantities were measured: Test run time, T (s); Weight of shelled seeds per unit time at main outlet, W_b (kg/h);

Weight of shelled seeds per unit time at chaff outlet, W_c (kg/h); Weight of unshelled seed per unit time at all outlets, W_d (kg/h); Weight of damaged seeds per unit time at all outlets, W_e (kg/h) and electric power consumed during the test, P (kWh).

Performance of the sheller was determined by following formulae:

Total throughput per unit time;

$$W_a = W_b + W_c + W_d \text{ (kg/h)} \tag{3.3}$$

Shelling efficiency;

$$E_S = \left(1 - \frac{W_d}{W_a}\right) * 100\% \tag{3.4}$$

Kernel Mechanical damage;

$$E_D = \left(\frac{W_e}{W_a} * 100\right)\% \tag{3.5}$$

Throughput per unit power consumed;

$$E_P = \frac{W_a}{P} \text{ (Kg/h)/kWh} \tag{3.6}$$

3.3. Experiment Design and Setup

3.3.1. Influence of Groundnut Characteristics on Sheller Performance

(i). Moisture Content

Each variety of groundnuts was divided into five equal portions. To obtain nuts at moisture contents of about 18%, 15%, 12%, 9% and 6%, four batches were soaked in water for a period of 20, 15, 10 and 5 minutes respectively and spread out in natural air for 8 hours before storing them in labeled polybags ready for subsequent experiments. The fifth batch was not soaked in water, thus retaining a moisture content of 6%. Moisture content levels were determined using a moisture meter.

A known weight of groundnuts at 18% moisture content was shelled at pre-set levels of variety, feed rate, shelling speed, concave clearance and sheller blade type. The time taken for the shelling process was determined by a stop clock. The weights of shelled, unshelled and damaged seeds at various machine outlets were measured by means of a weighing balance. Electric power consumed by the motor was determined by a watt meter connected to its cable. The experiment was repeated to obtain three replicates. The procedure for the next four levels of experiments was similar to the one already described except that the moisture contents of the nuts used were 15%, 12%, 9% and 6% respectively.

(ii). Variety

Five different varieties of groundnuts were chosen to represent the range of sizes available in the research site. The following were selected for this purpose: ICGV 99658, ICGV 9991, CG 7, Homa bay local and Valencia red. In the first set of experiments, ICGV 99658 of known weight was shelled at pre-set levels of moisture content, feed rate, concave clearance, shelling speed and shelling blades type. The remaining four varieties were used in the next four sets of experiments respectively. A replication of three was applied for all the experiments.

(iii). Feed Rate

A sliding gate in the form of a rectangular plate fitted on one of the slanting surfaces of the trapezoidal hopper was used to regulate feed rate. A fixed weight of groundnuts was shelled at various gate positions and the corresponding feed rates in kg/h recorded. The positions on the rectangular plate at which the feed rates of 400 kg/h, 800 kg/h, 1200 kg/h, 1600 kg/h and 2000 kg/h were attained were identified by use of a marker pen. The five levels of feed rates were then used in experiments for this section.

In the first set of experiments, a fixed quantity of groundnuts was shelled at a feed rate of 400 kg/h at pre-set levels of moisture content, variety, shelling speed, concave clearance and shelling blades type. Computation of sheller performance was done as described in section 3.2. The second, third, fourth and fifth sets of experiments were carried out in a similar manner to the first one but at feed rates of 800 kg/h, 1200 kg/h, 1600 kg/h and 2000 kg/h respectively.

3.3.2. Influence of Machine Characteristics on Sheller Performance

(i). Shelling Speed

From literature review, motorised shellers are commonly run at shaft speeds of between 160 rpm and 400 rpm. Five speed levels for experimental purposes were chosen as 150 rpm, 250 rpm, 350 rpm, 480 rpm and 580 rpm. The selected shaft speeds were attained by mounting pulleys available on the Kenyan market with a diameter range of 100 mm to 250 mm interchangeably on the two ends of the fan shaft. Belts of appropriate lengths were utilized to transmit power from the fan shaft to the shelling shaft. Velocity ratio and belt length formulae were used to calculate the diameter and lengths of the required pulleys and belts for experiments in this section.

Actual speeds during operation were measured by use of a tachometer. Five levels of experiments were carried out in this section with a replication of three for each. In the first level, a specified weight of groundnuts were shelled at a shaft speed of 150 rpm and at selected levels of moisture content, variety, feed rate, concave clearance and shelling blades type. In the second, third, fourth, and fifth levels of experiments, shaft speeds of 250 rpm, 350 rpm, 480 rpm and 580 rpm were applied respectively.

Tangential velocity changes proportionally with radius of the shelling blades for a given constant angular speed. Hence, there is need to determine the corresponding tangential velocities for the shaft speeds to obtain the shelling speeds for the blades. The following formulae were used;

$$v = wr \tag{3.7}$$

And

$$w = \frac{2\pi N}{60} \tag{3.8}$$

Where:
v = tangential velocity
w = angular velocity
r = shelling blade radius
N = shaft speed in revolutions per minute
Using equations 3.7 and 3.8, Table 3.2 was obtained.

Table 3.2. Shelling speeds.

Shaft speed, N (rpm)	Tangential shelling speed, v (m/s)
150	3.2
250	5.3
350	7.4
480	10.1
580	12.2

(ii). Concave Clearance

Concave clearance was determined by measuring the distance between the shelling blades and the concave sieve at the point where the clearance was at a minimum.

(iii). Shelling Blade Type

Five types of blades were employed in experiments under this section. The first type was made of iron paddles having a curved shape of radius 200 mm, thickness of 2 mm, length of 420 mm and a distance of 32 mm along the circumference. The second type was similar to the first but with the paddles covered with strips of rubber. Thirdly; steel pipes acted as the shelling blades. The thickness of the pipes was 2 mm with a diameter of 10 mm. The fourth type was similar to the third but with an extra circumscribed pipe free to roll around its axis. The fifth type consisted of steel pipes covered with rubber strips.

3.3.3. Factor Level Combinations for Optimum Sheller Performance

The characteristics whose influence on sheller performance were investigated together with their selected factor levels are shown in Table 3.3

Table 3.3. Selected levels of nut and machine characteristic.

Level number	Nut and machine characteristics					
	Moisture content (%)	Groundnut variety	Feed rate (Kg/h)	Shelling speed (m/s)	Concave clearance (mm)	Shelling blades type
1	6	ICGV 99658	400	3.2	10	Steel paddle
2	9	ICGV 9991	600	5.3	15	Rubber paddle
3	12	CG 7	1200	7.4	20	Fixed steel pipe
4	15	Homa bay local	1600	10.1	25	Rolling rubber pipe
5	18	Valencia Red	2000	12.2	30	Rolling steel pipe

In this section determination of combined influence of all the six charateristics on sheller performance at different combination levels was carried out. A Taguchi experiment design with an orthogonal array L25(56) was set up to obtain results for all experiments outlined in section 3.3 and is depicted in Table 3.4.

Table 3.4. Taguchi experiment design with an orthogonal array L25(56).

Experiment No.	Shelling speed (m/s)	Shelling blade type	Groundnut variety	Moisture content (%)	Feed rate (kg/h)	Concave clearance (mm)
1	3.2	Steel paddle	99658	6	400	10
2	3.2	Rubber paddle	9991	9	800	15
3	3.2	Steel pipe	CG 7	12	1200	20
4	3.2	Rubber pipe	Homa bay	15	1600	25
5	3.2	Rolling pipe	Valencia	18	2000	30
6	5.3	Steel paddle	9991	12	1600	30
7	5.3	Rubber paddle	CG 7	15	2000	10

Experiment No.	Shelling speed (m/s)	Shelling blade type	Groundnut variety	Moisture content (%)	Feed rate (kg/h)	Concave clearance (mm)
8	5.3	Steel pipe	Homa bay	18	400	15
9	5.3	Rubber pipe	Valencia	6	800	20
10	5.3	Rolling pipe	99658	9	1200	25
11	7.4	Steel paddle	CG 7	18	800	25
12	7.4	Rubber paddle	Homa bay	6	1200	30
13	7.4	Steel pipe	Valencia	9	1600	10
14	7.4	Rubber pipe	99658	12	2000	15
15	7.4	Rolling pipe	9991	15	400	20
16	10.1	Steel paddle	Homa bay	9	2000	20
17	10.1	Rubber paddle	Valencia	12	400	25
18	10.1	Steel pipe	99658	15	800	30
19	10.1	Rubber pipe	9991	18	1200	10
20	10.1	Rolling pipe	CG 7	6	1600	15
21	12.2	Steel paddle	Valencia	15	1200	15
22	12.2	Rubber paddle	99658	18	1600	20
23	12.2	Steel pipe	9991	6	2000	25
24	12.2	Rubber pipe	CG 7	9	400	30
25	12.2	Rolling pipe	Homa bay	12	800	10

Optimization of performance was achieved at the combination levels that gave the highest values of throughput per unit power consumption and shelling efficiency; and lowest value of kernel mechanical damage.

3.4. Data Analysis

The influence of a single factor on a performance parameter was determined by computing the average output values at each factor level and the results obtained analyzed graphically. Analysis for combined factor influences was done using signal-to-noise (S/N) ratio techniques. S/N values were calculated for all the experiment trials using mathematical expressions according to Wysk *et al.* (2000). The following formula was applied for calculating S/N for kernel mechanical damage (the smaller, the better);

$$S/N = -10 log \left[\frac{1}{n} \sum_{i=1}^{n} y_i^2 \right] \quad (3.9)$$

S/N values for throughput per unit power consumption and shelling efficiency (the larger, the better), were determined as;

$$S/N = -10 log \left[\frac{1}{n} \sum_{i=1}^{n} \frac{1}{y_i^2} \right] \quad (3.10)$$

Where;
n = number of experiment replications in a trial and
y_i = i^{th} measured output value for the trial.

Mean S/N values were determined for each factor level with the highest value corresponding to the optimum desired output. Output response graphs were plotted from these results to show the influence of the factors under investigation on performance of the sheller. In addition, significance of influence of factors on performance was determined by carrying out an analysis of variance (ANOVA) on the S/N ratios obtained. Confirmation experiments were then conducted to verify the performance of the optimum conditions as determined by the matrix experiment.

4. Research Findings

4.1. Moisture Content

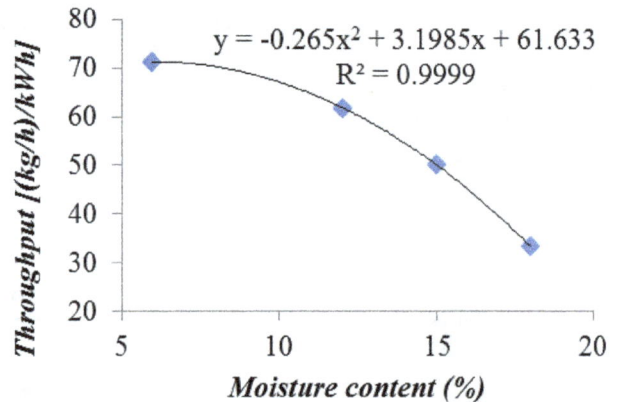

$$y = -0.265x^2 + 3.1985x + 61.633$$
$$R^2 = 0.9999$$

Figure 4.1. Throughput and moisture content.

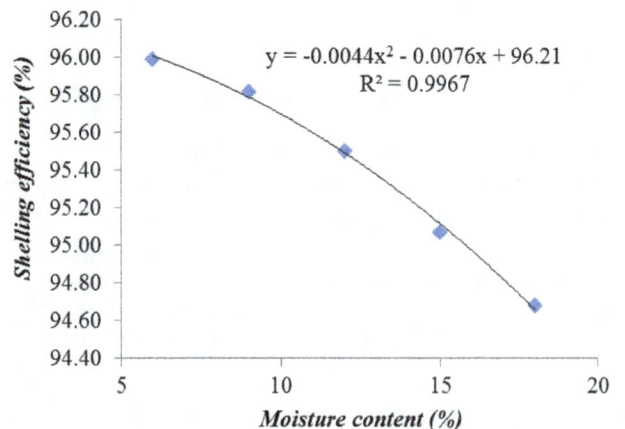

$$y = -0.0044x^2 - 0.0076x + 96.21$$
$$R^2 = 0.9967$$

Figure 4.2. Shelling efficiency with moisture content.

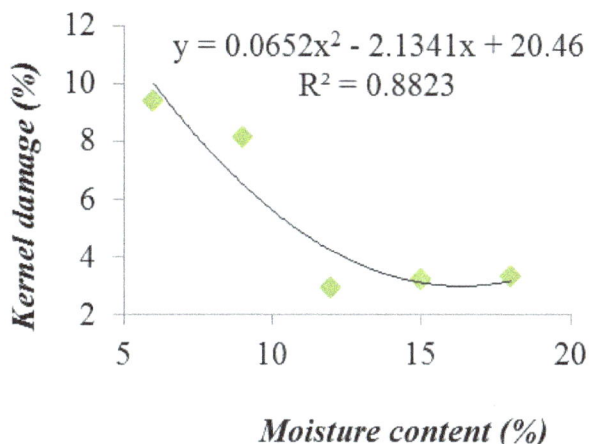

Figure 4.3. Kernel damage and moisture content.

Figure 4.1, 4.2 and 4.3 show how groundnut moisture content influenced throughput per unit power consumption, shelling efficiency and kernel mechanical damage respectively. The graphs were plotted from results for various experiments as shown in Appendix C.

The results show that machine throughput per unit power consumption increased with decrease in groundnut moisture content. This could be explained by the fact that the dry pods were more brittle than the wet ones, hence, they fractured faster upon being subjected to impact and frictional forces during the shelling process. As such, fewer motor revolutions were required to achieve complete shelling of a given quantity of groundnut pods with less moistutre content. The highest throughput per unit power consumption was achieved at 6% moisture content.

Shelling efficiency was also found to increase with reduction in moisture content with the highest efficiency being realised at 6% moisture content. The explanation for influence of moisture content on throughput per unit power consumption explained above also holds true for shelling efficiency. According to Nyaanga *et al.* (2007), pods with higher moisture content tend to flex instead of cracking and breaking hence leading to a higher percentage of unshelled groundnuts.

It was observed that kernel mechanical damage was highest at the lowest moisture content of about 6%, minimum at moisture contents between 15% and 18% and then increased marginally with further increase in moisture content. At a lower moisture content, the high brittleness of the kernels lends them to increased breakage. On the other hand, damage of kernels with very high moisture content was observed to occur by way of splitting along their middle axis. This could be attributed to a decrease in seed mechanical strength as explained by Gamal *et al.* (2008).

4.2. Groundnut Vareiety

The following varieties of groundnuts were used to carry out tests under this section; ICGV 99658, ICGV 9991, CG 7, Homa bay local and Valencia red and are shown in plate 4.2.

They were chosen to represent a wide spectrum of pod and

kernel physical characteristics such as size, density and repose angle. The results of their influence on sheller performance are shown graphically in Figures 4.4 to 4.6.

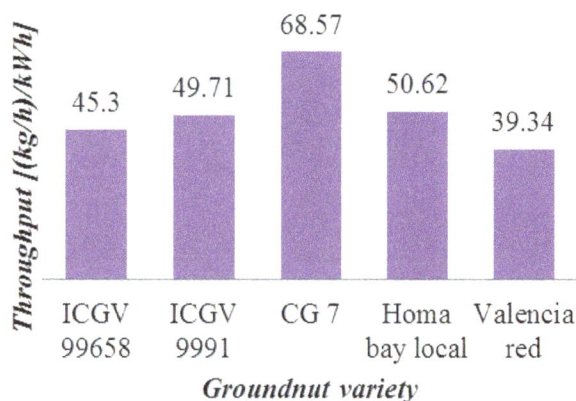

Figure 4.4. Throughput and nut variety.

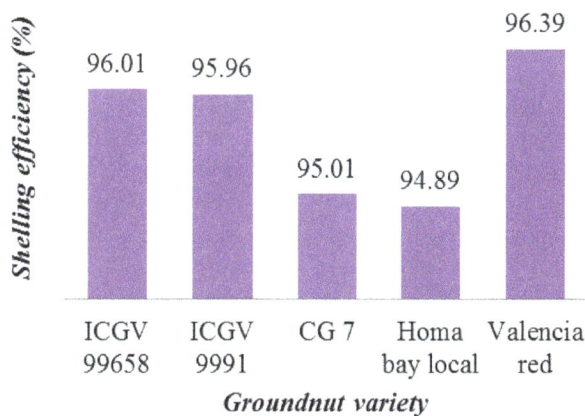

Figure 4.5. Shelling efficiency with nut variety.

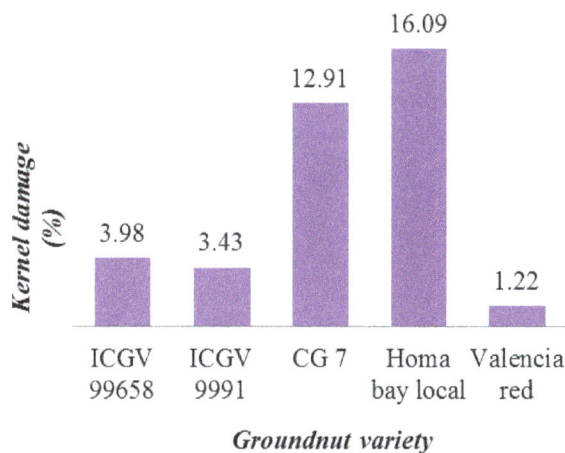

Figure 4.6. Kernel damage and nut variety.

The results show that variety CG 7 yielded the highest throughput per unit power consumption while Valencia red the lowest. It can also be seen from Figure 4.5 and 4.6 that the influence of the varieties under investigation on shelling efficiency and kernel mechanical damage follow the same pattern. Thus, Valencia red, ICGV 9991 and ICGV 99658 resulted in both high shelling efficiency and low kernel mechanical damage while CG 7 and Homa bay local led to low shelling efficiency and high kernel mechanical damage when shelled.

Experiments were carried out on several physical characteristics of groundnut varieties under study in a bid to explain the observations observed above. The results are shown in Table 4.1.

Table 4.1. Physical characteristics for selected groundnut varieties.

Variety	Pod bulk density (kg/m³)	Kernel bulk density (kg/m³)	Pod size (major diameter) (mm)	Kernel size (major diameter) (mm)	Kernel:Pod diameter ratio
CG 7	301.16	616.11	12.54	9.91	0.7903
Homa bay local	297.53	615.54	13.46	9.87	0.7332
ICGV 99658	212.43	639.46	12.65	9.27	0.7328
ICGV 9991	245.60	662.27	12.20	7.68	0.6295
Valencia red	224.56	673.61	13.00	7.46	0.5738

It can be infered from the results that throughput per unit power consumption increased with pod bulk density of the variety of groundnuts being shelled. Thus, variety CG 7 with the highest pod bulk density of 301.16 kg/m³ had the highest throughput per unit power consumption while ICGV 99658 with the lowest pod bulk density of 212.43 kg/m³ had the second last lowest throughput per unit power consumption.

Groundnut kernel to pod diameter ratio proved to be a vital characteristic as far as shelling efficiency and kernel mechanical damage are concerned. Results in Table 4.1 show that a high ratio translated into a low shelling efficiency and a high kernel mechanical damage. Following is a possible explanation for this scenario; A low kernel to pod diameter ratio corresponds to a wider air space between the husk and the kernel. This makes it relatively easier for the kernels to be released when the pods are fractured and they are less prone to impact and frictional forces occasioned by the rotating shelling blades. In addition, kernel mechanical damage rose with increase in kernel size. A large sized kernel, being heavier, collides with the fast moving shelling blades with greater momentum than a small one making it more vulnerable to cracking or splitting during the shelling process.

4.3. Shelling Speed

Shelling speeds ranging from 3.2 m/s to 12.2 m/s were employed in tests done under this section and the results obtained are as shown in the following figures;

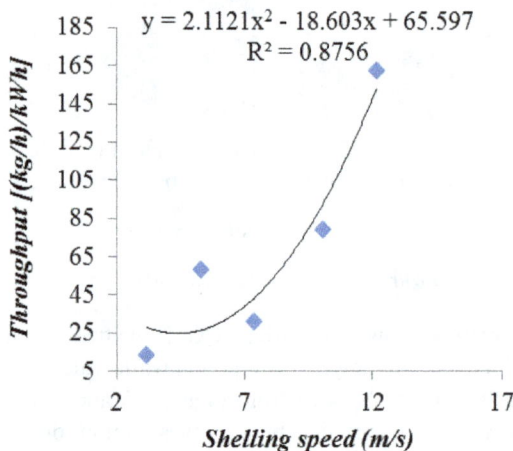

$$y = 0.0332x^2 + 0.0369x + 93.107$$
$$R^2 = 0.9475$$

Figure 4.8. Shelling efficiency and shelling speed.

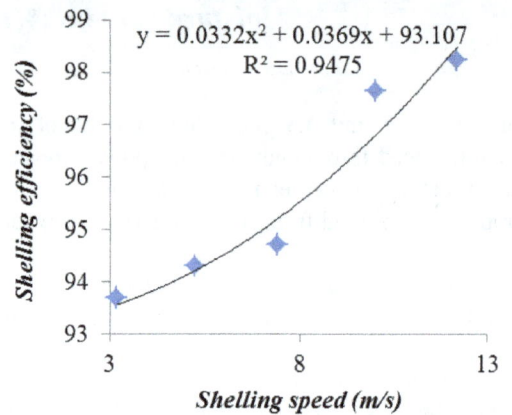

$$y = 0.1901x^2 - 2.1908x + 8.8097$$
$$R^2 = 0.9211$$

Figure 4.9. Kernel damage and shelling speed.

From Figure 4.10 and 4.11, both throughput per unit power consumption and shelling efficiency rose with increase in shelling speed with the highest values obtained around 12 m/s for this set of experiments. Figure 4.12 shows that kernel mechanical damage remained below 4 % for all speeds below 8 m/s then rose sharply with further increase in speed.

The collision and rubbing actions that generate the forces that result in the shelling of the groundnut pods; as well as the momentum of the shelling blades, increase with raise in shelling speed. This would lead to an increase in throughput per unit power consumption, shelling efficiency and kernel mechanical damage.

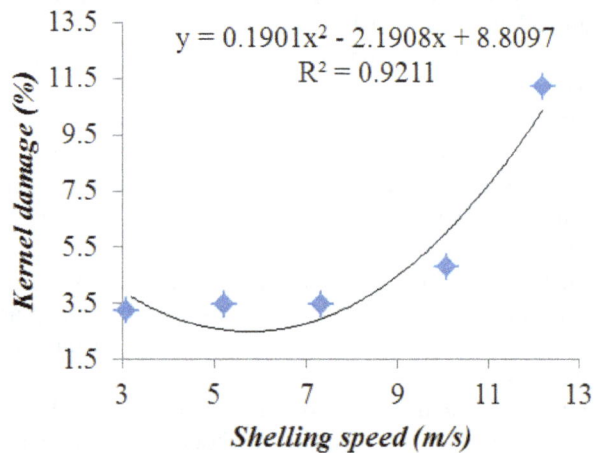

$$y = 2.1121x^2 - 18.603x + 65.597$$
$$R^2 = 0.8756$$

Figure 4.7. Throughput with shelling speed.

4.4. Concave Clearance

Five concave clearance levels of 10mm, 15mm, 20mm, 25mm and 30mm were used to obtain the results that generated the graphs shown in Figure 4.10 to 4.12.

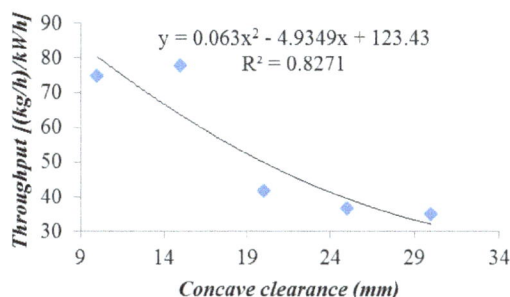

Figure 4.10. Throughput with concave clearance.

$$y = 0.063x^2 - 4.9349x + 123.43$$
$$R^2 = 0.8271$$

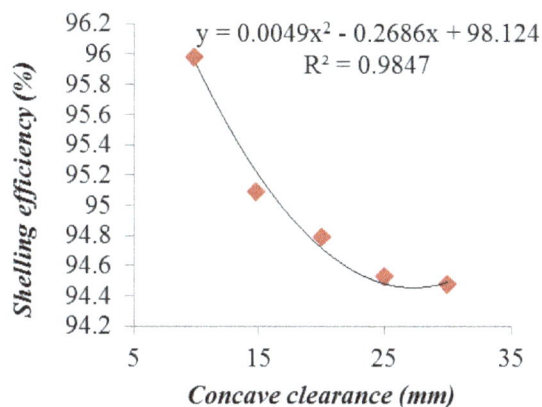

Figure 4.11. Shelling efficiency with concave clearance.

$$y = 0.0049x^2 - 0.2686x + 98.124$$
$$R^2 = 0.9847$$

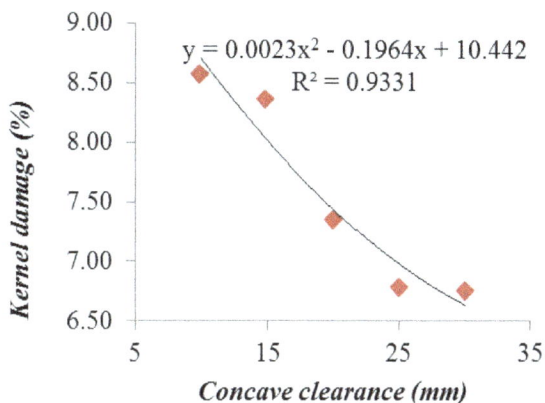

Figure 4.12. Kernel damage and concave clearance.

$$y = 0.0023x^2 - 0.1964x + 10.442$$
$$R^2 = 0.9331$$

From the figures, throughput per unit power consumption, shelling efficiency as well as kernel mechanical damage decreased with increase in concave clearance. This arose from the fact that at low concave clearance the groundnut pods are most compact and when subjected to impact and frictional forces during the shelling process, they are likely to fracture or open more easily leading to the high values of throughput per unit power consumption and shelling efficiency but this also resulted in high kernel mechanical damage.

The factor combinations that yielded the three optimum output performance, according to the mean S/N ratios and response graphs above, are as outlined in Table 4.2.

Table 4.2. Factor level combination for optimum groundnut sheller performance.

Groundnut sheller performance outputs	Optimum groundnut and machine factor combinations					
	Shelling speed (m/s)	Shelling blade type	Groundnut variety	Moisture content (%)	Feed rate (kg/h)	Concave clearance (mm)
Throughput per unit power consumption	12.2	Rubber paddle	CG 7	6	1600	10
Shelling efficiency	12.2	Rolling steel pipe	Valencia red	6	1600	10
Kernel mechanical damage	7.4	Fixed rubber pipe	Valencia red	12	800	15

4.4.1. Factor Significance

Significance of individual factors in influencing overall performance was determined by carrying out an analysis of variance (ANOVA) at 0.1 (α=0.1) level of significance on the S/N values of the three output parameters. The results are contained in Tables C8 to C10 in the Appendix section with a summary shown in Table 4.3

Table 4.3. Factor percentage contribution.

Output parameter	Factor contribution Factor contribution (%) (%)					
	Moisture content	Nut variety	Feed rate	Shelling speed	Concave clearance	Shelling blade type
Throughput per unit power consumed	6.94	3.16	6.90	67.94	11.64	3.42
Shelling efficiency Shelling efficiency	19.91	16.48	16.05	8.22	20.53	13.20
Kernel mechanical damage	31.02	36.56	5.08	6.24	12.11	3.31

Results from the three ANOVA tables led to several conclusions. First, shelling speed contributed most to the combined factor influence on throughput per unit power consumption, at 67.94% , while groundnut variety was the least significant at 3.16%. Secondly, the individual percentage contribution of the factors towards influence on shelling efficiency lay between 8.22% (shelling speed) and 20.53% (concave clearance). Hence, they were, approximately,

equally significant. Lastly, kernel mechanical damage was mostly influenced by two factors; groundnut variety (36.56%) and moisture content (31.02%), the rest of the factors posting contributions of 12.11% and below.

4.4.2. Optimal Outputs

Table 4.4. Expected optimum outputs.

Performance parameter	Optimum s/n ratio	Expected optimum outpts
Throughput per unit power consumption	63.13	985.14 (kg/h)/kWh
Shelling efficiency	40.36	104.23 %
Kernel mechanical damage	3.15	0.7 %

Actual outputs were determined by carrying out confirmation experiments at the optimal factor levels outlined in Table 4.4 above. Results obtained for the actual optimal outputs are shown in Table 4.5

Table 4.5. Actual optimum outputs.

Groundnut and machine factor combinationbination groundnut						Outputs		
Shelling speed (m/s)	Shelling blade type	Groundnut variety	Moisture content (%)	Feed rate (kg/h)	Concave clearance (mm)	Throughput per unit power consumption [(Kg/h)/kWh]	Shelling efficiency (%)	Kernel mechanical damage (%)
12.2	Rubber paddle	CG 7	6	1600	10	921.03*	97.38	52.15
12.2	Rolling steel pipe	Valencia red	6	2000	10	178.80	99.08*	8.08
7.4	Fixed rubber pipe	Valencia red	12	800	15	92.13	96.75	1.25*

*= Optimum values for each output parameter

A comparison of the theoretical and the actual optimum outputs shows that the two sets of values were close to each other. Experiments carried out under this study applied batch feeding mechanism, in which an average weight of 7 kg of in-pod groundnuts was shelled for each test run. Continuous feeding mechanism is often practiced under field conditions.

Hence, an extra experiment, employing continuous feeding mechanism was conducted at factor level setting for optimum kernel damage, in which 140 kg of in-pod groundnuts was fed into the sheller. Table 4.6 shows a comparison of the two set of results that were obtained.

Table 4.6. Output results for batch and continuous feeding mechanisms.

Feeding mechanism	Throughput per unit power consumption [(Kg/h)/kWh]	Shelling efficiency (%)	Kernel mechanical damage (%)
Continuous	495.23	96.57	2.02
Batch	92.13	96.75	1.25

It can be inferred from the results that continuous loading of in-pod groundnuts into the feed hopper yielded a much higher throughput per unit power consumption than batch feeding but with an accompanied slight increase in kernel damage. Shelling efficiency remained relatively constant.

5. Conclusions and Recommendations

Results obtained from this study show that the groundnut and machine characteristics considered influenced throughput per unit power consumption, shelling efficiency and kernel mechanical damage to various levels. Throughput per unit power consumption and shelling efficiency increased with reduction in % moisture content of the groundnut pods, with maximum outputs realized at 6% mc. Kernel mechanical damage decreased with increase in % mc up to a minimum at between 15% and 18% mc then increased marginally with further rise in moisture content.

Meanwhile, throughput per unit power consumption increased with bulk density of the groundnut variety being shelled. Thus,

variety CG 7 being heaviest yielded the greatest throughput per unit power consumed. In addition, kernel to pod diameter ratio had a significant influence on the outputs under study. Varieties with a higher kernel to pod diameter ratio like CG 7, resulted in lower shelling efficiency and higher kernel mechanical damage than those with a lower kernel to pod diameter ratio such as Valencia red.

All the three output parameters under review rose exponentially with increase in feed rate.

Throughput per unit power consumption and shelling efficiency rose steadily with increase in shelling speed with the highest values obtained at a shelling speed of about 12 m/s. On the other hand, kernel mechanical damage remained low (less than 4%) for speeds below 8 m/s, and then rose sharply with further increment in speed.

It was observed that all the output parameters increased with reduction in concave clearance with maximum values obtained at 10 mm clearance.

The conclusion concerning the influence of shelling blade type on output parameters was as follows; Steel and rubber paddles yielded the highest throughput per unit power consumption. At low shelling speeds (less than 8 m/s), rolling rubber and steel pipes resulted in lowest shelling efficiency and kernel mechanical damage but at higher speeds they resulted into both highest shelling efficiency and kernel mechanical damage.

The following are recommendations for further research: Influence of shelling blade type on performance outputs, yielded inconclusive results. The influence of machine factors on groundnut sheller performance can include sieve type and size. Research can be done to determine groundnut and machine factor combinations that result in optimum winnowing efficiency amongst motorised groundnut shellers

References

[1] Abubakar, M., and Abdulkadir, B. H. (2012). Design and Evaluation of a Motorised and Manually Operated Groundnut Shelling Machine. International Journal of Emerging Trends in Engineering and Development, 4(2): 673-682.

[2] Adedeji, O. S., and Ajuebor, F. N. (2002). Performance Evaluation of Motorised Groundnut Sheller. Journal of Agricultural Engineering, 39(2): 53-56.

[3] Akcali, I. D., Ince, A., and Guzel, E. (2006). Selected Physical Properties of Groundnuts. International Journal of Food Properties, 9(1): 25-37.

[4] Amoah, F. (2012). Modification and Evaluation of a Groundnut Cracker for Cracking Jatropha curcus Seeds. Unpublished Master's Thesis, Kwame Nkrumah University of Science and Technology, Department of Agricultural Engineering.

[5] Amadu, N. (2012). Development and Performance Evaluation of an Improved Soybean Thresher. Unpublished Master's Thesis, Ahmadu Bello University, Zaria Nigeria, Department of Agricultural Engineering.

[6] Anantachar, M., Maurya, N. L., and Navaravani, N. B. (1997). Development and Performance Evaluation of Pedal Operated Decorticator. Journal of Agricultural Sciences, 10(4): 1078-1081.

[7] Armitage, D., and Wontner-Smith, T. (2008). Grain moisture-Guidelines for Measurement. HGCA-Funded Project, Caledonia House 223 Petronville Road London NI9HY.

[8] Aslan, N., and Cebeci, Y. (2006). Application of Box-Behnken Design and Response Surface Methodology for Modeling of some Turkish Coals. Cumhuriyet University, Mining Engineering Department, Turkey.

[9] Atiku, A., Aviara, N., and Haque, M. (2004). Performance Evaluation of a Bambara Groundnut Sheller. Agricultural Engineering International, the CIGR journal of Scientific Research and Development, 6(Manuscript PM 04 002): 1-18.

[10] Balami, A. A., Adgidzi, D., Kenneth, C. A., and Lamuwa, G. (2012). Performance Evaluation of a Dehusking and Shelling Machine for Castor Fruits and Seeds. Journal of Engineering (IOSR JEN), 2(10): 44-48.

[11] Ballal, Y. P., Inamdar, K. H., and Patil, P. V. (2012). Application of Taguchi Method for Design of Experiments in Turning Gray Cast Iron. International journal of Engineering Research and Application (IJERA), 2(3)): 1391-1397.

[12] Bhatia, A. (2010). Pneumatic Conveying Systems.

[13] Unpublished lecture notes, Course No. M05-010 Continuous Education and Development, Inc; 9 Creyridge Farm Court, Story Point, NY 10980.

[14] Bobobee, E. (2002). No more Fingertip Shelling: The TEK groundnut Cracker to the Rescue; The International Journal of Small-scale Food Processing, 30: 12-15.

[15] Butts, C. L., Sorenseen, R. B., Nuti, R. C., Lamb, M. C., and Faircloth, W. H. (2009). Performance of Equipment for In-field Shelling of Groundnut for Biodiesel Production. American Society of Agricultural and Biological Engineers, 52(5): 1461-1469.

[16] Chinsuwan Winit (1983). Groundnut Shellers Project (Thailand) (File No. 3-80-0128, April, 1981-March, 1983). Submitted to International Development Research Centre (IDRC), Khon Kaen University, Department of Agriculture Engineering.

[17] Delhagen, W., Hussam, S., Mohdramli, R., and Alexander, Y. (2003). A Low-cost Groundnut Sheller for Use in Developing Nations.

[18] Duke, D., and Joyashree, R. (2011). Approach to Energy Efficiency among Micro, Small and Medium Enterprises in India. A field survey.

[19] Gamal, E., Radwan, S., ElAmir, M., and ElGamal, R. (2009). Investigating the Effects of Moisture Content on some Properties of Groundnut by aid of Digital Image Analysis. Food and Bio products Processing, 87: 273-281.

[20] Gitau, A. N., Mboya, P., Njoroge, B. K., and Mburu, M. (2013). Optimizing the Performance of a manually Operated Groundnut (Arachis hypogaea) Decorticator. Open Journal of Optimization, 2(1)): 26-32.

[21] Gitu, K. W., and Nzuma, J. M. (2003). Data Compendium for Kenya's Agricultural Sector; Kenya Institute of Public Policy Research and Analysis (KIPPRA)(Special Report, February 2003), Nairobi, Kenya.

[22] Helmy, M. A., Mitrroi, A., Abdallah, S. E., and Basioury, M. A. (2007). Modification and Evaluation of a Reciprocating Machine for Shelling Groundnut. Misr Journal of Agricultural Engineering, 24(2): 283-298.

[23] Henderson, S. M., and Perry, R. L. (1976). Agricultural Process Engineering. Westport, Connecticut: The AVI Publishing Company.

[24] Ikechukwu, C. U., Olawale, J. O., and Ibukun, B. I. (2014). Design and Fabrication of Groundnut Shelling and Separating Machine. International Journal of Engineering Science Invention, 3(4): 60-66.

[25] Ishola, T. A., Oni, K. C., Yahya, A., and Abubakar, M. S. (2011). Development and Testing of a Prosopis Africana Pod Thresher. Australian Journal of Basic and Applied Sciences, 5(5): 759-767.

[26] Kamboj, P., Singh, A., Kumar, M., and Din, S. (2012). Design and Development of Small Scale Pea Depoding Machine by using CAD Software. Agricultural Engineering International: CIGR Journal, 14(2): 40-48.

[27] Karuga, S., and Alfred, K. (2010). Staple Foods Value Chain Analysis- Country Report-Kenya.

[28] Khurmi, R. S., and Gupta, J. K. (2009). A Textbook of Machine Design (14th ed.). Ram Nagar, New Delhi: Eurasia Publishing House.

[29] Palomar, M. K. (1998). Groundnut in the Philippines Food System (A Macro Study, Groundnut in Local and Global Food Systems Series Report No. 1), University of Georgia, Department of Anthropology.

[30] Madhusudhana, B. (2013). A Survey on Area, Productivity of Groundnut Crop in India. IOSR Journal of Economics and Finance, 1(3): 1-7.

[31] Mezarcioz, S. M., and Ogulata, T. R. (2011). The Use of Taguchi Design of Experiment Method in Optimizing Spirality Angle of Single Jersey Fabrics. Texstil ve Konfeksiyon, 4(2011): 374-380.

[32] Nyaanga, D. M., Chemeli, M. C., and Wambua, R. M. (2007). Development and Testing of a Portable Hand-Operated Groundnut Sheller. Egerton Journal, 7(S): 117-130.

[33] Park, H. J., Cho, Y. K., Hong, S. G., and Song, C. J. (1990). Study on the development of a groundnut sheller for seed. Journal: Research Reports of the Rural Development Administration, Farm Management, Agricultural Engineering and Sericulture, 32(1): 61-67.

[34] Rai, A. K., Kottayi, S., and Murty, S. N. (2005). A low Cost Field Usable Portable Grain Moisture Meter with Direct Display of Moisture (%). African Journal of Science and Technology (AJST), Science and Engineering Series, 6(1): 97-104.

[35] Rostami, M. A., Azadshahraki, F., and Najafinezhad, H. (2009). Design, Development and Evaluation of a Groundnut Sheller. Journal: AMA, Agricultural Mechanization in Asia, Africa and Latin America, 40(2): 47-49.

[36] Siebenmorgan, T. J., Jia, C., Qin, G., and Schluterman. (2006). Evaluation of Selected Rice Laboratory Shelling Equipment. American Society of Agricultural and Biological Engineers, 22(3): 427-430.

[37] Simonyan, K. J., and Yiljep, Y. D. (2008). Investigating Grain Separation and Cleaning Efficiency Distribution of a Conventional Stationary Rasp-bar Sorghum Thresher. Agricultural Engineering International: the CIGR Ejournal, 10(Manuscript PM 07 028): 1-13.

[38] Tetsuro, T., and Shozo, K. (2010). Using IT to Eliminate Energy Waste in Production Lines. (Yokogawa Technical Report, English Edition, Vol.53, No.1).

[39] Thompson, M. (2009). Experimental Design and Optimization (3): Some Fractional Factorial Designs; AMCTB No. 36.

[40] Wysk, R. A., Niebel, B. W., Cohen, P. H., and Simpson, T. W. (2000). Manufacturing Processes: Integrated Product and Process Design. McGraw Hill, New York.

Permissions

List of Contributors

Md. Farhad, Md. Abdul Hakim and Md. Ashraful Alam
Wheat Research Centre, BARI, Nashipur, Dinajpur-5200, Bangladesh

N. C. D. Barma
Regional Wheat Research Centre, BARI, Joydebpur, Gazipur-1701, Bangladesh

Kodjovi Sotomè Detchinli and Jean Mianikpo Sogbedji
Ecole Supérieure d'Agronomie, Université de Lomé, Lomé, Togo

Omotoso Solomon Olusegun
Department of Crop, Soil and Environmental Sciences, Faculty of Agricultural Sciences, Ekiti State University, Ado-Ekiti, Nigeria

Muhammad Asif
Scientific Officer/AAE, CAEWRI NARC/Pakistan Agricultural Research Council (PARC) Islamabad

Col Islam-ul-Haque
Chairman, Ecological Sustainability through Environmental Services (Eco Steps), Islamabad, Pakistan

Laye Djouba Conde
State Key Laboratory for Conservation and Utilization of Subtropical Agro-Bioresources, Root Biology Center, South China Agricultural University, Guangzhou, 510642, People's Republic of China
National Department of Agriculture, Ministry of Agriculture and Livestock, Conakry BP: 576, Republic of Guinea

Zhijian Chen, Hongkao Chen and Hong Liao
State Key Laboratory for Conservation and Utilization of Subtropical Agro-Bioresources, Root Biology Center, South China Agricultural University, Guangzhou, 510642, People's Republic of China

Tekle Yoseph and Wondewosen Shiferaw
Southern Agricultural Research Institute, Jinka Agricultural Research Center, Department of Crop Science Research Process, Jinka, Ethiopia

Zemach Sorsa, Tibebu Simon and Abraham Shumbullo
Department of Plant Sciences and Horticulture, Wolaita Sodo University, Wolaita Sodo, Ethiopia

Woineshet Solomon
Southern Agricultural Research Institute, Hawassa Agricultural Research Center, Department of Crop Science Research Process, Hawassa, Ethiopia

Awotide Diran Olawale
Department of Agricultural Economics and Farm Management, Olabisi Onabanjo University, Yewa Campus, Ayetoro, Ogun State, Nigeria

Mafouasson Hortense Noelle Tontsa
Institut De Recherche Agricole Pour Le Developpement (IRAD), Yaounde, Cameroon

Arvind Bijalwan
Indian Institute of Forest management (IIFM), Nehru Nagar, Bhopal-462 003, M.P., India

Manmohan J. R. Dobriyal
Department of Silviculture and Agroforestry, ASPEE College of Horticulture and Forestry, Navsari Agricultural University, Navsari – 396 450, Gujarat, India

Bhartiya J. K.
Saranda Forest Division, Chaiwasa, Jharkhand, India

Anwar A. Alsanabani
Department of Horticulture and Forestry, Faculty of Agriculture, Sana'a University, Sana'a, Yemen

Sarwan Kumar
Department of Plant Breeding and Genetics, Punjab Agricultural University, Ludhiana-141 004, India

Alie Kamara and Patrick Andrew Sawyerr
Soil Science Department, School of Agriculture, Njala Campus, Njala University, Sierra Leone

Abibatu Kamara and Mary Mankutu Mansaray
Extension Division, Ministry of Agriculture, Forestry and Food Security, Sierra Leone

Muhammad Ashraf, Fayyaz Ahmad Tahir and Farah Umer
Soil and Water Testing Laboratory Layyah, Punjab Pakistan

Muhammad Nasir
Soil and Water Testing Laboratory Multan, Punjab Pakistan

Muhammad Bilal Khan
Soil and Water Testing Laboratory Muzaffar Garh, Punjab Pakistan

Tura Bareke Kifle, Admassu Addi Merti and Kibebew Wakjira Hora
Holeta Bee Research Centre, Oromia Agriculture Research Institute, Holeta, Ethiopia

Dilip Kumar Roy, Sujit Kumar Biswas, Abdur Razzaque Akanda, Khokan Kumer Sarker and Abeda Khatun
Irrigation and Water Management (IWM) Division, Bangladesh Agricultural Research Institute (BARI), Gazipur, Bangladesh

Kishor Chand Kumhar, Azariah Babu, Mitali Bordoloi and Ashif Ali
Tea Research Association, North Bengal Regional Research & Development Centre, Nagrakata, District – Jalpaiguri, West Bengal – 735 225, India

Gader Ghaffari
Department of Agricultural Engineering, Payame Noor University, East Azarbaijan Province, Iran

Mahmoud Toorchi, Saeid Aharizad and Mohammad-Reza Shakiba
Department of Crop Production and Breeding, Faculty of Agriculture, University of Tabriz, Iran

Bikal Koirala, Jay Prakash Dutta and Shiva Chandra Dhakal
1Department of Agricultural Economics, Institute of Agriculture and Animal Sciences, Rampur, Chitwan, Nepal

Krishna Kumar Pant
Department of Environmental Science, Institute of Agriculture and Animal Sciences, Rampur, Chitwan, Nepal

Siboniso M. Mavuso, Absalom M. Manyatsi and Bruce R. T. Vilane
University of Swaziland, Department of Agricultural and Biosystems Engineering, Manzini, Swaziland

Premanand Balkrishna Meshram, Nahar Singh Mawai and Ramkumar Malviya
Forest Entomology Division, Tropical Forest Research Institute (ICFRE), Jabalpur, India

Yacob Alemayehu Ademe
Department of Plant Science, College Agriculture and Natural Resource, Dilla University, Dilla, Ethiopia

Edson Edain González-Arredondo, Juan Carlos González-Hernández, Jorge Rodríguez-López and Christian Omar Martínez-Cámara
Laboratory of Biochemistry, Department of Biochemical Engineering, Technological Institute of Morelia, Lomas de Santiaguito, Morelia, Mexico

Md. Nehal Hasnain, Md. Elias Hossain and Md. Khairul Islam
Department of Economics, University of Rajshahi, Rajshahi-6205, Bangladesh

Absalom Mganu Manyatsi and Ntobeko Zwane, Musa Dlamini
University of Swaziland, Department of Agricultural and Biosystems Engineering, Manzini, Swaziland

James Lwelamira, John Safari and Patrick Wambura
Institute of Rural Development Planning (IRDP), Dodoma, Tanzania

Getahun Kassa and Eskinder Yigezu
Department of Agricultural Economics, Mizan-Tepi University, Mizan Teferi, Ethiopia

Azmi Elhag Aydrous
Department of Agricultural Engineering, Faculty of Agriculture, Omdurman Islamic University, Omdurman, Sudan

Abdel Moneim Elamin Mohamed
Department of Agricultural Engineering, Faculty of Agriculture, University of Khartoum, Khartoum, Sudan

Abdelbagi Ahmed Abdelbagi
Department of Crops sciences, Faculty of Agriculture, Omdurman Islamic University, Omdurman, Sudan

Md. Afzal Hossain
Ministry of Environment & Forests, Bangladesh Secretariat, Bangladesh

Md. Abu Kawochar and Md. Altaf Hossain
Tuber crops research Centre (TCRC), Bangladesh Agricultural Research Institute, Joydebpur, Gazipur, Bangladesh

Abdullah-Al-Mahmud and Ebna Habib Md. Shofiur Rahaman
International Potato Centre (CIP), SWCA, Banani, Dhaka, Bangladesh

Khondoker Md. Nasiruddin
Department of Biotechnology, Bangladesh Agricultural University, Bangladesh

Ufele Angela Nwogor, Ogbu Anthonia Uche, Ebenebe Cordelia Ifeyinwa and Akunne Chidi Emmanuel
Zoology Department, Nnamdi Azikiwe University, Awka, Anambra State, Nigeria

Md. Altaf Hossain
Tuber Crops Research Centre (TCRC), Bangladesh Agricultural Research Institute (BARI), Joydebpur, Gazipur, Bangladesh

Abdullah-Al-Mahmud
International Potato Center (CIP), USAID Horticulture Project, Bangladesh

Md. Abdullah-Al-Mamun
Department of Agriculture Extension (DAE), Rangpur, Bangladesh

Md. Shamimuzzaman
Department of Crop Sciences, University of Illinois at Urbana-Champaign, Urbana, Illinois 61801, USA

Md. Mizanur Rahman
Department of Horticulture, Bangabandhu Sheikh Mujibur Rahman Agricultural University, Joydebpur, Gazipur, Bangladesh

Philip Worlanyo Dugbley and Basuki Wasis
Department of Silviculture, Faculty of Forestry, Bogor Agricultural University, Bogor-Indonesia

Irdika Mansur
Department of Silviculture, Faculty of Forestry, Bogor Agricultural University, Bogor-Indonesia
SEAMEO Biotrop, Jalan Raya Tajur 6K, Bogor, Indonesia

Seu Jonathan Gogbeu
Laboratory of Plant Physiology and Pathology, Department of Agroforestry, University Jean Lorougnon Guede, Daloa, Cote d'Ivoire

Koffi Mathurin Okoma
Central Laboratory of Biotechnology, National Agronomic Research Centre (CNRA), Abidjan, Cote d'Ivoire

Koua Serge Beranger N'Goran
National Centre of Floristic, Department of Biosciences, University Felix Houphouet-Boigny, Abidjan, Cote d'Ivoire

Dénézon Odette Dogbo
Laboratory of Biology and Plant Production Improvement, Department of Sciences and Nature, University Nangui-Abrogoua, Abidjan, Cote d'Ivoire

Wangette Isaac S., Nyaanga Daudi M. and Njue Musa R.
Department of Agricultural Engineering, Egerton University, Njoro, Kenya

Index

www.ingramcontent.com/pod-product-compliance
Lightning Source LLC
Chambersburg PA
CBHW080629200326
41458CB00013B/4561